International Symposium on History of Machines and Mechanisms

A C.I.P. Catalogue record for this book is available from the Library of Congress.

ISBN 1-4020-2203-4 (HB)
ISBN 1-4020-2204-2 (e-book)

Published by Kluwer Academic Publishers,
P.O. Box 17, 3300 AA Dordrecht, The Netherlands.

Sold and distributed in North, Central and South America
by Kluwer Academic Publishers,
101 Philip Drive, Norwell, MA 02061, U.S.A.

In all other countries, sold and distributed
by Kluwer Academic Publishers,
P.O. Box 322, 3300 AH Dordrecht, The Netherlands.

Printed on acid-free paper

International Symposium on History of Machines and Mechanisms

Proceedings HMM2004

Edited by

MARCO CECCARELLI

Laboratorio di Robotica e Meccatronica,
Dipartimento di Meccanica, Strutture, Ambiente e Territorio,
Università di Cassino, Cassino, Italy

KLUWER ACADEMIC PUBLISHERS
DORDRECHT / BOSTON / LONDON

Table of Contents

8. Automata and Robots

Preface

The HMM2004 International Symposium on History of Machines and Mechanisms is the second event of a series that has been started in 2000 as main activity of the IFToMM Permanent Commission for History of MMS, Mechanism and Machine Science. The aim of the HMM Symposium is to be a forum to exchange views, opinions, and experiences on History of MMS from technical viewpoints in order to track the past but also to look at future developments in MMS.

The HMM Symposium Series is devoted to the technical aspects of historical developments and therefore it has been addressed mainly to the IFToMM Community. In fact, most the authors of the contributed papers are experts in MMS and related topics.

This year HMM Symposium came back to Cassino, after the challenging first event in 2000. The HMM2004 International Symposium on History of Machines and Mechanisms was held at the University of Cassino, Italy, from 12 to 15 May 2004.

These Proceedings contain 29 papers by authors from all around the world. These papers cover the wide field of the History of Mechanical Engineering and particularly the History of MMS. The contributions address mainly technical aspects of historical developments of Machines and Mechanisms. History of IFToMM, the International Federation for the Promotion of Mechanism and Machine Science is also outlined through the historical activities of some of its Commissions.

Original sources of technical achievements and works are referenced in the papers so that these Proceedings can be also considered a kind of handbook on History of Machines and Mechanisms. This book is of interest to researchers, graduate students and engineers specializing or addressing attention to History of Science and Technology, and particularly on History of MMS.

The figure of the cover has been extracted from the Table of the millstone work for modern TMM (now renamed as MMS because of the enlargement of the discipline): Lanz J.M. and Betancourt A., 1808, "Essai sur la composition des machines", Paris. The cover of the Proceedings of HMM2000 was selected from the book: Allievi L, 1895, "Cinematica della biella piana", Regia Tipografia Francesco Giannini & Figli, Napoli. For HMM2000 the figure was selected as a figure giving the idea of man walking ahead from the past by using an old scheme of kinematic properties of a mechanism that are still of great interest. In the case of the figure for the cover of the Proceedings of HMM2004 the table with several mechanisms has been selected to attract people with the richness of mechanism designs from the past that are still of current practical interest.

I believe that a reader will take advantage of the papers in these Proceedings with further satisfaction and motivation for her or his work (historical or not).

I would like to express my grateful thanks to the members of the International Organizing Committee of the Symposium: prof. Alexander Golovin (Russia), prof. Teun Koetsier (The Netherlands), prof. Carlos Lopez-Cajùn (Mexico), prof. Jammi S. Rao (India), prof. Junichi Takeno (Japan), prof. Lu Zhen (China-Beijing), prof. Hong-Sen Yan (China-Taipei) and to the members of the IFToMM Commission for History of MMS for co-operating enthusiastically for the success of the HMM initiative. I am grateful to the authors of the articles for their valuable contributions and for preparing their manuscripts on time, and to the reviewers for the time and effort they spent evaluating the papers. I would like to thank the sponsors of the Symposium: IFToMM and the University of Cassino, as well as my colleagues and assistants Dr Erika Ottaviano and Dr Giuseppe Carbone at the University of Cassino for their help.

I am grateful to my wife Brunella, my daughters Elisa and Sofia, and my young son Raffaele. Without their patience and comprehension it would not have been possible for me to organise HMM2004 International Symposium on History of Machines and Mechanisms.

Cassino,   May 2004

Marco Ceccarelli

Chairman for The International Symposium on History of Machines and Mechanisms

LIST of Contributors

Prof. Marco Ceccarelli
(Chairman of the IFToMM Permanent Commission for
History of MMS)
LARM: Laboratory of Robotics e Mechatronics
DiMSAT, University of Cassino
Via Di Biasio 43, 03043 Cassino (Fr)
Italy
Phone +39-0776-2993663 Fax +39-0776-2993711
email: ceccarelli@unicas.it
http://webuser.unicas.it/weblarm/larmindex.html

Prof. Vigen Arakelian
Department of Mechanics and Automatics
Institut National des Sciences Appliquées (INSA)
20 avenue des Buttes de Coësmes - CS 14315
F-35043 Rennes Cedex
France
Fax: + (33) 02 23 23 86 22
e-mail: vigen.arakelyan@insa-rennes.fr
INSA webpage : http://www.insa-rennes.fr/

Dr. Abdouslam Bashir
School of Mechanical & Systems Engineering,
Stephenson Building,
University of Newcastle upon Tyne, NE1 7RU,
Newcastle upon Tyne, NE1 7RU
UK

Prof. Emilio Bautista Paz
Laboratory of Machines & Mechanisms
Universidad Politécnica de Madrid
C/ José Gutierrez Abascal,
2 E-28006 Madrid
Spain
Phone +34-9133643119 Fax +34-913363118
email: ebautisa@etsii.upm.es

.Richard S. Berkof, Ph.D., P.E.
Professor of Mechanical Engineering
& Director of Pharmaceutical Manufacturing
Engineering Program
Stevens Institute of Technology
Castle Point on Hudson, Hoboken, NJ 07030,
USA
Phone: (201) 216-5538,   Fax: (201) 216-8315
E-mail: rberkof@stevens.edu

Hendrik Blauwendraat
Azaleastraat 29
NL-1214 CG Hilversum
The Netherlands
E-mail: hblauwe@few.vu.nl

Dr Robert (Bob) Bicker
School of Mechanical & Systems Eng,
University of Newcastle upon Tyne
Newcastle upon Tyne
England, NE1 7RU,
UNITED KINGDOM
Tel: ++44 191 222 6219
Fax: ++44 191 222 8600
e-mail : robert.bicker@ncl.ac.uk

Prof.em. Dr. Dr. h.c. Gerhard Bögelsack
Technische Universität Ilmenau
Fakultät für Maschinenbau
PF 100565
D-98684 Ilmenau (Thür.)
Germany
Telefon: 49 (3677) 69 2478 Telefax: 49 (3677) 69 1701
E-mail: gerhard.boegelsack@mb.tu-ilmenau.de

P. Gr. St-t Borisov Alexander
Material Plastic Treatment Machines and Technologies
Department,
Bauman Moscow State Technical University,
Ave. Enthusiastov, 100-1-24,
111531 Moscow
Russia
Phone +7 (095) 3077726
E-mail: borisanet@aha.ru

Dr Kevin Burn
School of Computing, Engineering and Technology,
University of Sunderland,
Sunderland SR1 3SD,
UK
e-mail: kevin.burn@sunderland.ac.uk

Prof. I-Ming Chen
School of Mechanical and Production Engineering
Nanyang Technological University
50 Nanyang Ave
Singapore 639798
Phone +65-6790-6203 Fax +65-6791-1859
email: michen@ntu.edu.sg
http://www.ntu.edu.sg/mpe/Research/Groups/Mod_Robot
ics/index.htm

Thomas G. Chondros PhD, Assistant Professor
University of Patras, School of Engineering,
Mechanical Engineering and Aeronautics Department,
265 00 Patras,
Greece.
Phone 30 2610 997263, 30 2610 997264, Fax. 30 2610
997744.
E-mail: chondros@mech.upatras.gr
http://www.mech.upatras.gr\~chondros

Prof. Dr. Carlos André Dias Bezerra
Dept. of Mechanical Engineering
Federal University of Ceara
Campus do Pici. Bloco 714.
60455-760. Fortaleza-Ceará.
Brazil.
E-mail: cadbe@ufc.br

Dr. Evert Dijksman
Mechanism Design Consultant
Luikersteenweg 538,
3920 Lommel
Belgium
E-mail: evert .dijksman @ belgacom.net

Prof. Javier Echavarri Otero
Laboratory of Machines & Mechanisms
Universidad Politécnica de Madrid
C/ José Gutierrez Abascal
2 E-28006 Madrid
Spain
Phone +34-913364217 Fax +34-913363118
email: jechavarri@etsii.upm.es

St-t Anastasiya Ermakova
Dept. Material Plastic Treatment Machines and
Technologies
Bauman Moscow State Technical University,
Brateevskaya Str., 25-1-142,
115612 Moscow
Russia
Phone +7 (095) 3423484
E-mail: eav_nastya@list.ru

Professor, D.Sc.Golovin Alexandre
Bauman Moscow State Technical University
Department "Theory of Mechanisms and Machines"
Adress (home): Koroleva Str., 4-1-379
129515 Moscow
 Russia
Phone: +7 (095) 2168918
e-mail: aalgol@mail.ru

Univ. Prof. Dr.-Ing. Klaus Hoffmann
Institute for Engineering Design and for Transport,
Handling and Conveying Systems
Vienna University of Technology
Getreidemarkt 9, A-1060 Vienna
Austria
Phone +43-1-58801-30746, Fax +43-1-58801-30799
email: hoffmann@ft.tuwien.ac.at
webpage: http://www.ft.tuwien.ac.at/

Dr. Zhongxu Hu
School of Mechanical & Systems Engineering,
Stephenson Building,
University of Newcastle upon Tyne, NE1 7RU,
Newcastle upon Tyne, NE1 7RU
UK

Hsing-Hui Huang, Graduate Student
Department of Mechanical Engineering,
National Cheng Kung University
Tainan 70101,
TAIWAN
phone: 886(6)208-2703 fax: 886(6)208-4972
email: hui@ms34.url.com.tw

Univ. Prof. Dr. Manfred L. Husty
Inst. for Engineering Mathematics, Geometry and
Computer Science
University of Innsbruck
Technikerstrasse 13, 6020 Innsbruck
Austria
Phone +43-512-507-6830 Fax +43-512-507-2709
email: manfred.husty@uibk.ac.at
http://techmath.uibk.ac.at/geometrie/

Dr. Eng. Theodor Ionescu
Romanian State Passenger Railways
Bd. Dinicu Golescu 38,
 010873 Bucharest
Romania
Tel. / Fax: + 40 (0) 21 224 93 06
E-mail: theodor.ionescu@cfr.ro

Prof. Sergey Jatsun
KSTU: Kursk State Technical University
Head of the Department of Mechanic and Mechatronic
50 let Oktyabrya 94, room 218
Kursk, 305040
Russia
Phone +007 0712 52 38 07 Fax +007 0712 52 38 07
 email: jatsun@kursknet.ru
http://kstugate.kursk.ru/kstu/kstu/mech/index6.htm

Svetlana Jatsun
KSU: Kursk State University
Head of the Department of Medical and Safety
Radischeva st.33,
Kursk, 305000
Russia
Phone +007 0712 56 30 72 Fax +007 0712 56 30 72
 email: jatsun@kursknet.ru
KSU webpage http://www.kgpu.ru

Dr. Hanfried Kerle
Peterskamp 12,
D-38108 Braunschweig
Germany
Phone: +49-531-351777
e-mail: h.kerle@t-online.de

prof. Teun Koetsier
Dept. of Mathematics and Computers Sciences
De Boelelaan, Vrije University of Amsterdam
Postbus 7161,
1081 Amsterdam
The Netherlands
tel +31-20-4447684; fax +31-20-6427705
e-mail: teun@cs.vu.nl

Prof. Pilar Leal Wiña
Laboratory of Machines & Mechanisms
Universidad Politécnica de Madrid
C/ José Gutierrez Abascal,
2 E-28006 Madrid
Spain
Phone +34-9133643119 Fax +34-913363118
 email: pleal@etsii.upm.es

Professor em. Dr.Sc.Tech. Tatu Leinonen
P.O.Box 4200
FIN-90014 University of Oulu
Oulu
Finland
Tel. +358 8 553 2050 Fax.+358 8 2026
e-mail: tatu@me.oulu.fi

Prof. Dr.-Ing. habil. Kurt Luck
TU Dresden
Mommsentrabe 13
D-30167 Dresden
Germany
e-mail: luck@mfk.mw.tu-dresden.de

Dr. Klaus Mauersberger
Technische Universität Dresden
Kustodie
Mommsenstrasse 13,
D-01062 Dresden
Germany
Phone +49-0351-46333992 Fax +49-0351-46337229
Email: Klaus.Mauersberger@mailbox.tu-dresden.de
Kustodie webpage: http://www.tu-dresden.de/kustodie

Dr. Jean-Pierre Merlet
COPRIN project, INRIA
2004 Route des Lucioles
06902 Sophia-Antipolis Cedex
France
Phone +33 4 92 387761 Fax +33 4 9238 7643
email: Jean-Pierre.Merlet@sophia.inria.fr
http://www-sop.inria.fr/coprin/index_english.html

Professor Francis C. Moon
J. Ford Profesor of Mechanical Engineering
204 Upson Hall
School of Mechanical and Aerospace Engineering
Cornell University
Ithaca, New York, 14853,
USA
e-mail: fcm3@cornell.edu

Prof. José Luis Muñoz Sanz
Laboratory of Machines & Mechanisms
Universidad Politécnica de Madrid
C/ José Gutierrez Abascal, 2 E-28006
Madrid España (E)
Phone +34-913364216 Fax +34-913363118
email: jlmunoz@etsii.upm.es

prof. Agamenon R.E. Oliveira
Applied Mechanics and Structures Department
UFRJ Federal University of Rio de Janeiro
P.O. Box 68536-21941 Ilha do Fondao
Rio de Janeiro
Brazil
Phone +21-2562-7977
e-mail: agamenon.oliveira@globo.com

Assoc. Prof. Viktor Strizhak
Department of Mechatronics
Tallinn Technical University
Ehitajate tee 5,
19086 Tallinn
Estonia
Phone +3726203303 Fax +3726203203
email: strizhak@staff.ttu.ee
webpage http://www.ttu.ee

Mr. Raymond Tay
School of Mechanical and Production Engineering
Nanyang Technological University
50 Nanyang Ave
Singapore 639798
Phone +65-6790-6317 Fax +65-6791-1859

Prof. Ivan Zaharov
KSTU: Kursk State Technical University
Rector of the Kursk State Technical University
50 let Oktyabrya 94,
Kursk, 305040
Russia
Phone+007 0712 -2 2-57-43 Fax +007 8-071-2 22-57-43.
email: zakharov@kstu.kursk.ru
KSTU webpage http://www.kstu.kstu.ru

Dr. Zampieri D.E.
Prof. Dr. Carlos André Dias Bezerra
Dept. of Mechanical Engineering
Federal University of Ceara
Campus do Pici. Bloco 714.
60455-760. Fortaleza-Ceará.
Brazil.
cadbe@ufc.br

Prof. Zhen Lu
Dept. of Mechanical Engineering
Beijing Universityv of Aero. & Astro.
Beijing, 100083
P.R. China
TEL+86-10-82317741 FAX+86-10-82316100
e-mail: zhen_luh@ yahoo.com

prof. Teresa Zielinska
Warsaw University of Technology
Institute of Aircraft Engineering and Applied Mechanics
ul. Nowowiejska 22,
 00-665 Warsaw
Poland
e-mail: teresaz@meil.pw.edu.pl

Univ. Ass. Nenad Zrni ụ Dipl.-Ing., Can. Tech. Sc.
Department of Mechanization
Faculty of Mechanical Engineering, University of
Belgrade
27 Marta 80, 11000 Belgrade
Serbia & Montenegro
Phone +381-11-3302454 Fax +381-11-3370364
email: nzrnic@mas.bg.ac.yu
Faculty of ME webpage http://www.mas.bg.ac.yu

Mr. Shusong Xing
School of Mechanical and Production Engineering
Nanyang Technological University
50 Nanyang Ave
Singapore 639798
Phone +65-6790-6317 Fax +65-6791-1859

Gao Xuan
Institute for History of Science and Technology &
Ancient Texts
Tsinghua University
Beijing,100084
P.R.China
Tel: +86(10)62794734
E-mail: gaoxuan@lib.tsinghua.edu.cn

Prof. Hong-Sen Yan
Department of Mechanical Engineering,
National Cheng Kung University
Tainan 70101,
TAIWAN
phone: 886(6)208-2703 fax: 886(6)208-4972
email: hsyan@mail.ncku.edu.tw
webpage http://msdlab.me.ncku.edu.tw

Prof. Song Huat Yeo
School of Mechanical and Production Engineering
Nanyang Technological University
50 Nanyang Ave
Singapore 639798
Phone +65-6790-4701 Fax +65-6791-1859
email: myeosh@ntu.edu.sg
Webpage:
http://www.ntu.edu.sg/mpe/Research/Groups/Mod_Robot
ics/index.htm

1. History of IFToMM

Ceccarelli M.: *A Short Introduction on IFToMM officers over time*

Ceccarelli M., Koetsier T.: *On the IFToMM Permanent Commission for History of MMS*

Ionescu T., Bogelsack G., Leinonen T.: *Past, Present and Future in Mechanism and Machine Science Terminology*

Rieger N. F.: *Development and Activities of the IFToMM Rotordynamic Committee, 1977-2004*

A SHORT INTRODUCTION ON IFTOMM OFFICERS OVER TIME

Prof. Marco Ceccarelli
(Chairman of the IFToMM Permanent Commission for History of MMS)
LARM: Laboratory of Robotics e Mechatronics
DiMSAT, University of Cassino
Via Di Biasio 43, 03043 Cassino (Fr), Italy
email: ceccarelli@unicas.it

IFToMM was founded in 1969 by persons who believed that collaborations among engineers and scientists from many countries would have improved the knowledge and effectiveness of Mechanism and Machine Science (MMS) for enhancement of life and society.

Activities and results in IFToMM have been achieved because of persons who formed IFToMM and worked for IFToMM. Memory of their names is not only memory of personalities in MMS but even keeping track of the evolution of IFToMM through the persons who have been involved in its activity.

The following tables give the names of people who have been IFToMM officers but many others have contributed with their activity to the success of IFToMM and they should remembered too because IFToMM is a large worldwide Community. Those other names can be found in the list of the Commission/Committee members, list of participants in IFToMM Conferences, and list of authors of scientific works presented and published in IFToMM Proceedings and Journals.

The following tables illustrate also the organization of IFToMM as a world well-established organism.

Fig. 1. IFToMM Presidents at HMM2000 (from left to right): Giovanni Bianchi (1984-1987 and 1988-1991), ArcadyBessonov in substitution of Ivan I. Artobolevsky (1969-1971 and 1972-1975), Bernard Roth (1980-1983), JorgeAngeles (1996-2000), Kenneth J. Waldron (2000-2003 and 2004-2007), Leonard Maunder (1976±1979), Adam Morecki (1992-1995), and Marco Ceccarelli (Chairman of HMM200 Symposium). (The years in the brackets indicate the term of Presidentmandate.)

Table A-a: IFToMM Executive Council

Term	President	Vice-President	Secretary General	Treasurer	Members
1969-1971	Ivan I. Artobolevskii (USSR)	F.R. Erskine Crossley (USA)	M.S. Konstantinov (Bulgaria)	Werner Thomas (West Germany)	B.M. Belgaumkar (India) K.H. Hunt (Australia) Jan Oderfeld (Poland)
1971-1975	Ivan I. Artobolevskii (USSR)	F.R. Erskine Crossley (USA)	M.S. Konstantinov Emil Stanchev (acting) (Bulgaria)	Werner Thomas (West Germany)	Giovanni Bianchi (Italy) Leonard Maunder (England) Jan Oderfeld (Poland) Theodor Pantelic (Yugoslavia) Jack Phillips (Australia)
1976-1979	Leonard Maunder (United Kingdom)	C.S. Pelecudi (Romania)	Adam Morecki (Poland)	H. Rankers (The Netherlands)	Arcady P. Bessonov (USSR) Giovanni Bianchi (Italy) Kurt Luck (GDR) T Pantelic (Yugoslavia) Jack Phillips (Australia) Bernard Roth (USA)
1980-1983	Bernard Roth (USA)	Arcady P. Bessonov (USSR)	Adam Morecki (Poland)	H. Rankers (The Netherlands)	Elizabeth Filemon (Hungary) Mark S. Konstantinov (Bulgaria) Kurt Luck (GDR) M.O.M. Osman (Cxanada) J.M Prentis (UK) Z. Zlvkovic (Yugoslavia)
1984-1987	Giovanni Bianchi (Italy)	Mark S. Konstantinov (Bulgaria)	Elizabeth Filemon (Hungary)	J.M Prentis (UK)	Gerhard Bogelsack (GDR) K.V. Frolov (USSR) T. Hayashi (Japan) Jammi S. Rao (India) A. A. Seireg (USA) Jean Vertut (France)
1988-1991	Giovanni Bianchi (Italy)	Gerhard Bogelsack (GDR)	L. Pust (CSSR)	J.N Fawcett (UK)	G. Dittrich (DDR) K.V. Frolov (USSR) T. Hayashi (Japan) Jammi S. Rao (India) Ali A.. Seireg (USA) Justo Nieto (Spain)
1992-1995	Adam Morecki (Poland)	Terry E. Shoup (USA)	L. Pust (CSSR)	J.N Fawcett (UK)	Jorge Angeles (Canada) Y. Chen (P.R. China) G. Dittrich (DDR) Yuko Hori (Japan) Justo Nieto (Spain) Yuri L. Sarkisyan (Armenia)

Table A-a: IFToMM Executive Council

Term	President	Vice-President	Secretary General	Treasurer	Members
1996-1999	Jorge Angeles (Canada)	Yuko Hori (Japan)	Tatu Leinonen (Finland)	Robert Bicker (UK)	Arcady P. Bessonov (USSR) Joe K. Davidson (USA) JeanClaude Guinot (France) Alberto Rovetta (Italy) Gabor Stefan (Hungary) Karl Wohlhart (Austria)
2000-2003	Kenneth J. Waldron (USA)	JeanClaude Guinot (France)	Tatu Leinonen (Finland)	Robert Bicker (UK)	Tian Huang (P.R. China) Manfred Hiller (Germany) Gabor Stefan (Hungary) Alberto Rovetta (Italy) Hirofumi Miura (Japan) Kristoff Kedzior (Poland)
2004-2007	Kenneth J. Waldron (USA)	Tian Huang (P.R. China)	Marco Ceccarelli (Italy)	Joseph Rooney (UK)	Manfred Hiller (Germany) Theodor Ionescu (Romania) Carlos Lòpez-Cajun (Mexico) Hirofumi Miura (Japan) Kristoff Kedzior (Poland) James Trevelyan (Australia)

Table B1: Chairmen of the IFToMM Permanent Commissions (°)

Term / Commissions	1969-1971	1971-1975	1975-1979	1979-1981	1982-1985	1986-1989	1990-1993	1994-1997	1998-2001	2002-2005
Communications	T. Pantelic (Yugoslavia)	T. Pantelic (Yugoslavia)	unknown	unknown	unknown	J.S. Rao (India)	J.S. Rao (India)	J.S. Rao (India)	G. Stepan (Hungary)	Constantinos Mavroidis (USA)
Education	M.S. Konstantinov (Bulgaria)	R. Bogdan (Romania)	M.S. Konstantinov (Bulgaria)	unknown	unknown	unknown	Aldo Rossi (Italy)	Aldo Rossi (Italy)	Kenneth J. Waldron (USA)	Kenneth J. Waldron (USA)
History of MMS	Established in 1973	Jack Phillips (Australia)	Jack Phillips (Australia)	Jack Phillips (Australia)	Elisabeth Filemon (Hungary)	Elisabeth Filemon (Hungary)	Teun Koetsier (The Netherlands)	Teun Koetsier (The Netherlands)	Marco Ceccarelli (Italy)	Marco Ceccarelli (Italy)
Publications	E.R.J. Crossley (USA)	E.R.J. Crossley (USA) & A. Bessonov (USSR)	E.R.J. Crossley (USA) & A. Bessonov (USSR)	E.R.J. Crossley (USA) & A. Bessonov (USSR)	Leonard Maunder (UK)	Leonard Maunder (UK)	Terry E. Shoup (USA)	Terry E. Shoup (USA)	Terry E. Shoup (USA)	Vincenzo Parenti-Castelli (Italy)
Standardization of Terminology	D. Muster (USA)	D. Muster (USA)	D. Muster (USA) G. Bogelsack (Germany)	G. Bogelsack (Germany)	J.M. Prentis (UK)	J.M. Prentis (UK)	Tatu E. Leinonen (Finland)	Tatu E. Leinonen (Finland)	Teohdor Ionescu (Romania)	Tehodor Ionescu (Romania)

(°): Data are taken from Minutes of the IFToMM Executive Council meetings and from information by current IFToMM officers.

Table B2-a: Chairmen of the IFToMM Technical Committees (°)

Term/Commissions	1969-1971	1971-1975	1975-1979	1979-1981	1982-1985	1986-1989	1990-1993	1994-1997	1998-2001	2002-2005
Computational Kinematics						Started in 1991	Barham Ravani (USA)	Barham Ravani (USA)	JeanPierre Merlet (France)	JeanPierre Merlet (France)
Gearing		Started in 1976	Darle W. Dudley (USA)	Darle W. Dudley (USA)	Darle W. Dudley (USA)	Karl Stolzle (Germany)	Karl Stolzle (Germany)	Aizon Kubo (Japan)	Veniamin I. Goldfarb (Russia)	Veniamin I. Goldfarb (Russia
Human-Machine Systems					Started in 1986	K.V. Frolov (U.S.S.R.)	Kreystof Kedzior (Poland)	Kreystof Kedzior (Poland)	Kreystof Kedzior (Poland)	V.P. Tregoubov (Russia)
Linkages and Cams		Started in 1975	K. Kunad (Germany)	K. Kunad (Germany)	Fl. Duditza (Romania)	Fl. Duditza (Romania)	K. Luck (Germany)	K. Luck (Germany)	M. Vaclavik (Checz Republik)	M. Vaclavik (Checz Republik)
Mechatronics							Started in 1994	Manfred Hiller (Germany)	Manfred Hiller (Germany)	Henrik Van Brussel (Belgium)
Micromachines							Started in1994	T. Hayashi (Japan)	T. Hayashi (Japan)	Alberto Rovetta (Italy)
Nonlinear Oscillations								Started in 1998	L. Pust (Checz Republik)	Gabor Stepan (Hungary)

v(°): Data are taken from Minutes of the IFToMM Executive Council meetings and from information by current IFToMM officers.

Table B2-b: Chairmen of the IFToMM Technical Committees (°)

Term/Commissions	1969-1971	1971-1975	1975-1979	1979-1981	1982-1985	1986-1989	1990-1993	1994-1997	1998-2001	2002-2005
Reliability								Started in 1998	O.V. Berestnev (Belarus)	O.V. Berestnev (Belarus)
Robotics		Started in 1971	Bernard Roth (USA)	Giovanni Biannchi (Italy)	Adam Morecki (Poland)	Adam Morecki (Poland)	Kenneth J. Waldron (USA)	Kenneth J. Waldron (USA)	Jean Claude Guinot (France)	Jean Claude Guinot (France) Bodo Heinman (Germany) 2003
Rotordynamics			Started in 1977	Z. Parszewski (Poland)	J.S. Rao (India)	Yukio Hori (Japan)	G. Diana (Italy)	G. Diana (Italy)	Neville Rieger (USA)	Neville Rieger (USA)
Transportation Machinery								Started in 1998	Barham Ravani (USA)	Barham Ravani (USA)

(°): Data are taken from Minutes of the IFToMM Executive Council meetings and from information by current IFToMM officers.

Table B2: Editors of IFToMM Journals

Term / Journal	1969-1971	1971-1975	1975-1979	1979-1981	1982-1985	1986-1989	1990-1993	1994-1997	1998-2001	2002-2005
Mechanism and Machine Theory (Editor-in-Chief) (*)	John J. Uicker Jr. (USA)	John J. Uicker Jr. (USA)	John J. Uicker Jr. & Terry E. Shoup (USA)	Terry E. Shoup (USA)	Terry E. Shoup (USA)	Terry E. Shoup (USA)	Terry E. Shoup (USA)	Terry E. Shoup (USA)	Terry E. Shoup (USA)	Terry E. Shoup (USA) Andrés Kecskeméthy (Germany)
Journal of Applied Mechanics (Editor)									Started 2000	Nodar Davitashvili (Georgia)
Journal of Gearing and Transmissions (Editor)									Started in 2002	Veniamin I. Goldfarb (Russia)
Electronic Journal on Computational Kinematics (Editor)									Started in 2002	JeanPierre Merlet (France)

(*) MMT started in 1966 with E.R.J. Crossley (USA) as Editor; in 1972 it became IFToMM journal.

Table B3: Chairmen of the General Assembly Commissions (°)

Term / Commissions	1969-1971	1971-1975	1975-1979	1979-1981	1982-1985	1986-1989	1990-1993	1994-1997	1999-2003	2002-2005
Constitution		Established in 1975	E.R.J. Crossley (USA)	Jan Oderfeld (Poland)	Jan Oderfeld (Poland)	Bernard Roth (U.S.A.)	Bernard Roth (U.S.A.)	Giovanni Bianchi (Italy	Adam Morecki (Poland)	Jorge Angeles (Canada)
Nominating		A. Morecki (Poland)	E.R.J. Crossley (USA)	unknown	Leonard Maunder (UK)	Bernard Roth (USA)	Bernard Roth (USA)	Adam Morecki (Poland)	Jorge Angeles (Canada)	Jorge Angeles (Canada)
Honors and Awards								Started in 1998	Terry E. Shoup (USA)	Terry E. Shoup (USA)

(°): Data are taken from Minutes of the IFToMM Executive Council meetings and from information by current IFToMM officers

ON THE IFToMM PERMANENT COMMISSION FOR HISTORY OF MMS

Marco Ceccarelli (1) and Teun Koetsier (2)

(1) Laboratory of Robotics and Mechatronics
DiMSAT, University of Cassino, Via Di Biasio 43, 03043 Cassino (Fr), Italy
e-mail: ceccarelli@unicas.it

(2) Department of Mathematics,
Vrije Universiteit, De Boelelaan 1081, NL-1081HV Amsterdam, The Netherlands
e-mail: teun@cs.vu.nl

ABSTRACT- In this paper we have outlined the historical development of the IFToMM Permanent Commission for History of MMS (Mechanism and Machine Science) by also looking at the recently established field of History of MMS with technical perspectives. The activity of the PC for History of MMS has been overviewed thought facts and efforts of the many members over the time. Subjects of History of MMS have been presented by using published results of historical investigations with modern technical reformulation.

KEYWORDS: MMS, History of TMM, History of IFToMM, History of PC for History of MMS

INTRODUCTION

The history of a discipline is usually considered to be a part of that discipline. Why is that so? One could argue that the history of a discipline is a source of entertaining stories. That is correct, but there is more. The history of a discipline is also a source of information with respect to who invented what and when. It enables us to give the credit for an invention to the right person. Moreover, the history of a discipline teaches us a certain modesty; it shows that what is considered to be a perfect solution today could be completely outdated tomorrow. The history of a discipline is finally a source of pedagogical ideas and it may suggest other and better solutions to the problems of the present.

Yet, there is a more important reason for the fact that the history of a discipline is considered to be part of the discipline. The reason is that the history of a discipline concerns the identity of the discipline. It is characteristic of man that he can think about

11

12

his own existence. That is why inevitably human beings ask themselves one day or another: Who am I? Where am I from? This is done by individuals but by groups and institutions as well. People who do not know the identity of their parents in the course of time often start a search for their parents. Obviously often there are much more urgent problems than the question: Where am I from? Survival always has priority. Yet, the fact that the history of a discipline concerns the identity of the discipline explains why the history of the discipline is a part of the discipline.

In this way the history of IFToMM is part of the identity of IFToMM, The International Federation for the Promotion of Mechanism and Machine Science (MMS) and the History of the Theory of Mechanisms and History of the Science of Mechanisms and Machines is part of this Science. To the engineers participating in IFToMM or merely working in the Theory of Machines and Mechanisms it can be put in a very personal way: We can say to them: the History of MMS and the History of IFToMM concerns your identity. Moreover, there are very good reasons to be proud of what was done in MMS in the IFToMM member countries since the foundation of IFToMM. It deserves to be studied.

Essentially this is why the IFToMM Permanent Commission for History of MMS was established in 1973 by strong support of the first IFToMM President, and because of the great motivation of the first PC chairman Jack Philips.

Between 1973 and the present two periods can be distinguished in the existence of the Commission for History. The first period is between 1973 and 1998. In that period Jack Philips (1973-1981), Elisabeth Filemon (1982-1989) and Teun Koetsier (1990-1997) chaired the Commission. In the second period Marco Ceccarelli (1998-present) chaired the Commission. The Chairpersons are shown in Fig.1.

In the first period there was a slow growth in the activities of the Commission. In this first period of the Commission's existence, after 1983 meetings of the Commission were held at IFToMM World Congresses.

When Marco Ceccarelli took over the chair of the Commission and at the same time the new millennium was approaching it was decided that a whole new start should be made in order to turn the Commission into body functioning satisfactorily.

Fig.1: Chairmen of IFToMM Permanent Commission for History of MMS at HMM00 Symposium in Cassino in May 2000 (from left to right): Teun Koetsier (1990-'97), Jack Phillips (1973-'81), Elisabeth Filemon (1982-'89), and Marco Ceccarelli (1998-2005). (The years in the brackets indicate the term of Chair mandate).

THE FOUNDATION OF IFTOMM

In 1965 the first World Congress on the Theory of Machines and Mechanisms was organized in Varna in Bulgaria. At that congress the Bulgarian delegation proposed the establishment of an international federation for the Theory of Machines and Mechanisms. Four years later, in 1969, in Zakopane in Poland IFToMM, the International Federation for the Theory of Machines and Mechanisms was established, Figs.2 and 3.

Fig.2: I. I. Artobolewskij (second from the right), first president of IFToMM, together with. L. Maunder, G. Bianchi at his left and A. Bessonov at his right, at the founding congress in Zakopane, Poland in 1969. (Courtesy of Kurt Luck).

Fig.3: F. R. E. Crossley, first vice-president of IFToMM, addresses the founding congress of IFToMM in 1969 in Zakopane in Poland. (Courtesy of Kurt Luck).

In the middle of the cold war IFToMM represented a highly remarkable example of successful cooperation between the East and the West. The first president of IFToMM was I. I. Artobolewskij (USSR) and the first vice-president was F. R. E. Crossley

14

(USA). M. S. Konstantinov (Bulgaria) became the first secretary-general.
Since 1969 the field of the Theory of Machines and Mechanisms TMM has undergone a spectacular development into what we now prefer to call Mechanism and Machine Science (MMS). In this development IFToMM has been the major forum for international scientific cooperation in MMS.

THE EARLY HISTORY OF THE COMMISSION
The first period of the Commission is between 1973 and 1998. In that period Jack Philips (1973-1981), Elisabeth Filemon (1982-1989) and Teun Koetsier (1990-1997) chaired the Commission with a slow growth of the activities. In this first period of the Commission's existence after 1983 formal meetings of the Commission started to take place at IFToMM World Congresses.
In April, 26-29, 1983 Elisabeth Filemon organized the first IFToMM Symposium on the History of TMM in Miskolc, Hungary, Fig.4. Representatives from different countries lectured on the History of TMM in their country. The lectures were not published. At the same time the first formal meeting of the Commission was held, also in Miskolc. The Commission consisted at the time of few members as listed in the Appendix A.2.

Fig.4: The first IFToMM Symposium on the History of TMM in 1983 in Miskolc, Hungary: Branko Gligoric (Yugoslavia) next to Teun Koetsier (on the right).

At that time slowly sessions on the History of TMM started to be organized on IFToMM World Congresses.
At the Seventh World Congress in Sevilla, Spain 1987, there was a session on History, pp.1887-1898 of (IFToMM1987), with two lectures:
-    T. Koetsier: Some remarks on and around Burmester's work on the occasion of the centenary of his "Lehrbuch der Kinematik";
-    M. K. Uskov & A.A. Parkhomenko: History of machines and mechanisms theory in the work of Soviet scientists.
At the Eighth World Congress in Prague, Czechoslovakia 1991 there was an invited lecture on History of IFToMM that can be considered the first historical overview of IFToMM over time:
-    F. E. Crossley, The Early Days of IFToMM, pp. 3-8 of (IFToMM1991).
Moreover, at the same congress in Prague there was one full Session on History of

TMM, pp. 791-806 of (IFToMM1991), with the papers:
- T. Koetsier: A note on the early history of the Euler-Savary Theorem;
- T. Nakada, S. Nishijma: Applying Roller-Disc Integrator, the Ancient Chinese "South Pointing Chariot" can be converted to the Vehicle with Route-Log Recorder;
- Popescu: Des Mecanismes a Cames dans la Technique Populaire Roumaine;
- J. Muller & K. Mauersberger: CAM Mechanisms in Hydraulic Pumps - Technical-Historical Review.

At the Ninth World Congress on the Theory of Machines and Mechanisms 1995 in Milan there was again one full Session on History of TMM. Th following papers, (pp. 3187-3204 of [IFToMM1995]), were presented with an increased number of attendees:
- M. Ceccarelli: Screw Axis Defined by Giulio Mozzi in 1763;
- M. Cigola & M. Ceccarelli: On the Evolution of Mechanism Drawing;
- K. Luck: Some Remarks on the History and Future of the Theory of Mechanisms;
- Zeno Terplan: Die Auswirkung der industriellen Revolution auf die Maschinenbau-Wissenschaft in Ungarn.

In October 10, 1997 in Amsterdam T. Koetsier organized a small Symposium on "The Role of Mathematics in the Formation of Mechanical Engineering as an Independent Discipline" with the following presentations that were not published.:
- " The Role of Mathematics in Applied Mechanics Teaching to French Engineers in XIXth Century" by Bruno Belhoste (Paris);
- " On the Historical Development of Mechanisms Drawing" by Marco Ceccarelli (Cassino);
- " On the Mathematization of Kinematics of Mechanisms" by Teun Koetsier (Amsterdam);
- " Die Entwicklung des wissenschaftlichen Maschinenbaus in Deutschland im Spannungsfeld von visuellem Denken und mathematisvche Abstraktion" by Klaus Mauersberger (Dresden);
- " What is so Applied about Applied Mechanics?" by Gerard Alberts (Nijmegen)
- "On the Increasing Role of Fluid Mechanics in Engineering: J.M. Burgers and his Work on Pumps" by Fons Alkemade (Amsterdam)

Yet, although the interest in History of MMS within IFToMM was growing, the work of the Commission did not gain enough momentum. The members did not meet between IFToMM World Congresses. Plans suggested by the chair or during world congresses often did not work out.

DEVELOPMENTS IN 1998--2003

In January 1998 Marco Ceccarelli started his work by appointing as new members in the new Commission only three persons, namely Teun Koetsier (the Netherlands), last Andreas Dimarogonas (USA), and Hong-Sen Yan (China-Taipei). The idea was to have members, with great interest in History of MMS, although not necessarily experts on it, although those first three members have repute on History of MMS. But the idea was to have who members would be available to promote activity in the field of History of MMS. Indeed this first group was very active and rapidly the Commission received requests for additional nominations. Even past members have asked to be involved in the re-new activity of the Commission, like Elizabeth Filemon (Hungary) and the late

Joseph Duffy (USA). Thus, on July 1999 at IFToMM World Congress in Oulu the members were 23 and year by year the number of members has increased with more and more interested persons up to the current number of 48, as listed in Appendix A.4. The increased attention and interest on the PC for History of MMS has brought also an increase of activity so that it has been thought convenient to have sub- commissions for different activities. The chairmen (or women) of the sub-commissions are qualitate as the vice-chairmen of the PC for History of MMS.

Thus, in 2001 the following sub-commissions have been established and are currently formed as:

- Sub-Commission for Africa; aim: to promote and organize activity (publishing of papers, workshops, lectures, local meetings) in the field of History of MMS in African countries; members: not enough available members;

- Sub-Commission for Asia; aim: to promote and organize activity (publishing of papers, workshops, lectures, local meetings) in the field of History of MMS in Asian countries; members: Zhen (Vice Chairman), Rao, Seo, Yan, Ang;

- Sub-Commission for Oceania; aim: to promote and organize activity (publishing of papers, workshops, lectures, local meetings) in the field of History of MMS in Oceania countries; members: not enough available members;

- Sub-Commission for Europe; aim: to promote and organize activity (publishing of papers, workshops, lectures, local meetings) in the field of History of MMS in European countries; members: Cuadrado (Vice Chairman), Golovin, Havlik, Kerle, Popescu, Tolocka;

- Sub-Commission for America; aim: to promote and organize activity (publishing of papers, workshops, lectures, local meetings) in the field of History of MMS in American countries; members: Gosselin (Vice Chairman), Lopez-Cajùn, Mavroidis, Carvalho;

- Sub-Commission for IFToMM Archive; aim: to promote and facilitate the collection, and circulation of material and information in the field of TMM for the IFToMM Archive at CISM, Udine, Italy; members (a member will be the Chairman of the PC for History of MMS): Koetsier (Vice Chairman), Shoham, Golovin, Mavroidis, Kerle, Ceccarelli;

- Sub-Commission for the Web page; aim: to define, accept and update the contents of a Web Page on the History of MMS; members: Merlet (Vice Chairman), Cuadrado, Havlik, Kopey, Mavroidis;

- Sub-Commission for the HMM Symposium; aim: to promote, organize and hold HMM International Symposia on the History of Machines and Mechanisms ; members: (the Chairman of the PC for History of MMS chairs this Sub_Commision): Ceccarelli (Vice Chairman), Koetsier, Zhen, Rao, Yan, Takeno, Golovin, Lopez-Cajùn;

- Sub-Commission for Relations; aim: to promote, establish, and work cooperation with other IFToMM Commissions and other Institutions working on History of MMS; members: Yan (Vice Chairman), Zhou, Dai, Merlet, Carvalho.

The activity of the PC started in 1998 with small activities like lectures but main efforts were spent to promote interest and future works in subjects of the History of MMS.

First positive results have been obtained with the submission of 12 papers at the IFToMM World Congress 1999 in Oulu, Finland. There it was possible to have a full oral session on History of MMS and even an additional poster session with the accepted papers, as published in pp.37-85 of (IFToMM 1999). The oral session was well attended with more than 80 attendees, who have expressed great satisfaction for the presented papers. This positive result has been confirmed with the submission of 13 papers at the IFToMM World Congress 2003 in which we have had two sessions with the 12 accepted papers, (IFToMM2003).

Summarizing, in the period 1998-2003 the following important results have been obtained:

- having a Commission with several active members;
- reinvigorating interest History of MMS with papers in the IFToMM World Congress and in the IFToMM journal Mechanism and Machine Theory (MMT), and even elsewhere;
- re-establishing the IFToMM Archives, (IFToMM Archives);
- establishing the HMM Symposium Series on History of Machines and Mechanisms;
- intense lecture activity, even within the frame of small Workshops;
- a continuous investigation activity on History of MMS;
- periodical meetings and annual meetings of the Commission.

Several meetings have been organized at which the members could meet each other, exchange opinions and experiences, establish collaborations, and program further activities. Unfortunately it was impossible to have all the members always present, as shown in the photos of Figs.5, 6 and 7, due to the fact that most of the members serve on personal economic capability or in combination with programs of technical projects. This is because funds are not yet available in the IFToMM Community for working on History of MMS.

Details on the above-mentioned activity that has been carried out by members of the PC in the period 1998-2003 can be found in the PC reports that are available at IFToMM Archives in CISM, Udine, (IFToMM Archives 2003), but also in the PC web page (PC WebPages 2002).

Fig.5: Members of IFToMM Permanent Commission for History of MMS at the meeting 1999 in Oulu during the IFToMM World Congress: From left to right: Hanfried Kerle, Zhen Lu, Moshe Shoham, Bohdan Kopey, Lihua Zhou,Teun Koetsier, Marco Ceccarelli, Marek Kujath, Elisabeth Filemon, Joseph Duffy, Hong-Sen Yan and Jian S. Dai.

18

Fig.6: Members of IFToMM Permanent Commission for History of TMM at the meeting in Cassino on 12 May 2000 (from left to right): Carlos Lopez-Cajun, Marco Ceccarelli, Ignacio Cuadrado, Hanfried Kerle, Jean-Pierre Merlet, Elisabeth Filemon, Teun Koetsier, Yuri Soliterman, Manfred Husty, Jammi S. Rao, Alexander Golovin, Lu Zhen.

Fig.7: Members of IFToMM Permanent Commission for History of MMS at the meeting in Seoul on 21 May 2001 (from left to right): Moon Hwo Seo, Moshe Shoham, Joao Carlos Mendes Carvalho, Jae-Kyung Shim, Hsing-Hui Huang (representingHong-Sen Yan), Marco Ceccarelli, JeanPierre Merlet, Manfred Husty, Vicente Mata (representing Ignacio Cuadrado), Marcelo Ang.

The so-called IFToMM field of History of MMS (or TMM) has been established in technical sense by stating in the Introduction of HMM2000, (HMM2000): "the goal was to stimulate experts in TMM (MMS) with some feeling for history, people who can understand, appreciate, and refresh past works in TMM (MMS) to write a historical technical paper".

a)   *b)*

Fig.8: Members of IFToMM Permanent Commission for History of MMS at the meetings in 2002 (from left to right): a) at Admont Worshop in October 20-26: Librarian, *Manfred Husty, Marco Ceccarelli, Ignacio Cuadrado Teun Koetsier, Austrian attendee, Atsuo Takanishi, Hanfired Kerle; b) at ASME DETC02 in September 2002:* Jae-Kyung *Shim, Moshe Shoham, Manfred Husty, Marco Ceccarelli, Carlos Lopez-Cajùn, Clement Gosselin.*

Indeed, the History of MMS should be understood as an independent discipline both within MMS and the History of Science, with a specific character consisting in treating past technical developments and personalities with the aim to give historical track of Techniques but understand at the same time the technical details with an eye on modern applications. The IFToMM field of History of MMS differs from History of Science since the goal of History of MMS is to investigate in depth the technical details in order to refresh, reformulate, and reuse the knowledge and achievements of the past in modern terms.

Because of the above-mentioned identification of the field, History of MMS has attracted more and more attention, mainly in the IFToMM technical Community.

Thus, since 2000 History of MMS has been recognized as a suitable technical topic also for regular publication in the IFToMM Journal Mechanism and Machine Theory.

In the IFToMM World Congress 1999 in Oulu, lectures and papers have been presented in different subjects regarding also different periods as in the following topics, (IFToMM 1999): review of past works and formulations; historical developments of topics of MMS; ancient mechanism designs; past personalities and their works.

Again, in the IFToMM World Congress 2003 in Tianjin, lectures and papers have been presented (IFToMM 2003) as dealing with: past machines and mechanisms; people and their work; recent past mechanism designs; re-formulation of past theories; historical overviews.

In addition, a relevant impulse in the development of the field of History of MMS has been achieved through HMM2000, the International Symposium on History of Machines and Mechanisms. Because of its success it has established itself as a Conference series for presentation and discussion of more and more subjects belonging to the History of MMS. In fact, at HMM2000 interesting papers have been presented in the following topics, (HMM2000): history of IFToMM by past Presidents; past mechanical designs; ancient machines; people in TMM and their work; historical national overview; history of teaching; history of mechanical Engineering; automata and robots.

In many of the published papers the historical content has been organized in sections with historical information of a humanistic type on life and environment of past authors. But most of the content of these papers is devoted to the technical arguments that have been revised and even reformulated and discussed for modern interpretation and even practical current applications. Significant examples of this kind of papers are the papers (Ceccarelli and Vinciguerra 2000) and (Shoham 1999).

In (Ceccarelli and Vinciguerra 2000) the Chebyshev approach for designing approximate circle-tracing mechanisms has been reformulated into current terminology and mathematical means that have permitted even an extension of the design procedure to more general design problems by reducing the model constraints.

In (Shoham 1999) Clifford's formulation for bi-quaternion has been re-formulated in a modern form not only to show the current feasibility but even to deduce modern results through Clifford's derivation when bi-quaternions are used in their rotational sense rather then as ratio of two vectors.

Even TMM personalities have been investigated to stress their technical contributions, like in (Husty 2002) in which the life of prof. Walter Wunderlich is illustrated by looking at his achievements in Mechanism Kinematics.

Another interesting result of the increasing interest can be considered the study of past designs and machinery with attempts of interpretation and formulating design procedures that have been used for them in the past. Significant examples are the papers that are referenced as (Yan 1999) and (Bautista Paz et al. 2000).

In (Yan 1999) a legendary ancient Chinese walking machine has been investigated by presenting a possible design and discussing its functions though historical documents and a modern mechanism analysis.

In (Bautista Paz et al. 2000) the so-called dancing machine for water pumping up hill, is re-discovered as a mechanism design of Spanish Engineering of XVIth century by looking at historical technical documents that illustrate its design, construction, installation, and use. The mechanics of the complex mechanism was astonishing for that time and it is examined with a modern view to stress the current significance yet.

Strongly related to a technical historical interest can be considered the renewed attention for the collection of past models for basic mechanisms or scaled machinery, not only for museum purposes but even for technical teaching and investigation purposes. Since always, teachers of machinery have helped the understanding of mechanism design and operation by showing scaled models of mechanisms to the engineering students. The models were made of wood, iron, or bronze and recently they are made of plastic and light material alloy. Even rapid prototyping is used to make them. Usually, those models are built and used for one-teacher's own purposes, as one can find mechanism models in any office of University professors working in mechanisms. But, there are also collection of mechanisms that were built for the market and they are sometimes available in the Universities and even in some Science Museums. Relevant is the mechanism collection (800 mechanism models of which 300 were available also for the market) that was prepared by Franz Reuleaux and built by the firm Gustav Voigt MechanischeWerkstatt of Berlin in 1876. Part of this successful collection is still available in many Universities, like for examples in Dresden (Germany), in Ithaca (USA), and in Turin (Italy). Indeed, this collection has been also a stimulus for enlarging and completing the collection with new mechanisms, like for example in the Bauman Moscow University. Scaled models of mechanisms, built in the past, can be seen also in Science Museums, like for example at the British Museum in London and the Science Museum in Milan. Past models are shown in the hall of many Universities to attract interest and deserve memory of past technical achievements, like for example at the Technical University of Catalunya in Barcelona, the Technical University in Turin, and the Bauman Moscow University, as shown in Fig.9. Even recent past modern systems are shown as pieces of History of MMS, like for example in Tokyo at Waseda University and in Palo Alto at Stanford University where pioneer robots are shown in a proper corners of building entrance halls, as shown in Fig.10.

Another development can be considered the attention to the evolution of technical knowledge in the past with attempts to discover motivations, correct and incorrect reasoning that can be still of great importance for today's advances. Significant examples are the papers (Zhen and Xuan 2000) and (Ceccarelli 2000).

In (Zhen and Xuan 2000) the authors have studied water powered machines in ancient China in order to discuss properties, capabilities, and working principles for a correct understanding of those mechanisms that are an important example of the evolution of Chinese Technology of mechanisms.

In (Ceccarelli 2000) a survey is given of the History of Theory of Screw by looking at the early works since the XVth century. Giulio Mozzi is credited to have defined first in 1763 the screw axis and motions by treating its kinematics and dynamics in a rigorous mathematical form in his work (Mozzi 1763). Other fundamental works are revised to remark the evolution to a modern approach and to stress that beside Michel Chasles also Gaetano Giorgini in 1831 formulated in a modern way the Screw axis and its fundamental kinematics. The paper (Ceccarelli 2000) has stimulated somehow further attention on these early works, like for example the republishing of the papers by Chasles and Rodriguez that have been translated in (Baker and Parkin 2003) as example of cooperation of a PC-member, Ian A. Parkin, with another kinematician.

a)   b)

Fig.9: Past mechanism models shown at: a) Technical University in Turin; b) Bauman Moscow University.

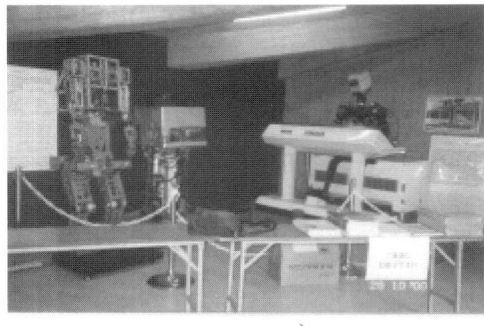 

a)   b)

Fig.10: Recent past modern systems shown as pieces of History of MMS at: a) Wabot robots of the 70's and 80's at Waseda University in Tokyo (Courtesy of Waseda University); b) Stanford arm of 70's at Stanford University in Palo Alto (Courtesy of Stanford University).

A novelty can be also considered the study of the very recent past with a historical view in order to fix the historical significance of recent results that are promising for future technical developments and achievements. This attention has been addressed both to technical fields and teaching aspects, as outlined for examples in the references (Bansevicius and Tolocka 2000) and (Crossley 1988).

In (Bansevicius and Tolocka 2000) the use of piezoactive materials is illustrated in the evolution of their application in mechanisms and in integrated machine design, by referring to national experiences in Lithuania.

In (Crossley 1988) Crossley has written his memories in teaching mechanisms since 1943 by illustrating conditions, subjects, and evolutions in the field of University curricula as he experienced personally. He also recognizes several personalities who met as visitors or in conferences or in meetings for organizing events and Journal of Mechanisms (now MMT). He also gives his view of how IFToMM was started being one of the main promoters together with Artobolevski and Konstantinov.

The recent past has been also studied by evolving one-view interpretation or memory to exhaustive examination. Regarding this aspect, of relevant significance is the overview of the History of IFToMM that has been attempted sometimes by singular personal views. Relevant are the cases by Ereskin F. Crossely and Adam Morecki in keynote speeches of IFToMM World Congresses in 1991 and 1999 respectively, that are reported in pp.3-8 of (IFToMM 1991) and in pp.30-36 of (IFToMM 1999). Crossley has described the activity in the first meetings in which IFToMM was founded by recognizing the founding fathers. Morecki has presented general remarks concerning early days of IFToMM and development of its activity and tracks the near future of IFToMM. But then, History of IFToMM has been outlined and discussed in its many aspects by all the IFToMM Past Presidents in a common frame for a specific Chapter of the HMMM2000 Proceedings, (HMM2000). Indeed, the participation of the Past IFToMM Presidents at HMM2000 can be recognized as recognition of the established significance of both History of MMS and History of IFToMM in the IFToMM

Community yet.

The study of several historical aspects has also stimulated attempts of overviews of the evolution of Mechanical Engineering at large. Interesting examples can be considered the papers (Koetsier 2000) and (Ceccarelli 2001). In (Koetsier 2000) the author has presented an outline of the pre-twentieth history of investigations on the Kinematics of machines in order to identify the main developments but the identity of the modern Mechanism and Machine Science as well. In (Ceccarelli 2001) the historical developments of MMS are overviewed by illustrating and discussing through few very significant examples the basic concepts and facts that have enhanced Mechanism design over the time.

The above-mentioned outline of the evolution of interests in the newly established field of History of MMS has been limited to few main aspects that the authors consider significant to characterize the work done by the PC for History of MMS but even to stimulate further activity. Therefore, the references have been restricted to the few examples that have been published within the frame of IFToMM or even more specifically within the sphere of activity of the PC for History of MMS.

FUTURE ACTIVITY AND CHALLENGES FOR THE COMMISSION

What have we learned from the past for the future of the PC and the field of History of MMS?

The answer to this question can be argued from different viewpoints and even with different motivations from person to person. Indeed, this is because the History of MMS can be even be viewed as a personal background that can motivate awareness, activity, and future actions not only in the field of History of MMS.

In particular, the future activity for the PC for History of MMS can be outlined as an extension and improvement of the past and current results.

A very important aspect is clearly the need for a wider awareness of the past, in the IFToMM Community both in terms of technical aspects and IFToMM History.

Specific activity for the PC can be expected:
- to manitain an active Commission with members from all over the world
- to solicit papers on History of MMS for Conferences and Journals
- to circulate information on the Commission
- to exchange opinions among the members of the Commission, mainly by e-mail
- to organize lectures and/or workshops on History of MMS
- to establish collaboration for historical investigations
- to maintain the IFToMM Archive
- to re-publish and re-consider past works and theories on MMS.

The future developments of the activity of the PC for History of MMS can be considered as a challenge since the great specialization of expertise in technical fields allows less and less interest or availability for the History of MMS.

CONCLUSIONS

In every discipline its history, dealing the maintenance of awareness and knowledge of past activity, can be recognized as a part of that discipline, the aim being not only to keep memory of past achievements, efforts, and personalities alive, but also to guide future developments.. These goals have been recognized as fundamental in the IFToMM Community since the beginning of its activity. In fact, a Permanent Commission for History of MMS (Mechanism and Machine Science) has been established in 1973 to serve as the historical memory of IFToMM but also to work for tracking the technical developments of MMS over the time. Indeed, only recently the PC for History of MMS has achieved its maturity with a promising future through a well-organized activity of several representatives of the IFToMM members, after having had a quite long starting process. The growing interest and participation in the activity of the PC are strongly related to the technical study and re-consideration of past achievements and personalities in the field of MMS and Mechanical Engineering at large. Results in terms of republishing and research papers can and will be obtained not only in the IFToMM Community by so-called Historicians (that is coined as fusion of the words Historian and Kinematician). They will address attention mainly to technical contents and developments in past activity in fields that are related or included in MMS.

REFERENCES

Baker J.E. and Parkin I.A. 2003, Fundamentals of Screw Motion: Seminal papers by Michel Chasles and Olinde Rodrigues (A Translation), The University of Sydney.

Bansevicius R. and Tolocka R.T. 2000, Modern history of piezomechanics in Lithuania, HMM2000, pp.225-231.

Bautista Paz et al. 2000, The dancing machine: the secret of raising water uphill, HMM2000, pp.119-125.

Ceccarelli M., 2000, "Screw Axis defined by Giulio Mozzi in 1763 and Early Studies on Helicoidal Motion", Mechanism and Machine Theory, Vol.35, pp.761-770.

Ceccarelli M., Vinciguerra A. 2000, Approximate Four-Bar Circle-Tracing Mechanisms: Classical and New Synthesis, Mechanism and Machine Theory, 2000, Vol.35, n.11, pp.1579-1599.

Ceccarelli M. 2001, The Challenges for Machine and Mechanism Design at the Beginning of the Third Millenium as Viewed from the Past, Proceedings of COBEM2001 Brazilian Congress on Mechanical Engineering, Uberlandia, 2001, Invited Lecture, Vol.20, pp.132-151.

Crossley F.E. 1988, Recollection from forty years of teaching mechanisms, ASME Jnl. of Mechanisms, Trans., and Auto. In Design, Vol.110, pp.232-242

Koetsier T. 2000, Mechanism and Machine Science: its history and its identity, HMM2000, pp.5-24.

HMM2000, Proceedings of HMM2000 International Symposium on History of Machines and Mechanisms, Kluwer Academic Press, Dordrecht, 2000.

Husty M. 2002, prof. Wunderlich and his work, CD Presentation Volume of Admont Workshop on History of Mechanical Engineering, IFToMM Archives, Udine.

IFToMM 1987, Proceedings of 7th IFToMM World Congress on the Theory of Machines and Mechanisms, Bautista E. et al.. (Eds.), Seville, September 17-22, 1987, Pergamon Press, Oxford.

IFToMM 1991, Proceedings of 8th IFToMM World Congress on the Theory of Machines and Mechanisms, M.Okroulik and L. Pust (Eds.), Prague, August 26-31, 1991, Society of Czechoslovak Mathematicians and Physicians, Prague.

IFToMM 1995, Proceedings of 9th IFToMM World Congress on the Theory of

Machines and Mechanisms, Rovetta A. (Ed.), August 29 - September 2, 1995, Politecnico di Milano, Milan.

IFToMM 1999, Proceedings of 10-th IFToMM World Congress on the Theory of Machines and Mechanisms, Leinonen T. (Ed.), June 20-24, 1999, University of Oulu. Oulu.

IFToMM 2003, Proceedings of 11-th IFToMM World Congress on the Theory of Machines and Mechanisms, Huan T. (Ed.), April 1-4, 2004, University of Tianjin, Tianjin.

IFToMM Archive 2003, IFToMM documents over time, Archives at CISM, Udine.

Mozzi G. , 1763, "Discorso matematico sopra il rotamento momentaneo dei corpi", Stamp. di Donato Campo, Napoli.

PC Webpage 2003, http://www-sop.inria.fr/coprin/IFTOMM/History/.

Shoham M. 1999, A note on Clifford's derivation of Bi-Quaternions, IFToMM 1999, pp.43-47.

Yan H.-S. 1999, A design of ancient China's cattle machine Chu-Ko Liang's wooden ox and glinding horse, IFToMM1999, pp.57-62.

Zhen L. and Xuan G. 2002, The development of water-powered machines of China in 10-14-the century, HMM2000, pp.129-134.

APPENDICES

A.1 LIST OF MEMBERS IN 1973

List not available in official IFToMM documents.

A.2 LIST OF MEMBERS ON APRIL 1983 BY ELIZABETH FILEMON

A.P. Bessonov, USSR; G. Dizioglu, Germany; G.F.R., J. Duffy, U.S.A.,; E. Filemon, Hungary; B. Gligoric, Yugoslavia; P. Genova, Bulgaria; A. Jakubovich, Poland; K. Luck, G. D. R.; J. Muller, GDR; J. Philips, Australia; J. Rees-Jones, UK; A. Vinciguerra, Italy.

A.3 LIST OF MEMBERS ON 1997 BY TEUN KOETSIER

A.P. Bessonov, USSR; Fl. Dudita, Romania; G. Dizioglu, G.F.R.; J. Duffy, U.S.A.; E. Filemon, Hungary; B. Gligoric, Yugoslavia; P. Genova, Bulgaria; A. Jakubovich, Poland; T. Koetsier, The Netherlands (Chair); K. Luck, G. D. R; J. Muller, GDR; J. Philips, Australia; A. Vinciguerra, Italy.

A.4 LIST OF MEMBERS ON JULY 2003 BY MARCO CECCARELLI

Juan Ignacio Cuadrado Iglesias (Vice Chairman for Europe), Spain; Clement M. Gosselin (Vice Chairman for America), Canada; Jean Pierre Merlet (Vice Chairman for Web page), France; Teun Koetsier (Vice Chairman for IFToMM Archive), The Netherland; Zhen Lu (Vice Chairman for Asia), China-Bejing; Hong-Sen Yan (Vice Chairman for Relations), China-Taipei; Mario Acevedo, Mexico; Jorge A. C. Ambrosio (observer), Portugal; Marcelo H. Ang Jr, Singapore; Roque Calero Perez, Spain; Thomas G. Chondros, Greece; Jian S. Dai, U.K.; G. Reg Dunlop (observer), New Zealand; Elisabeth Filemon, Hungary; John A. Gal , Australia; Alexander A. Golovin, Russia; Sergey Jatsun, Russia; Stefan Havlik, Slovakia; Manfred Husty, Austria; Hanfried Kerle, Germany; Yaroslav T. Kinitsky, Ukraina; José Ignacio López Soria (observer), Perù; Franz Otto Kopp, (obsrever)Germany; Bohdan Kopey, Ukraine; Marek Kujath, Canada; Carlos S. Lopez Cajùn, Mexico; Francis Moon (obsrever), U.S.A.; Klaus Mauersberger , Germany; Constantinos Mavroidis, U.S.A.; João Carlos Mendes Carvalho, Brazil; Agamenon R.E. Oliveira, Brasil; Ben Fathi Ouezdou, France; Evangelos Papadopoulos, Greece; Ian Parkin, Australia; Iulian Popescu, Romania; Jammi S. Rao , India; Bahram Ravani, U. S. A.; Alberto Rovetta, Italy; Moon Hwo Seo, Korea; Moshe Shoham, Israel; Jae-Kyung Shim, Korea; Yuri Soliterman, Republic of Belarus; Atsuo Takanishi, Japan; Junichi Takeno, Japan; Rymmantas Tadas Tolocka, Lithuania; Ching-Huan Tseng, China-Taipei; Janusz Wawrzecki, Poland; Teresa Zielinska, Poland; Lihua Zhou, China-Bejing.

PAST, PRESENT AND FUTURE IN MECHANISM AND MACHINE SCIENCE TERMINOLOGY

Theodor Ionescu, Romanian State Passenger Railways, Department of Technical Documentation,
Bd. Dinicu Golescu 38, 010873, Bucharest 1, Romania
E-mail: theodor.ionescu@cfr.ro
Gerhard Bögelsack, Technische Universität Ilmenau, Germany
E-mail: gerhard.boegelsack@mb.tu-ilmenau.de
Tatu Leinonen, University of Oulu, Finland
P.O. Box 4200, FIN-90014 University of Oulu,Finland
E-mail: tatu@me.oulu.fi

ABSTRACT – The paper deals with the historical and methodological aspects of the MMS terminology work done within IFToMM for over 32 years. It provides an overview beginning with the early days, pinpointing the past and present achievements and suggesting tasks for the future.

KEYWORDS: Machines and Mechanisms, Terminology, History, Methodology

INTRODUCTION

IFToMM, formerly the *International Federation for the Theory of Machines and Mechanisms*, at present the *International Federation for the Promotion of Mechanism and Machine Science*, is celebrating its 35th anniversary in 2004.

The Federation, formally founded on 27 September 1969 at the 2nd *International Congress for the Theory of Machines and Mechanisms*, in Zakopane/PL, is the world's premier scientific and engineering body in the field of – Theory of Machines and Mechanisms (TMM) / Mechanism and Machine Science (MMS). Australia, Bulgaria, Czechoslovakia, German Democratic Republic (GDR), German Federal Republic (GFR), Hungary, India, Italy, The Netherlands, Norway, Poland, Romania, UK, USA, URSS and Yugoslavia were represented at the Inaugural Assembly. IFToMM National or Territorial Member Committees have since been established in 45 other countries.

The work of the Federation is conducted by Permanent Commissions and Technical Committees; the Commissions for Standardization of Terminology, Education, Communications, History of TMM, and Publications; the Committees for Computational Kinematics, Gearing, Linkages and Cams, Human – Machine Systems, Mechatronics, Micromechanisms (later Micromachines and nowadays Micro – and Nanomechanisms), Nonlinear Oscillations, Reliability, Robotics, Rotordynamics, and Transportation Machinery. [1]

The official journal of IFToMM is *Mechanism and Machine Theory*, the leading journal in the field, published initially by Pergamon Press and now by Elsevier Science Ltd. It recently celebrated its 35th anniversary. [2]
More information about the history of IFToMM can be found in [3].

ESTABLISHMENT OF THE COMMISSION FOR STANDARDIZATION OF TERMINOLOGY

The constituent meeting of the commission for "Standardization of Terminology" was held on 18 September 1971, during the 3rd World Congress on the Theory of Machines and Mechanisms in Kupari, Yugoslavia. There were five participants: Professors Baticle (France), Bazjanac (Yugoslavia), Bögelsack (GDR), Davies (United Kingdom) and Keller (GFR).

As Professors D. Muster (USA) and N. I. Levitskyi, prenominated by the Executive Council as chairman and vice-chairman, respectively, of the Commission did not attend the Congress, the constitutive meeting was chaired by Prof. T. Davies. Initial lists of terms were already submitted by United Kingdom and GDR during the meeting, and a provisional program with responsibilities and a set of rules were established [4].

From the very beginning, the objectives of the commission were to establish a specific and unitary terminology for MMS. In the past, several national and international groups had successfully endeavoured to compile dictionaries and glossaries in this field. Some related publications can be mentioned as examples: In [5] are listed 90 terms in Russian, English, French and German, but defined in the Russian language only. A preceding academic bulletin was published in 1938. The German glossary [6] contains 221 terms and definitions illustrated by sketches and drawings. Dictionary [7] includes 610 terms in German, English, French, Russian and Bulgarian without definitions.

These very useful results had to be compiled and complemented significantly. Above all, definitions should appear in all of the four official IFToMM languages. The IFToMM Constitution & By Laws stipulates: *"The Permanent Commission for the Standardization of Terminology has to write a standard terminology (with definitions) for mechanism and machine science, including terms pertaining to the realization of motion in machines and its control, associated problems in dynamics and kinematics, and in machine components; to cooperate with other international and national bodies concerned with terminology and related matters.*

METHODOLOGY

In view of the demand that a well-made definition should distinguish by identifying and identify by distinguishing, the commission reached an agreement on the following rules to be observed in the methodology of defining:

- In each context, it must be possible to replace the term to be defined (*definiendum*) by the definition (*definiens*).
- A definition may neither contain nor cause logic contradiction.
- The term to be defined may not appear in the definition either openly or as concealed (circular definition).
- The predicate of a definition should not be negative.
- *Definiendum* and *definiens* must be identical in extent.

- A term should be neither overdefined (more characteristics in the definition than in the term) nor underdefined.

Some more guidelines were proposed by Prof. J. M. Prentis [8]:

- Terms should be elegantly defined in the simplest possible language.
- Definitions should be concise.
- Terms should not be needlessly multiplied, e.g. (common adjective) + (old term) = (new term)
- Terms should not be included (or, even worse, invented) simply to provide a counter-point to other terms.
- A term that is easier to understand than the definition should be deleted unless a simpler definition can be found.
- When in doubt, leave it out!

ON THE HISTORY OF THE COMMISSION, 1971 – 1997

As a rule, Commission working meetings have been organized, with very rare exceptions, every two years:

1971	Kupari/YU	1980	Miskolc/HU	1990	Oulu/SF	2000	Kaunas/LT
1973	Nieborów/PL	1982	Cambridge/GB	1992	Poitiers/FR	2002	Mezötur/HU
1975	Warszawa/PL	1984	Brasov/RO	1994	Delft/NL		
1976	Oberhof/GDR	1986	Siegen/GFR	1996	Warszawa/PL		
1978	Genova/IT	1988	Niš/YU	1998	Brno/SK		
		1989	Smolenice/CSSR				

The next meeting is scheduled for 2004 in Kosice, Slovakia. Additional brief meeting were held during the following World Congresses: 1987 Seville / ES, 1991 Praha / CSSR, 1995 Milan / IT, 1999 Oulu / FI and SYROM Conferences: 1977, 1981, 1985, 1997, 2001 Bucuresti / RO.

The chairmen of the commission have been D. Muster (1972 / 1976), G. Bögelsack (1976 / 1986), J. Prentis (1986 / 1990), T. Leinonen (1990 / 1998), T. Ionescu (since 1998).

Initially, a master version in English was set up and published in the IFToMM Journal, Mechanism and Machine Theory Vol. 18, No. 6, 1983 by a team chaired by Prof. G. Bögelsack. [9] Useful comments were received and incorporated into the subsequent draft. A four–language glossary in English, French, German and Russian was then published in the same journal, Vol. 26, No. 5, 1991 by a team chaired by Prof. T. Leinonen. [10] It contained 763 terms and their definitions and was divided into the following chapters and subchapters:

- Structure of Machines and Mechanisms (Components, Sub-assemblies; Mechanisms)
- Kinematics (General; Motion; Kinematic Geometry)

- Dynamics (General; Force and Moment; Momentum, Energy, Work and Power, Principles; Structural Behaviour and Characteristics; Structural Concepts; Dynamical Concepts; Dynamical Systems and Characteristics; Vibrations)
- Machine Control and Measurements (Signals and Functions; Accuracy and Errors; Devices and Components)
- Robotics (Systems; Components; Motion; Control; Miscellaneous)
- Appendix (General Terms Used in TMM).

As a continuation, a standard set of "Abbreviations / Symbols for Terms in TMM" and "Graphical Symbols for the Representation of Mechanisms" were finalized and published in MMT Journal Vol. 32, No. 6, 1997. [11]

1998 – 2003

The activity on the enhancement of MMS terminology continued in the years 1998 – 2003 with the intention to publish a new edition in the early 2000. Following a declaration of intentions and two invitations made to the IFToMM community at large to contribute to the forthcoming issue, the Commission worked on two dimensions: updating and enriching the available chapters as well as adding new specialized sections, such as non – linear oscillations, biomechanisms, mechatronics, rotordynamics, gearing, and stability with the anticipated science and technology developments. The entire activity aimed at:

- facilitating the adoption of new terms and their insertion among the existing ones, with minimum reworking of numbering and indexes;
- permitting independent, modular enhancement of any section/chapter of terminology;
- increasing reader accessibility to terms and definitions;
- making our terminology more user-friendly and usable for extensive application.

Beginning with the 18th Commission meeting (Brno, 1998), the work on terminology expanded, some working principles and new tasks were agreed on, and the bases of the first cooperative actions with the TC for "Nonlinear Oscillations" were established. Important decisions were made at the 19th Commission meeting (Kaunas, 2000) concerning the new edition of TMM terminology, the basic character of our terminology, the need for a proper working instrument, the voluntary nature of our activity and its (individual) real costs, the personnel politics (considering meeting participation, effective interest in activity and the new related scientific fields), as well as the aspects of cooperation with IFToMM TCs and PCs. A new project of cooperation, with the "Gearing" TC was established. We also decided upon the premises of adopting a new, more flexible and reliable system of organizing terms.

At the 20th Commission meeting (Mezötur, 2002), the draft of the new printed version of the MMS terminology was completed.

Subsequently, this was approved by the IFToMM Executive Council at its annual meeting in 2002 (Udine), and eventually, the 3rd edition of the MMS terminology appeared in the Mechanism and Machine Theory journal Vol. 38, Nos. 7 – 10, 2003. This present edition contains [12]:

- a "basic core" of 771 terms and definitions in the English, French, German, Russian languages obtained by revising / updating, completing (with System & Model and Robotics terms) and partly rearranging the 1990 edition. There are seven chapters: Generalities, Structure of Machines and Mechanisms, Kinematics, Dynamics, Machine Control and Measurements, Robotics, General Terms used in MMS, which are partly consistent with the natural structure and divisions of MMS;
- a "supplement" of 823 terms and definitions in English was obtained by developing the existing and new MMS subdomains as Dynamics, Rotordynamics & Measurement, Vibrations & Oscillations, Stability, Biomechanics, Gearing, Mechatronics. There are seven new chapters, alphabetically arranged to facilitate future development. These chapters express the actual trends and challenges in MMS as well as IFToMM Technical Committee's preoccupations.

The entire documentation is an outcome of dedicated personal and team activity by the following persons: Prof. A. J. Klein Breteler, Prof. C. W. Stammers (responsible for the English language), Prof. J. P. Lallemand (responsible for the French language), Prof. G. Bögelsack (responsible for the German language), Prof. V. Plakhtin (responsible for the Russian language), Dr. T. Ionescu, Prof. T. Leinonen, Prof. A. Morecki (†), Prof. I. Biro, Prof. J. Novotny, Dr. V. Oravsky, Prof. D. Re mond, Prof. I. Salyi(†), Prof. S. Segla, Prof. T. E. Shoup, Dr. H. Socha, Prof. V. E. Starzhinsky, Prof. R. T. Tolocka. Many specialists inside and outside IFToMM were also consulted and contributed to this terminology. The fruitful cooperation with the IFToMM Technical Committees for Nonlinear Oscillations and for Gearing should be mentioned. Now that the document is made available to the MMS community at large for extensive instrumental use, on the 30th anniversary of IFToMM Permanent Commission for Standardization of Terminology, we hope that this work will become a useful stage and reference for future terminological developments as well.

Based on re-organization, eight new subcommissions start operating today within our PC: "Structure and Kinematics", "Dynamics", "Robotics, Machine Control and Measurements", "Biomechanisms", "Mechatronics", "Gearing", "Micro – and Nanomechanisms ", "Electronic Dictionary".

AIMS AND TASKS FOR THE FUTURE

Some important directions of activity are:

- translation of the new chapters (terms & definitions) of the 3rd edition from English into French, German, Russian;
- continuous enhancement, enrichment and updating of the MMS terminology; inclusion of new related terminological domains;
- elaboration of an alphabetical, data-based and computer-aided terminology ("Electronic Dictionary ");
- consideration of CD – ROM and/or the IFToMM Internet site as preferred media for the future edition of the terminology;
- creating a web site of our PC and linking it to (the main) IFToMM site;

- amplification of exchange and cooperation on terminological matters with IFToMM Technical Committees and Permanent Commission and with other qualified institutions;
- enrolment of new active members for the quantitative and qualitative growth of the Commission's capacity and work.

CONCLUSIONS

As far as the basic character of our terminology is concerned, a choice between a general version for widespread use and a strictly limited, specific version must be made. Although there has been some recent criticism of the present (and future) development of our terminology, which takes over terms from other sciences and technical fields, the criterion of utility must be respected. It is our belief that most professionals related to and working in the MMS discipline do need proper reference documentation covering, in terms and definitions, every notion and conceptual category in use. If necessary, a differentiation mark between "own" and "borrowed" terms can be included and introduced in the next edition.

Regarding the need for a proper working instrument, the Commission must be able to focus and develope its activity in a continuous and efficient way and with good results. This implies organizational and structural measures promoting creative work, an activity plan with tasks for each subcommission and for every member, responsibilities, schedules, cooperation, etc. Being voluntary in character, our project is entirely based on the members' goodwill and commitment, and moreover, this activity costs the members both time and money. Obviously, the consequences affect the Commission's power and outcomes.

It is true that the last few years were marked by a decline in the "personnel" domain due to political and institutional changes at international and national macro levels, natural causes, retirement and lack of interest. At the same time, we tried to counteract this phenomenon and to involve in the Commission's activities competent professionals from within and outside IFToMM. New members were proposed and nominated.

As the TMM terminology must express globally our federation's concerns and keep in touch with the latest developments and achievements in science and technology (see also the discussions on TMM 21 topic), it seems evident, on the one hand, to ask for a full cooperation with the IFToMM TCs and PCs (which dealing with specific problems, use and develop specific terminologies). On the other hand, a timing to suit the actual circumstances and enlargement of our Commission are inevitable in order to cope with the present and future terminological tasks. In this views, delegated representatives of the IFToMM TCs / PCs are most welcome to our Commission.

The current members of the Commission are: Prof. Paun Antonescu, Prof. István Biro, Prof. Gerhard Bögelsack, Prof. Mikio Horie, Prof. A. J. Klein Breteler, Dr. Theodor Ionescu, Prof. Tatu Leinonen, Prof. Dmitry N. Levitsky, Prof. Jen-Yu Liu, Prof. Jiri Novotny, Dr. Vladimir Oravsky, Prof. Vladimir Plakhtin, Prof. Didier Remond, Prof. Aldo Rossi, Prof. Yuri L. Sarkissyan, Prof. Stefan Segl'a, Prof. Terry E. Shoup, Prof. Charles W. Stammers, Prof. Victor E. Starzhinsky, Prof. Rymantas T. Tolocka, Prof. Lan Zhaohui, Prof. Zivota Zivkovic.

The MMS terminology is historical in character, similarly to MMS itself; it reflects a certain stage of existence and technical & scientific development. But terminology must be also viewed as a science, "La terminologie est la science des terms: elle etudie leurs caracteristiques, notamment les lois linguistiques et socio – psychologiques de leur naissance et de leur evolution", with a marked historical pattern. [13] However, a stable equilibrium of *Past, Present and Future* on the cycle of evolution is characteristic of all worthwhile terminological activities.

Wishing to pay homage to all those professionals involved, with mind and heart, in MMS terminology, we express here our firm belief that IFToMM PC on "Standardization of Terminology" (also known as Commission A) will continue its highly appreciated activity to address and serve the engineering and scientific world audience. [14]

REFERENCES

1. Bianchi, G.and Pust, L., Twenty years of IFToMM, Mechanism and Machine Theory, Vol. 25, No. 1, pp II – IV, 1990.
2. Shoup, T. E., Happy Birthday Mechanism and Machine Theory and Happy Birthday IFToMM, Mechanism and Machine Theory, Vol. 25, No. 1, pp. I, 1990.
3. Morecki, A., Past, Present and Future of IFToMM, Mechanism and Machine Theory, Vol. 30, No. 1, pp. 1 – 9, 1995.
4. Bögelsack, G. Twenty – five Years IFToMM Commission A "Standardization of Terminology" – History, Methodology, Results and Future Work, Mechanisms and Machine Theory, Vol. 33, No. 1/2, pp. 1 – 5, 1998.
5. Levitskyi, N. I. et al. Teorija Mechanizmow - Terminologija. Isdatelstwo Nauka Moskwa 1964.
6. VDI Richtlinie 2127 Getriebetechnische Grundlagen - Begriffsbestimmungen der Getriebe. VDI Verlag Düsseldorf 1962.
7. Konstantinov, M.S., Artobolewskyi, I.I., Hartenberg, R.S. et al. Concise Terminological Dictionary on Kinematics and Dynamics of Machines. Sofija 1965
8. Prentis, J.M., Guidelines for Terms and Definitions. Protocol Minutes of the Eleventh Meeting of IFToMM Commission A, Smolenice 1989.
9. IFToMM Commission A, Terminology for the Theory of Machines and Mechanisms, Mechanism and Machine Theory, Vol. 18, No. 6, pp. 379 – 407, 1983.
10. IFToMM Commission A, Terminology for the Theory of Machines and Mechanisms, Mechanism and Machine Theory, Vol. 26, No. 5, pp. 435 – 539, 1991.
11. IFToMM Commission A, Abbreviations / Symbols for Terms in TMM, Mechanism and Machine Theory, Vol. 32, No. 6, pp. 641 – 666, 1997.
12. IFToMM Commission A, Terminology for the Mechanism and Machine Science, Mechanism and Machine Theory,Vol. 38, Nos. 7-10, pp.598 – 1111, 2003.
13. Hermans, A., La Definition des Terms Scientifiques, Centre de Terminologie de Bruxelles, Meta XXXIV, 3, 1989.
14. Ionescu, T., Past, Present and Future in TMM Terminology, Proceedings of the Scientific Seminar of the 19th Meeting of IFToMM PC on "Standardization of Terminology", pp. 7 – 12, Kaunas, 2000.

DEVELOPMENT AND ACTIVITIES OF THE IFToMM ROTORDYNAMIC COMMITTEE, 1977-2004

Neville F. Rieger, Chief Scientist
(Chairman of the IFToMM TC for Rotordynamics)
STI Technologies, Inc.
1800 Brighton-Henrietta TL Rd.
Rochester, NY 14623, USA
Nrieger@sti-tech.com

Abstract

This paper contains a summary of the objectives and achievements of the IFToMM Rotordynamics Committee since its inception in 1977 to the present. Future plans and activities are also mentioned.

Introduction

The IFToMM Sub-Committee on Rotor Dynamics (or the RC) grew from a suggestion by Professor J.S. Rao of IIT, New Delhi, and Professor Z. Parziewski of the University of Lodz, Poland, at the IFToMM Congress at Lodz, in 1975. The creation of a Sub-Committee on Rotordynamics was subsequently proposed to the EC by Prof. Giovanni Bianchi of the Polytechnico di Milano. Professor Zizlaw Parziewski was elected Foundation Chairman of this Committee. One of the earliest activities of the Committee was to organize an international meeting to discuss and present new activities in rotor-bearing dynamics. It was decided that the Committee would sponsor such a meeting once every four years, in a major city of an IFToMM member country. The following are the recollections of the author of this paper. He has been a member of the RC since 1979, and is currently its Chairman.

It is the policy of the RC to:

a. Promote the science and practice of rotor dynamic knowledge through the coordination and dissemination of knowledge of this subject.

b. To make and to actively seek a membership which truly is international in its representation, so that the Committee may serve as THE International Society for Rotor Dynamics, worldwide

c. To hold the International Conference on Rotordynamics in a different country on each occasion so that knowledge may be disseminated widely and more generally.

The main activity of the Committee is the organization of the International Conference on Rotordynamics. This meeting is held once every four years at a major city in a member country. To date there have been six International Conferences since 1982. Committee meetings are held every two years, at the IFToMM Rotordynamics Conference and at the IMechE Vibrations of Rotating Machinery Conference. There are now 42 member countries and 50 delegates (including six delegates-elect). The Committee is served by an elected Chairperson and an elected Secretary.

First International Conference on Rotordynamics, Rome 1982

The major event of this early period was preparation for the First International Conference on Rotordynamics, which was held in Rome in August 1982, under the chairmanship of Prof. Bianchi. About 30 papers were presented and about 40 delegates from Italy, the U.K., Switzerland, France, Germany, Japan and the U.S.A. were present. The Rotordynamics Committee held its first official meeting at this conference. Prof J S Rao was appointed the incoming Chairman of the Committee. It was also decided that the Second International Conference on Rotordynamics would be held in Tokyo, Japan, under the Chairmanship of Prof Yukio Hori, of Tokyo University The meeting in Rome was held in difficult economic times worldwide, which hurt the attendance somewhat.

IFToMM International Congress, New Delhi, 1983

A special session on problems of turbomachinery blading was held at the IFToMM Congress in New Delhi, in December 1983. A special volume of proceedings on blade problems was published and printed (separately) by STI. For this meeting. Chairman J.S. Rao of I.I.T, New Delhi, organized and chaired this session.

Tokyo, 1986

The second International Conference on Rotordynamics was organized by Chairman J S Rao and was held in Tokyo in 1986. About 50 delegates attended this meeting and 36 papers were presented. Professor Hori served as Chairman of the Conference. Lyon, France, was selected as the site for the Third Rotordynamics Conference, to be held in 1990. Prof Michel Lalanne was selected as the Third Conference Chairman.

IFToMM International Congress, Seville, 1987

A special session on Rotordynamics was held at the IFToMM Congress in Seville, in 1987. Rotodynamics papers on several topics (shaft dynamics, bearings, blades, disks) were presented. A special volume of Proceedings was published for this session, These Proceedings were printed (separately) by STI.

A meeting of the RC was held, and Chairman J S Rao called for a meeting of the RC in Lyon in September, 1988, to coordinate committee activities for the Lyon Conference in 1990.

Lyon, 1990

The Third International Conference on Rotordynamics was held in Lyon, France in 1990, with Professor Michel Lalanne of INSA serving as Conference Chairman. A total of 60 papers were presented and about 60 delegates attended from 25 countries. A meeting of the RC was held at this Conference. Official records of Committee activities date from this meeting. Professor Yukio Hori of the University of Tokyo was appointed as Committee Chairman, and Dr. N.F. Rieger of STI was appointed as Secretary.

It was decided that the Fourth Conference On Rotodynamics would be held in Chicago in 1994, and that Drs. N.F. Rieger of STI and R.L. Eshelman of the Vibration Institute would be the co-Chairman of this meeting. It was also decided that meetings of the Committee would be held bi-annually, alternately at the International Rotordynamics Conference (1994, 1998, and 2002...) and at the IMech 'Vibrations in Rotating Machinery' Conference (1992, 1996, and 2000...).

Bath, 1992

A meeting of the RC was held with Professor Hori as Chairperson. Professor Hori announced his resignation at this meeting, but no successor was appointed. This step was then considered to be the prerogative of the IFToMM Executive Committee. Eventually the EC appointed Professor Giorgio Diana to be the succeeding Chairman, although no vote was taken by the RC on this. At this meeting it was also decided that the Fifth Rotordynamcis Conference in 1998 would be held in Darmstadt, Germany, and that Professor R. Nordman of Darmstadt, and Professor H. Irretier of Kassel would serve as Co-Chairmen.

Chicago, 1994

This meeting attracted over 80 delegates and about 60 papers from about 25 nations. It was decided that the next RC meeting would be held at Milan, to coincide with the IFToMM World Congress in 1995. Professor Diana called this meeting and the first selection of a conference site by presentation was made. After discussions, the Sixth International Conference in 2002 was awarded to Sydney, Australia.

Oxford, 1996

The meeting was chaired by Professor G. Diana. Activity reports were presented by the Secretary, Dr. N.F. Rieger, concerning the 1994 Conference, and by Professor E.Hahn of the University of NSW concerning preparations for the 6th Conference.

Fifth International Conference on Rotordynamics, Darmstadt, 1998.

The Conference Chairmen were Professor Rainer Nordman of Darmstadt University and Professor Horst Irretier of Kassel University. There were about 75 attendees and 65 papers presented. The resignation of Professor G. Diana as Committee Chairman was announced, and following an election by email preceding the Committee meeting, Dr.

N.F. Rieger was elected Chairman (designate) of the RC. Dr. A.J. Smalley of SWRI, Texas, was elected Secretary.

IMechE Vibration of Rotating Machinery Conference, Nottingham, 2000

Activity reports were presented concerning Committee activities by Dr. A.J. Smalley, and on progress concerning preparations for the Sixth International Conference in 2002 (Professor Hahn). In view of the sad and untimely death of the great rotordynamicist, Professor J W Lund of Denmark, it was decided that a medal in his name would be awarded every four years at the Rotordynamics Conference, to commemorate his contributions to the field. The venue for the Seventh International Conference in 2006 was chosen to be Vienna, Austria. Professor Helmut Springer of Vienna Institute of Technology was elected as Conference Chairman.

EC Meeting in Milan, August 2001

Dr. N.F. Rieger attended, was confirmed incoming RC Chairman and presented the RC activities report. Approval was given by the award of a medal by the RC for best paper at each International Conference, to be known as the Lund Medal, in honor of Professor J.W. Lund (1930-2000) of Denmark, a pioneer of modern rotodynamics. Approval was also given by the EC for the award of up to six travel grants of $500 USD each, for travel to the IFToMM Congress in Beijing in 2003 (later postponed to April 2004). These travel awards would be selected for the best papers at the Rotordynamics Conference..

Sixth International Conference on Rotordynamics, Sydney, 2002

A total of 85 delegates attended this meeting. A total of 62 papers were presented. The RC accepted a total of 6 new members-elect, subject to confirmation by the EC. The First Lund medal was awarded to Professor R. Nordman and associates of Darmstadt for best paper of the Conference. The Conference organizers were commended for an exceptionally fine meeting by the delegates. The Lund Medal and Paper and five other prizewinning papers were selected and announced at the Conference Banquet, which was held in the N.S.W. State Parliament Banquet Hall, overlooking Sydney Harbour.

Conclusion

The Rotordynamics Committee notes with sorrow the passing in 2003 of Prof Giovanni Bianchi, a foundation member and past-president of IFToMM. The Committee extends its condolences to Signora Bianchi and her family on their sad loss.

The next International Conference on Rotordynamics will be held in Vienna, Austria, in 2006. Professor H Springer of Vienna Institute of Technology will be the Conference Chairman. The RC will meet again at the forthcoming of the IMech VRM Conference in 2004, in Swansea, U.K. The main agenda items will be the assignment of sub-committee activities and the appointment of sub-committee chairpersons. It is expected that a venue for the 2010 IRC will be selected at this meeting.

2. Ancient Machines

A STUDY ON WESTERN AND CHINESE LOCKS BASED ON ENCYCLOPEDIAS AND DICTIONARIES

Hong-Sen Yan   Hsing-Hui Huang
Professor   Graduate Student
Department of Mechanical Engineering
National Cheng Kung University
Tainan 70101, TAIWAN, R.O.C.
fax: 886(6) 208-4972   e-mail: hsyan@mail.ncku.edu.tw

ABSTRACT- This work collects and integrates locks-related information and literatures in English encyclopedias and dictionaries to explore their meanings and mechanisms in different periods. First, we classify the basic mechanisms of locks into three parts: the fastening, the opening, and the obstacle. Most of the Western locks were designed by applying pins just like the ancient Egyptian wooden locks. The Ancient Chinese locks used bolts or latches on the entrance gate and used the splitting-spring padlocks for personal securities in general. The mechanisms of Western locks have been improved from mechanical devices to electronics. The mechanisms also transform from the simply pin lock in origin to various mechanisms, and the components are gears, cams, linkages, and springs,…etc, in major. Comparatively, the evolvements of ancient Chinese locks are slow; this may result from the extended family life of ancient Chinese that finally bring about the elimination through the competition of Western locks.

KEYWORDS: Western lock, Chinese lock.

INTRODUCTION

Lock, a simple mechanical device at first sight, was broadly used to protect people's security and privacy. The development and advances made of locks have been changed significantly in the past; from a simple bolt to various components, from a crude, clumsy shape to ingenious and delicate one. And, the functions of the locks have been turned from merely protection to decoration and beautification.

Traditionally, locks were designed by specialized technicians in order to prevent the knowledge of the mechanisms of locks from lawless people to further ensure security. But that also blocked the way of the public's interests to understand the mechanisms of locks. Consequently, only very few people has been interested in tracing the origin, existence, and development of locks. Through the literatures about Chinese and Western locks, one can clearly see that in the West, people thoroughly recorded the history of the locks and as time progressed, there were numerous improvements documented. Through different eras, various newly-designed styles of Western locks emerged. On the other hand, information and development of Chinese locks were barely recorded. The word "lock" was often seen in documentation in conjunction with military security, and historical records of Chinese locks were still scant and incomplete. This paper introduces the basic components of locks, and explains the mechanisms of Western and Chinese locks made of these components first. The definitions and relative etymology, aided with the explanations of mechanisms and the developments of locks, are then investigated and studied. Also, the literatures of different periods in Chinese history are utilized to strengthen our understanding of the meaning and evolvements of ancient Chinese locks.

COMPONENTS OF LOCKS

In general, a lock consists of a fastening device, an opening device, and an obstacle component.

a. Fastening- Most mechanical locks use a sliding or rotary "bolt" as the fasten component. In the second half of the 20 century, electromagnetism began to be used as the fastening device.

b. Obstacle- There are numerous types of obstacles in locks to discriminate and obstruct wrong opening devices. Most western locks used warded and tumblers as the obstacles. Most Chinese locks, however, used splitting-springs as the obstacle.

c. Opening- The opening device is used to overcome the obstacle component and can be in the form of a key, numbers or a secret code.

Figure 1 shows the relationship of the fastening device, the opening device, and the obstacle components that construct the operation of a lock. When the locking device is in the locked/unlocked condition, the opening devices can be operated. Meanwhile, the opening devices are discriminated by the obstacle to ensure the validity. If the opening device is incorrect, the correct opening device should be renewed for proceeding the operation of a lock. Once the obstacle is overcome, the fastening device is released and the lock will get into the unlocked/locked condition.

Although the design, the shape, and even the mechanism of locks in various eras are somewhat different, the basic principle and intention behind a lock are identical. With this fact, locks can reflect the technological level in each period of the time we studied.

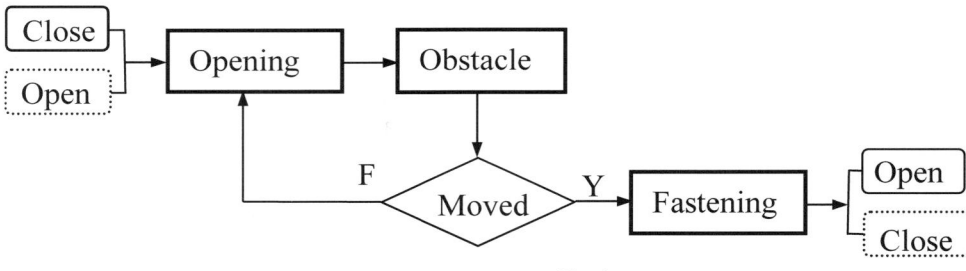

Fig. 1   Movement of locks

MECHANISMS OF LOCKS

The mechanisms of ancient Egyptian locks are quite the same as the modern everyday-used locks: the pin tumbler locks. In the following, we began with introducing the mechanisms of Egyptian locks for a better understanding of the mechanisms of locks, and then followed by the introduction of the modern western locks; and the special ancient Chinese splitting-spring padlock would be the last one mentioned.

The Ancient Egyptian Lock

The ancient Egyptian wooden lock has become an immemorial though mechanically-completed lock, Fig 2. The key can be inserted into the bolt, and the front part of the bolt is inserted into the square hole of the wall. There are some little holes on the bolt with pins inside that can move up and down thoroughly in the holes as the obstacle for the movement of the bolt. The key of such a lock is shaped as a big wood comb, and the location of wood treenails on the key can just fit the holes on the bolt. When the lock is locked, the pins in the hole of the bolt will fall down by its own weight to stop the movement of the bolt. Inserting the key and lift up the pins in the holes of the bolt by the treenails on the key will enable the movement of the bolt for opening the lock.

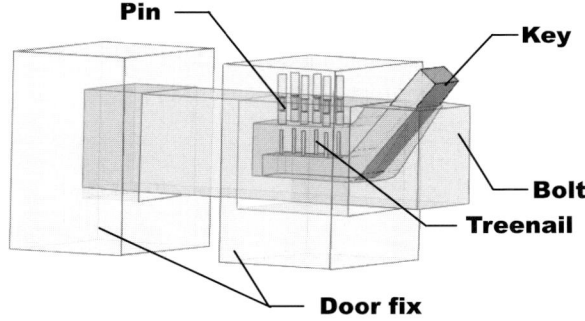

Fig. 2   Configuration of an ancient Egyptian lock

Modern Western Locks [1, 2]

The definition in 《Merit Students Encyclopedia》 classifies locks into pin-tumbler cylinder locks, warded locks, lever-tumbler locks, disc-tumbler cylinder locks and combinations locks.[3]

a. Pin-tumbler cylinder locks, Fig 3(a).- Pin-tumbler cylinder locks and the Egyptian wooden lock is quite the same. They both use a small pin to prevent the lock from opening. The difference between the two is that a pin-tumbler cylinder lock uses the power of springs to release the small pin and the treenail on the key is replaced by warded. With the improvements in the design, the locks became firmer and more secured. In addition, Pin-tumbler cylinder locks come in various styles and the applications are broader, such as the cabinet locks, padlocks, utility locks …etc.

b. Warded locks, Fig 3(b).- A warded lock has a dent, in which the key rests on when trying to unlock. In order for the lock to be opened or closed, the warded has to be moved to the right height. Warded locks are one of the older types of locks; its existence can be traced back to the Roman times. Due to the simplicity in mechanism, it is less secured and is fading out nowadays.

c. Lever-tumbler locks, Fig 3(c).- Lever-tumbler locks come in many different designs. Usually there are more than three tumblers to ensure better security. It also utilizes springs to push the tumblers to the position of locking. To open the lock, one needs to only insert the right key and turn, then move all the tumblers to a certain place to open.

d. Disc-tumbler cylinder locks, Fig 3(d).- Disc-tumbler cylinder locks use a series of thin, moveable, metal discs, and no bolts, for the tumblers. They may have a key with grooves on one or two sides for opening. With the design, disc-tumbler cylinder locks cannot vary the design as much as the way pin-tumbler cylinder locks do. As a result, there are only a few applications in the market at present.

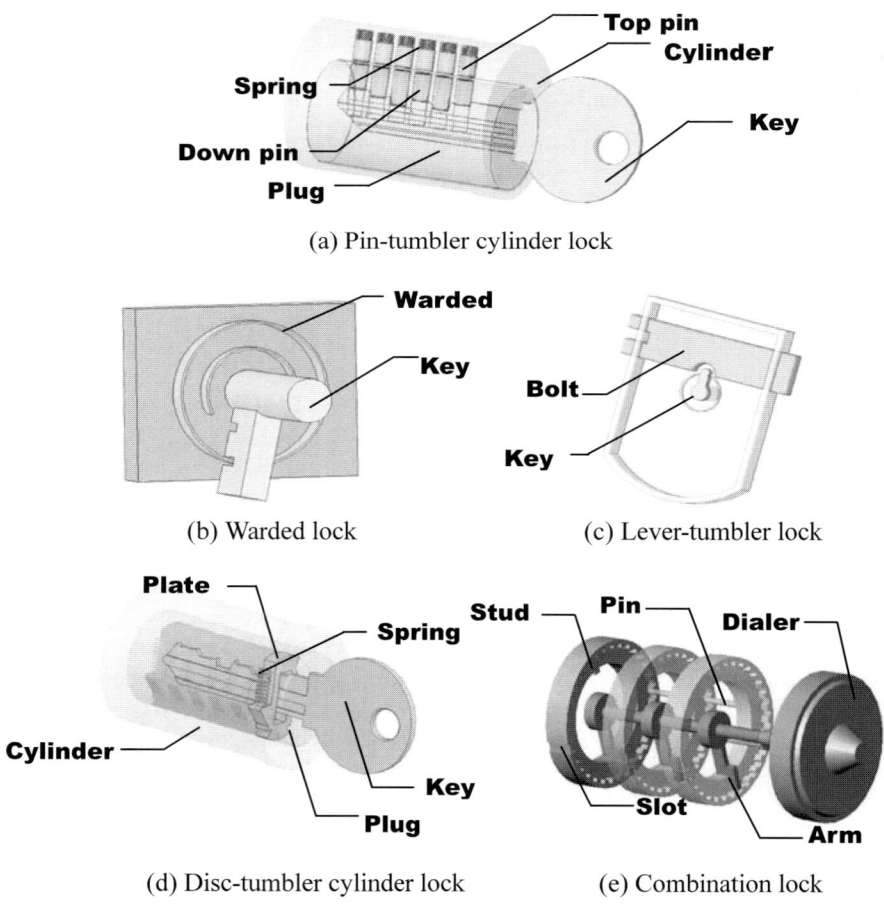

(a) Pin-tumbler cylinder lock

(b) Warded lock   (c) Lever-tumbler lock

(d) Disc-tumbler cylinder lock   (e) Combination lock

Fig. 3   Mechanisms of modern Western locks

e. Combination locks, Fig 3(e).- Combination locks need no keys for opening. This type of lock appeared in the 16th century. The earliest combination locks consisted of a series of tumblers (or turning wheels) that lined up closely. When a series of number or letters are arranged correctly, all the notches are aligned. And the main rod or lock-peg can be moved for the opening of the lock.

Ancient Chinese locks

Ancient Chinese locks can be classified into bolt locks and padlocks. Fig. 4 shows mechanism of a bolt lock in which the obstacles are horizontal and vertical timbers. Ancient Chinese padlocks can be classified into splitting-springs and combination locks. Locks that utilize springs can be further classified into broad locks and pattern locks. Broad locks usually adopt a rectangular shape, Fig. 5(a), and pattern locks are always varied in their appearances, Fig. 5(b). Combination locks have several turning wheels and use etching poems or phrases to be the secret codes, Fig. 6(a).

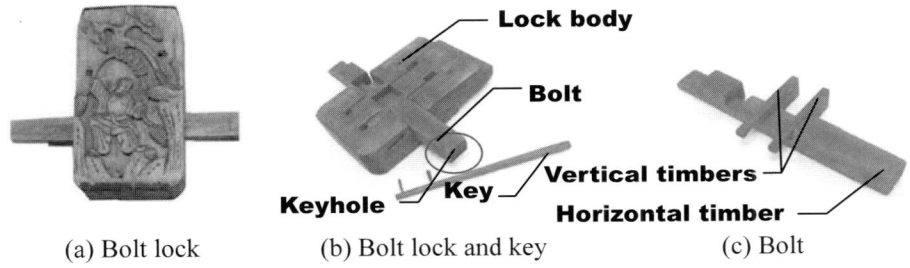

| (a) Bolt lock | (b) Bolt lock and key | (c) Bolt |

Fig. 4   Mechanism of an ancient Chinese bolt lock

Most splitting-spring locks consist of a lock body, a shackle (fastening mechanism), a key (opening mechanism), and splitting-springs (obstacle). The shackle and the splitting-springs make up the lock-peg, Fig. 5(c). The lock body provides a keyhole which allows the key to be inserted and guide the movement of lock-peg, and the splitting-springs is connected to the lock-peg by a lock-stem. When it is locked, the splitting-springs on the lock-peg are stuck in the inner walls of the lock body; when the key is inserted, the shape of the keyheads are designed to squeeze all the splitting-springs and enable the movement of the lock-peg to open the lock [4]. The statements in the book Zheng Zi Tong (正字通) written by Zi-lie Zhang (Zi-lie Zhang, Ming Dynasty) said: "Key (鍉, di), that can be used to open a lock (钥, yue) because of the splitting-springs (須, xu) in the lock body. When inserting the key, the splitting-springs can be squeezed for opening the lock, that we also called it key (匙, shi) " [5]. This explains the method for opening a splitting-spring lock accurately.

46

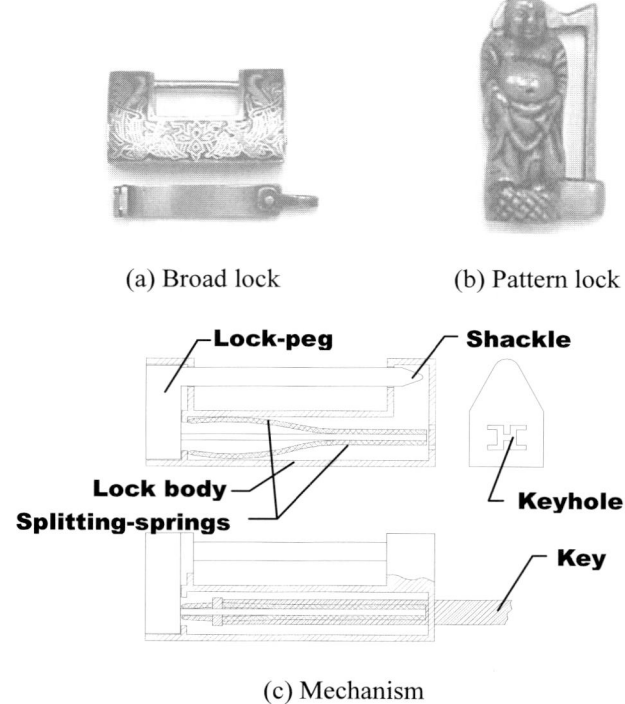

(a) Broad lock　　　　　　(b) Pattern lock

(c) Mechanism

Fig. 5　Ancient Chinese splitting-spring padlocks

Combination locks are consisted of a lock body, lock-peg, and turning wheels. For opening the locks, the turning wheels should be lined up to form the etching of characters in the correct arrangement for separating the lock body and lock-peg, Fig. 6(b).

DEFINITIONS OF LOCKS

As technology improves and the needs of people increase, the mechanism of locks becomes more intricate. It ranges from the simple, basic locks to the more precise electrical devices. A device can be called a "lock" as long as it achieves the goal of protecting, enclosing, and firming. Here, we study the developments of western locks based on the literatures of English encyclopedias and dictionaries of different era and editions.

Western Locks

The descriptions and definitions about Western locks are more integrated. From various encyclopedias in different era, we can inquire not only the definitions and classifications of these Western locks, but also the developments of the mechanisms.

(a) letter-combination padlock

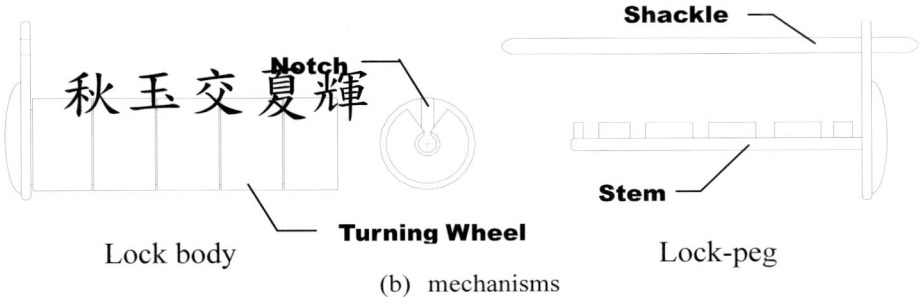

(b) mechanisms

Fig. 6   Ancient Chinese letter-combination padlock

《The New Encyclopedia Britannica, 1768》 defines [6]：

> *Lock, mechanical devices for securing a door or receptacle so that it cannot be opened except by a key or by a series of manipulations that can be carried out only by a person knowing the secret or code.*

At this time, locks are perceived as mechanical devices that can be classified into key-operated or combination locks.

《Funk & Wagnalls New Encyclopedia, 1876》 defines [7]：

> *Lock, mechanical device used for fastening doors, chests, and lids, consisting essentially of bolt guarded by a mechanism released by a key or a combination.*

As we can see, locks use bolts to guard intrusions, and take a key or combination codes to open.

《Columbia Encyclopedia, 1950》 defines [8]：

> *Locks and key, fastening consisting essentially of a sliding, pivoted, or rotary bolt guarded by an obstacle either fixed (warded) or movable (tumbler) and operated by a key. Usually the key has a shank that is flat or in the form of a pipe or a pin.*

The mechanical structure of locks device began to evolve into various types; and, bolts can now be sliding, pivoted or rotated.

《The Columbia-Viking Desk Encyclopedia, 1953》 defined [9]：

> *Lock usually consists of sliding, pivoted, or rotary bolt guarded by an obstacle either fixed (warded) as in most padlocks or movable (tumbler) and controlled by a key with a flat, pipe, or pin shank.*

Now, keys can come in different shapes, such as flat, pipe or pin shank. The intricacy of the keys can increase the complications for opening locks.

《The American Peoples Encyclopedia, 1962》defines [10]：

> *Lock, a mechanical device for fastening a cover or a door. It includes a bolt or other mechanism held in position by some obstacles that must be removed, usually by a key or a dial arrangement of letters or numbers, in order for the lock to be opened.*

Here, a lock is still a mechanical device and the codes of a combination lock can be letters or numbers.

《Encyclopedia International, 1963》defines[11]：

> *A lock is a fastening for a door, drawer, lid, or container to prevent unauthorized entry, seizure, or movement. Locks can be operated by a key or other mechanism such as a dial, set of push buttons, or electric circuit.*

At this time, electrical devices begin to apply in locks in addition to manipulate a lock by a key, dial, or button pushed.

《Merit Students Encyclopedia, 1967》defines [3]：

> *Lock (Lok), a mechanical device used to safeguard an enclosures, containers, or machinery. Generally a lock is connected to a fastening device, such as a bolt on a door, to prevent its movement. Some locks, such as automobile ignition locks, operate electrical contacts. A lock is operated by a key or, in the case of the combination lock, by dialing the correct sequence of positions. The different types of locks presently in use are warded locks, lever-tumbler locks, pin-tumbler cylinder locks, disc-tumbler cylinder locks, and combination locks.*

Although locks have begun to utilize electricity, it is still not widely used. And, here mechanical locks are clear classified. This reveals the sophistication of locks.

《The Encyclopedia American, 1980》defined[12]：

> *Lock, a device that secures such things as a door of a house or cabinet, a lid of briefcases or other luggage, and the action of an ignition system by means of a bolt or latch that can be released by a mechanical, hydraulic, or electrical actuator.*

Other than the original mechanism, there are hydraulic and electrical actuators being used for opening locks.

Besides the encyclopedias of different era and editions, the contemporary dictionaries and scholars also contributed to the definitions of locks.

《The New Century Dictionary, 1927》defines [13]：

> *A contrivance for fastening or securing something: specif., an*

> *appliance for securing a door, gate, lid, drawer, or the like in position when closed, and consisting of a bolt or system of bolts propelled and withdrawn by a mechanism operated by a key or other device.*

Specify the obstacle of locks as bolts, and the keys were used to move or rotate the bolts.

《The American College Dictionary, 1947》 defines [14] :

> *A device for securing a door, gate, lid, drawer, or the like, in position when closed, consisting of a bolt or system of bolts propelled and withdrawn by a mechanism operated by a key, dial, etc..*

Besides indicating opening by the keys, it also bringing up the usages of dials.

《The New Oxford Illustrated Dictionary, 1962》 defines [15] :

> *Appliance for fastening door, lid, etc., with bolt that requires key of particular shape to work it.*

The shapes of keys are various.

《Standard College Dictionary, 1963》 defines [16] :

> *A mechanical closing or fastening device having a bolt or combination of bolts secured or released by a key, dial, etc., and used to prevent unauthorized entry, access, or operation.*

Bolts are no longer simple mechanisms.

《The New American Heritage Dictionary Of the English Language, 1969》 defines [17] :

> *A device used to provide restraint: especially, a key. Or combination-operated mechanism used to fasten shut a door, lid, or the like. Such a device used to prevent unauthorized operation of a machine.*

From the definitions of locks in 1947-1969, the role of locks remains the same.

《The New Lexicon Webster's Dictionary of the English Language, 1972》defines [18] :

> *A fastening device for doors, lids, drawers, etc., in which a bolt is secured and released by a mechanism operated usually by a key acting on moving components(wards), or by some other means, such as a combination.*

Besides using keys to open the locks, here also indicates the usages of combination locks.

《Chambers 20th Century Dictionary, 1983》 defines [19] :

> *A fastening device, esp. one in which a bolt is moved by mechanism, with or without a key.*

The opening of locks is not restricted to keys or number utilized; the explanation here offers the improvement of locks an infinite space.

《Webster Dictionary, 1988》 defines [20] :

> *A mechanical device furnished with a bolt and usually, a spring,*

for fastening a door, strongbox, etc. by means of a key or combination.
Anything that fastens something else and prevents it from opening, turning, etc.

The perception of locks is quite general.   Anything that can prevent something from being opened or rotated can be considered a "lock".
《The New Oxford American Dictionary, 2001》defines [21] :

A mechanism for keeping a door, lid, etc., fastened, typically operated only by a key of a particular form; the key turned firmly in the lock.

The explanations of locks in the dictionaries focused on the function of locks; descriptions on the mechanisms are obviously less than the definitions.
Furthermore, Hopkins (1928) defined locks as [22] :

A lock is a fastening consisting of a bolt, or a substitute for a bolt, held in a definite position in order to prevent the opening of a door, cupboard, etc., until released by an authorized implement, or by arrangement if letters, numbers.

And, Ashley (1993) [23] :

A lock is a temporary fastening device operated by a key, dial or circuit that prevents something from being opened or removed.

From the above definitions, the function of Western locks is for safety and protection, and the mechanisms progress from mechanical to electronic came in gradually from the 17th century to the 20th century.

Chinese Locks
The following will conclude the piecemeal literature for introducing the definition of Chinese locks through observing the development, progress and speculating locks' influence on the daily life.   Also, in combination with the introduction of basic mechanisms of locks, we can further intensify the understanding of Chinese locks.
The earliest literature regarding the mechanism of locks and development process was "The locksmith's art" of Science and Civilization in China written by Needham [24]. It describes the method of studying the ancient Chinese locks as this: "One can only offer a few scattered observations. Conscious of the first difficulty which needs clearing up, namely the exact significance of the words employed in the early times."   Because it was so long ago and the locks are hard to maintain, the best way to understand the significance and influence of locks is to search the related statements and phraseology from the ancient literature.   Besides, the book The Beauty of Ancient Chinese locks [4] written by Yan, systematic introduces the characteristics of Chinese locks.
Because of the lacking of the paleography that records the origin and mechanisms of Chinese locks, we here list some statements in tool books explaining the definition of locks.   Then we locate relative words of locks to understand and conjecture the usages from the fragmentary of literatures.
《Tse Hai, 1941》defines them as [25]:

Lock, an enclosure device that needs to be opened by a key.

《Shui Wen, 1948》 described them as [26]：

> *Locks, made by metals, use for closing the doors.*

《Tse Yuan, 1948》 defines them as [27]:

> *Used for fastening a door or box. A device that prevents one from opening an object or devices, used to be called "jian (鍵)", now called a "suo (鎖, lock)".*

《Cheng Chun Shin Yin Yi Amalgamated Dictionary, 1977》 defines [28]:

> *"Men-jian (門鍵, lock)" is said as a lock. Any kind of devices that fasten a door, closet, or box that we can also called it "Men-jian (門鍵, lock)".*

Furthermore, we also excerpt from other literatures to explain "lock" from etymologies integrated, for more profound comprehension of Chinese locks.

The related Chinese characters about ancient Chinese lock are numerous, such as "quang (關)", "jiong (扃)", "bi (閉)", "shuan (閂)", "zhi (植)", "jian (楗)", "ju (鐻)", "suo (鏁)", and "yao (鑰)",…etc. From the shape and the definition of the character, we can textual research sketchy the meaning and usages of ancient Chinese locks.

Shui Wen (說文) states: "quang (關, lock), guard the door by a horizontal timber", and "jiong (扃, lock), that locks (關, Quang) the door from outside" [26]; both of the characters have the meaning of locking the door by a horizontal timber. And "bi (閉, lock)" also has similar meaning with "quang (關, lock)"; both characters refer to the locks inside the door. The difference between these two words is that "quang (關, lock)" always indicates the big entrance, such as the gate of a city or the door of a huge building, and "bi (閉, lock)" refers to the small door of a house. Besides, "zhi (植, lock)" is a vertical timber for guarding a door. No matter which characters, they all indicate the form of bolt lock [29].

Character "jian (楗, lock)" may indicate a wood-made lock that appeared in the Miu Xun of Huai Nan Zi (淮南子・繆訓) [30], and the narration of Er Ya Xu (爾雅序): "… lock (鈐, qian) and key (鍵, jian) of six art. "鈐 (qian, lock)" also mentioned the meanings of locks [31].

Statements in Qu Qie of Zhuang Zi (莊子・胠篋): "…afraid the lock (扃鐻, jiong- ju) are not firm enough."[32] in which jiong- ju (扃鐻, lock) also have the meaning of lock. Xianc Yi Bian (賢奕編) stated: "The Locks (鏁, suo) with round shape are earlier made, and locks with square shapes are lately made."[31] that explains not only the character of "鏁 (lock)", but also differentiate the shape from different generation. The Fifth of Fang Yan stated: "door locks (戶鑰, hu-yao), on the east from the gateway, between the country of Chen and Chu, are called "jian (鍵, lock)", and on the west from the gateway are called "you (鑰, lock)", that also poined out the different appellation in different area [33].

The shape of the padlocks began to show variety with specific significance on the periods of Han and Tang Dynasty (618~907). And the most remarkable shape should be the lock with the form of fish. Feng Su Tong (風俗通) written by Ying-shao (應紹, Han Dynasty) stated: " The lock should be made in the appearance of fish because of the membrane of the fisheyes make fish see thoroughly."[34] From this statement, we can presume that the form of fish should be used on the lock appearance earlier.　Also, Zhi Tian Lu (芝田錄) said: "The lock of the door should be made as a fish, for a fish never closes its eyes and therefore it can guard all night."[35,36]　Both statements also explained the reason of shaping the locks as a fish.　For the progresses of using and designing the locks, the appearance of a lock becomes diversified gradually. Furthermore, most of the keyholes of locks have been designed and embellished, and have also endowed with distinctive significant.　From the design of keyhole we may also differentiate the dignity of the user [37].

Besides the investigation of the definition and relative etymology, we also have further studied the usages from the literature.　The earliest Chinese locks should be the ones with wooden bolts that fasten the board-door on the wall by using a bolt, and may have appeared as early as in the Spring and Autumn period (770-550 BC). Jun Shou of IV Shi Chun Qiu · (呂氏春秋 · 君守) stated [38]:

> What that prevents one come out from inside, called "jiong (扃, lock)", and what that prevent one get in from outside, called "bi (閉, lock)".

From the words, the definition and usage of locks are quite complete.　As to the size and usage of the bolt may also tell sketchy from Mo Zi · Bei Cheng Men (墨子 · 備城門) [39]:

> The bolt set horizontal and vertical for fastening the door should be firmed by a piece of thin metal ring.　And where the across of the horizontal and vertical bolts may also be dominated by a wood lock which is about two feet in length.　Besides, it also adds a metal lock outer the wood lock for further firmly, and sealed at last.　Instruct one to guard the seal for more security, and also takes care the solid of the horizontal and vertical bolts.

We can see the way ancient Chinese people secure their gate, and we also noticed the names of these lock in the statement clearly.　The statement called the horizontal bolt "guan (關, lock)" and vertical bolt "zhi (植, lock)".　The addition wooden lock that called "shu-guan (梳關, lock) , can also be called "huang -guan (桄關, lock)", and the outer metal lock has another name called "xian (筧, lock)".　Moreover, according to Miu Chen of Huai Nan Zi (淮南子 · 繆稱)[30]：

> The lock should not be used to lock without the length more then one foot.

That shows the size of an ordinary lock of the house is about one foot, and the lock stated in Bei Cheng Men of Mo Zi (墨子 · 備城門) [40] is for the gate, that the size should be multiple.

From the various definitions and etymologies stated above, we know that, as the mechanisms and usages of both Chinese and Western locks keep changing over times, their functions for defense are also being improved continuously.　When integrated the definition, function, and mechanisms of the locks stated above, we may now define "lock" extensively as:

Any enclosure devices that serve the purpose of protecting, managing and embellishing are called "lock".

COMPARISONS BETWEEN WESTERN AND CHINESE LOCKS

The uses of Chinese locks did not go through much progress; they just sized down the horizontal and vertical bolts and place it into the lock body. Not until the design of splitting-spring locks and their mass usage, did locks start to use on locking the boxes and chest for protecting private properties popularly, which also opened a new era for Chinese locks.

From the literature, we can find the differences of the improvements and developments between Chinese and Western locks. In the West, people mainly used pin locks. The mechanical designs of the pin locks were improved and varied, and numerous components were applied on it, such as gears, cams, linkages, and springs. On the other hand, the development of Chinese locks was slower and was used mainly by rich families. The splitting-spring locks were used mainly to protect one's privacy. This difference should have been related to the Chinese culture. Due to the fact that Chinese people are more used to domestic life, when leaving their houses, they tend not to lock their doors. However, they might also be afraid that family members or servants might steal or rummage their personal valuables, so they used splitting-spring locks to guard their own property. Another benefit added to this was that it also acted as a form of decoration. Because of its frequent usage, the splitting-spring locks were better designed and more intricate, and the development of the bolt locks of the gates was much slower in comparison. Even until the end of Qing Dynasty (1911) we could still see bolt locks used on gate of the cities [41].

In appearance, the sumptuous western locks are used to show high social status and the Chinese splitting-spring locks are restricted to the social status of the users. For more dignity, the design of Chinese locks is more delicate and intricate, and also reveals the Chinese civilization.

Although there are no records, the Chinese splitting-spring locks should have come from a child's intellectual toy and was further developed. As a result, locks are also getting more ingenious and amusing. The western locks, in order to invent locks that are more difficult to unlock, the locksmiths spent a lot of time designing and also looking for lock openers in order to show the heightened security of the locks. As for the invention of combination locks, that creation really provided challenges for the locksmiths [6].

CONCLUSION

The fundamental mechanisms of the ancient Western and the Chinese locks, such as the Egyptian pin locks and the Chinese bolt locks are very similar. These old locks urge the development of more convenient and delicate of modern pin-tumbler locks. Although the historical evolvement of Western locks went through a dim period of time, but it can always improved step by step by a regular and basic ways. The improvements of the Chinese bolt locks focused only on the number and the size of the pins, not much on the evolvement of the mechanisms. This study of western and Chinese locks, based on encyclopedias and dictionaries, also provides a brief approach to understand the historical developments of ancient locks indirectly. Through the

54

understanding of ancient locks, especially Chinese locks, should be done by studying various literatures, e.g., from available poetry, in various period in detail. One milestone of the development of Chinese locks is the invention of the splitting-spring padlocks, and the usage of splitting-springs as obstacles. The splitting-spring locks not only demonstrated the intelligent and craft ship of the ancient Chinese people, but also reflected the level of technology in each period.

ACKNOWLEDGEMENT

The authors are grateful to The Ancient Chinese Machinery Cultural Foundation (Tainan, TAIWAN) for the financial support of this work.

REFERENCES

01. Hu, G.. C., *Lock*, Foundation of the Xu's Publishing Co., Taipei, 1992.
02. Rathjen, J. E., Locksmithing, McGraw-Hill, New York, 1995.
03. Merit Students Encyclopedia, Vol.13, Macmillan Educational Company, New York, 1967.
04. Yan, H. S., The Beauty of Ancient Chinese Locks (Chinese and English edition), Ancient Chinese Machinery Cultural Foundation, Tainan, TAIWAN, May 2003.
05. Zhang, Z. L. (Ming Dynasty), *Zheng Zi Tong* (正字通), China Worker Publishing Co., Beijing, 1686.
06. The New Encyclopedia Britannica, Vol. 7, Encyclopedia Britannica Inc., Chicago, 1768.
07. Funk & Wagnalls New Encyclopedia, Vol.16, Funk & Wagnalls Inc., USA, 1876.
08. Columbia Encyclopedia, Rockville House Publishers, USA, 1950.
09. The Columbia-Viking Desk Encyclopedia, Columbia University, USA, 1953.
10. The American Peoples Encyclopedia, Vol. 12, Grolier Inc., USA,, 1962.
11. Encyclopedia International, Grolier Inc., USA, 1963.
12. The Encyclopedia Americana, International Edition, Grolier Inc., U S A, 1980.
13. The New Century Dictionary, The Century Co., USA, 1927.
14. The American College Dictionary, USA, 1947.
15. The New Oxford Illustrated Dictionary, Oxford University Press and Bay Books Pty Ltd., USA, 1962.
16. Standard College Dictionary, Funk & Wagnalls, USA, 1963.
17. The New American Heritage Dictionary of the English Language, American Heritage Publishing Co. Inc., Taiwan, 1969.
18. The New Lexicon Webster's Dictionary of the English Language, Lexicon Publication, Inc., USA, 1972.
19. Chambers 20th Century Dictionary, W & R Chambers Ltd, Great Britain, 1983.
20. Neufeldt, V, Webster's Dictionary, Simon & Schuster Inc, Ohio, 1988.
21. The New Oxford American Dictionary, Oxford University Press Inc., USA, 2001.
22. Hopkins, A., The Lure of the Lock, The General Society of Mechanics and Tradesmen, New York, 1928, p.14.
23. Ashley, S., "Under Lock and Key", ASME Mechanical Engineering, Vol.135, August 1993, pp.62-67.
24. Needham, J., Science and Civilization in China, Vol. 4, Part II: Mechanical

Engineering, Cambridge University Press, 1965.

25. Editors of Zhong Hua Book Co., *Tse Hai* (辭海), Zhong Hua Book Co., Kunming, 1941.

26. Xu, S., *Shui Wen* (說文), Taiwan Shan Wu Publishing Co., Taipei, 1971.

27. Fang, Y., *Tse Yuan* (辭源), Shan Wu Book Co., Sanghai, 1933.

28. Gao, S. F., *Cheng Chun Shin Yin Yi Amalgamated Dictionary* (正中形音義綜合大字典), Cheng Chun Book Co., 1977.

29. Guo, P., *Er Ya Xu* (爾雅序), Zhong Hua Book Co., Beijing, 1985.

30. Liu, A. (Western Han), *Miu Xun* of *Huai Nan Zi* (淮南子‧繆訓), Taiwan Ancient Books Publishing Co., Taipei, 2000.

31. Guo, X. annotated, *Zhuang Zi* (莊子), Yi Wen Printing Co., Taipei, 1983.

32. Liu, Y. C. edited, *Xianc Yi Bian* (賢奕編), Zhong Hua Publishing Co., Beijing, 1985.

33. Yang, X. noted, Guo, P. annotated, *The Fifth* of *Fang Yan* (方言‧第五), Zhong Hua Publishing Co., Beijing, 1985.

34. Ying, S. (Han Dynasty), *Feng Su Tong* (風俗通), Zhong Hua Publishing Co., Beijing, 1985.

35. Wei, C. H., *Zhi Tian Lu* of *Fiction of Hang Dynasty* (全唐小說‧芝田錄), Shang Dong Wrn Yi Publishing Co., Tsinan, 1993, P. 2846.

36. Yan, H. S. and Huang, H. H., "Fish locks of Ancient China," Technology Museum Review, Vol. 7, No. 2, National science and technology museum, Kaoshung, 2003, pp.3-10.

37. Yan, H. S. and Huang, H. H., "On the spring configurations of ancient Chinese locks, " Proceedings of HMM2000, International Symposium on History of Machines and Mechanisms, Cassino, Italy, Kluwer Academic Publishers, May, 2000, pp.87-92.

38. Vincent, J. M. E., Locks and Keys throughout the Ages, Lips' Safe and Lock Manufacturing Company, London, 1957.

39. lV, B. W. (Qin Dynasty), *lV Shi Chun Qiu* (呂氏春秋), Taiwan Zhong Hua Book Co., Taipei, 1966.

40. Zhang, C. Y., *Mo Zi Ji Jie* (墨子集解), Wen Shi Zhe Publishing Co., Taipe, 1982.

41. Hommel, R. P., China at Work, The John Day Company, New York, 1937, pp.291-304.

A Simulating Experiment

on the Ancient Chinese Astronomical Clock Tower

Gao Xuan
Tsinghua University, Beijing 100084, P. R. China

Lu Zhen
Beihang University, Beijing 100083, P. R. China

Abstract: It is very helpful and illuminating to utilize a new method to reconstruct an ancient machine which does not exist now. This paper introduces a research experiment with the computer simulation on the astronomical clock tower built in ancient China. We have set a simulation model with computer software and tried to reconstruct the ancient machine and revealed its structure, working principle and target, etc. The technique will provide a new way that helps us to research the ancient machine and solve the mystery problems on it.

Key words: the astronomical clock tower, simulating experiment, UG model, ADAMS

1. Introduction

The astronomical clock tower was built in Kaifeng, the capital of ancient China in 1089. The very huge water-powered machine produced huge influence at that time. Many extant ancient literatures recorded it. Including the outside structure, it was about 12 meters high and had 3 layers in it. It could be used for time-keeping system, celestial observing and demonstrating. There were about 400 parts in the whole mechanism and 108 parts that had an indeed name of them. It was the most complicated and exact machine in ancient China. The astronomical clock tower only existed 34 years before it was destroyed by the war in north China at that time [1]. Fortunately, the officer Su Song which presided the setting engineering had written a book named *Xin Yi Xiang Fa Yao* that recorded particularly whole and parts of the astronomical clock tower, even the structure and sizes of most of parts. It left 47 types of mechanical drawings. They are the earliest mechanical drawings in ancient we can see now [2]. The valuable records and drawings provide the condition for our study. Then the astronomical clock tower becomes an important research object for the scholars in history of science and technology.

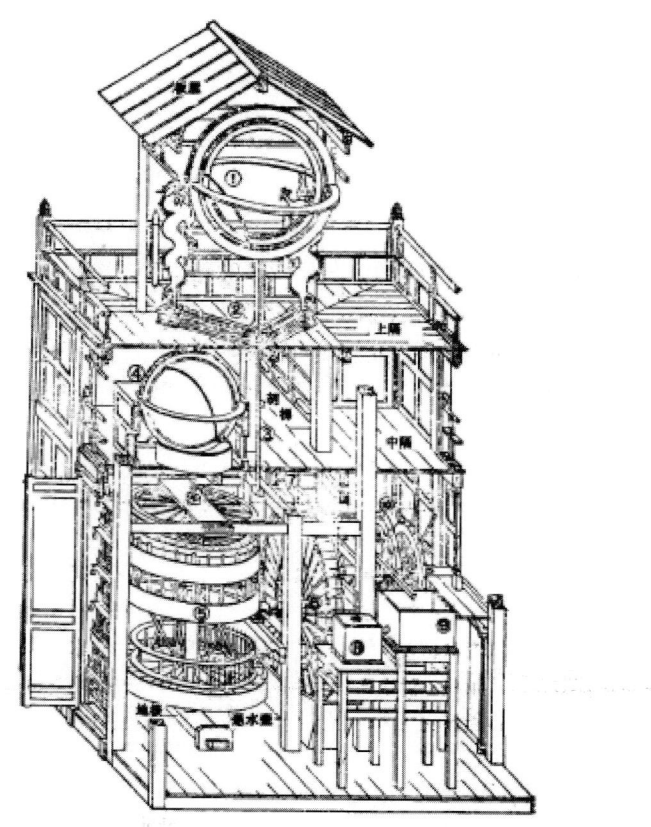

Fig.1 A reconstruction drawing of the astronomical clock tower, it was drawn by Zhang Cheng in 1958 for the work of setting up a contractible model taken charge by Prof. Wang Zhenduo [3]. This model is the first finished reconstruction of the ancient machine. It is now putting on the National Chinese Museum in Beijing.

Scholars usually did their research on the ancient machines just used records from extant ancient literatures before. They got their conclusions by analysis and guess according as some records from the extant ancient literatures and documents. The conclusions were not validated by any experiment for their dependability. Now we try to make use of the simulating method to do our research. We have set up a three-dimensional model of the astronomical clock tower by AUTOCAD software to simulate the ancient machine. We make our model of the machine partly consulting the data recording from the ancient book *Xin Yi Xiang Fa Yao*. It's based on the ADAMS—Automatic Dynamic Analysis of Mechanical System—with the virtual prototype technology to study eradicative ancient machine. Then we input the moving relationship of parts of the machine into the model to analyze their working. We wish to solve some problems on the machine in sides of dynamics and kinematics, and probe the feasibility of reconstructing the machine set up 900 years ago.

2. Preparative work of simulating experiment

Before we started to make our simulating model, we gathered most of the parameters of those parts coming from the huge ancient machine. It is more than 100 parts described their names, forms and sizes in the ancient book, *Xin Yi Xiang Fa Yao*. Because the data of sizes is ancient measure in originally records, so we must change them to modern measure according to the new research productions on the ancient metrology. From the book History of Science and Technology in China, we know the conversion proportion of length between Song Dynasty and now. It is that 1 unit of Song length is equal to 31.4 centimeters [4]. Then we classify all the parts in 4 different system classes: power unit, transmission unit, control unit and work unit, and list all the parameters in different parts of different units [5].

The astronomical clock tower would convert the water power to an intermissive rotation used a very big water-wheel. Then a control unit kept the water-wheel a uniform running velocity. And then a series of gears slowed down the velocity of rotation to make it was similar to the motion of the planets. Moreover, there was a transmission unit to transfer the rotation to work unit. So the whole machine could be used to do the work of time-keeping, celestial observing and demonstrating.

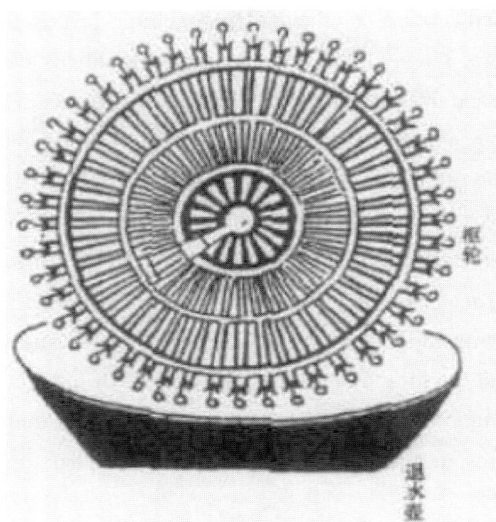

Fig.2 The water-wheel drawing in ancient book, we can't understand its work principle clearly because the different ancient representation.

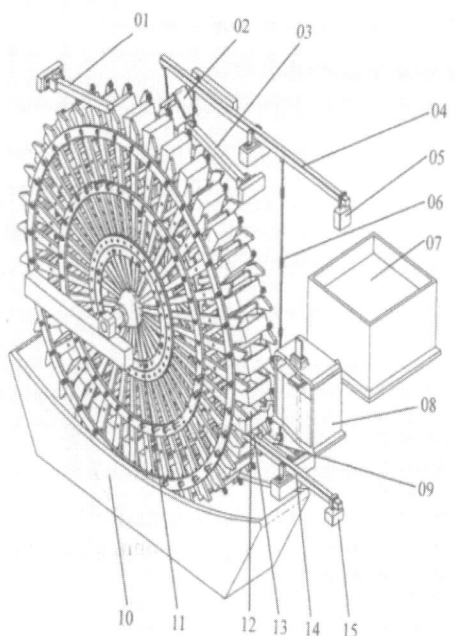

Fig.3 A modern description about the water-wheel and its work principle, the plan comes from The Chinese Encyclopedia, Vol. On Machine and Mechanism.[6]

3. Establishing and making use of the simulating model

Making a dynamics and kinematics model we need to set up a geometric model and put the restricting frameworks on it. In order to recur closely the original shape of the ancient machine, we have made a model with the computer software Epigraphic; thereinafter we call it shortened form UG, including all the main parts concerning our reconstruction. UG is excellent as it can help us to make a 3D model. It can show details of parts and make them looks like ancient machine. Otherwise, we can provide a series of necessary data of the parts in the machine by the experiment on our revisable model, just like the number of teeth of gears, and diameters of ratchet-wheels, etc. To the lacking data of the machine, we can get some experimental results by amending the parameters of the simulating model in spite of resetting a new one. It's the best of a digital model not as a practical model. After we get our UG model, we can input it to the platform of ADAMS by a kind of Para solid format. Then we can suppose some properties of parts whose lack of necessary data like their masses, original positions, first velocities, moving directions and reestablished loads with ADAMS. And can limit the relative motions between different components by a series of restrictions. And can link all the components to a mechanical system. It will achieve by a volume of restricting facilities.

We can get some important information on structure and parts of the ancient machine. But they are not clear enough for us to get the detailed materials to make the simulating model.

After we get all the useful parameters of the machine, we can do our experiment to calculate other parameters just like moments of inertia, transmitting power, etc. And then we can deduce its motion and working effect of the ancient machine.

At last, we will explain the UG model. Because not all the parts of the ancient machine are necessary for us to do the simulating experiment, so we ignored some parts which are independent of our simulating results, such as canteens, puppets for reporting time, etc. And we predigested some complicated parts as a simple rigid body.

4. Establishing and making use of the moving disciplinarian

In macroscopically, main motion of the astronomical clock tower is a rotation of water-wheel. It's a periodic and uniform velocity motion. We set it as the inputting motion of our whole mechanism of simulating model. In fact, rotation of the water-wheel is very complicated with some periodic liberation and collision, and some stochastic motion. According to the data on parts of gears, the ratio from water-wheel to work unit is:

$$i_{3.9} = \frac{\omega_3}{\omega_9} = \frac{n_3}{n_9} = \frac{Z_4 \times Z_7 \times Z_9}{Z_3 \times Z_6 \times Z_8} = \frac{96 \times 12 \times 480}{6 \times 80 \times 12} = 96$$

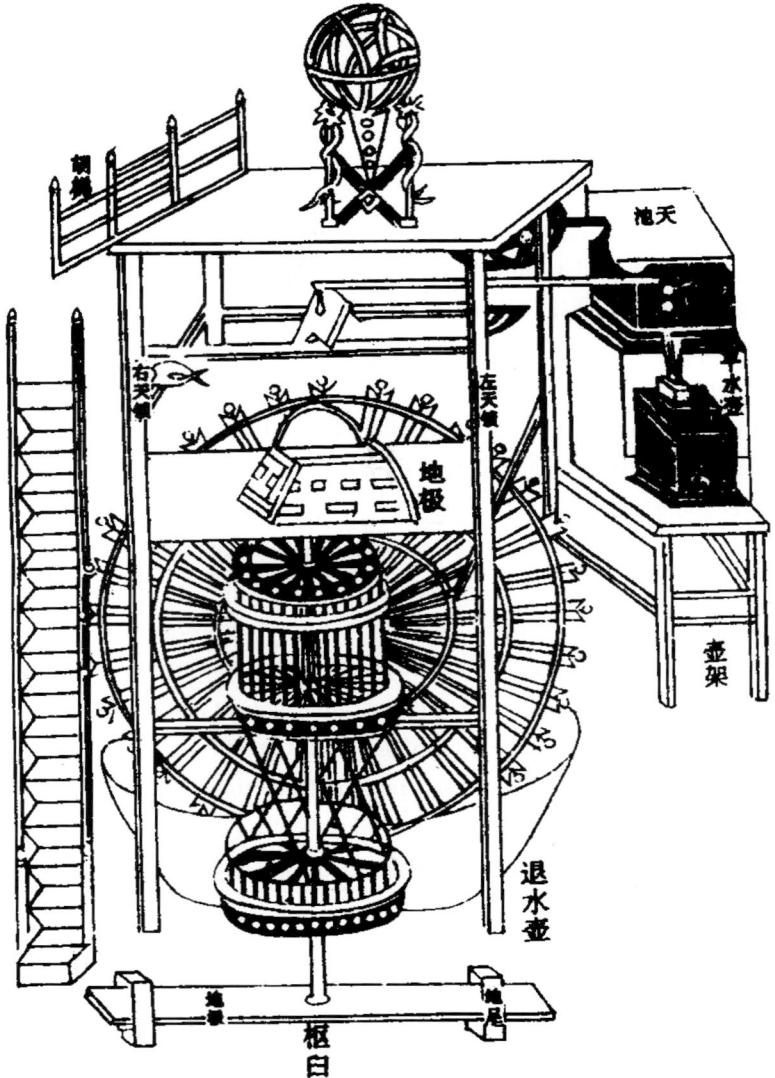

Fig 4. The main structure of the astronomical clock tower, including its water-wheel and
clepsydras of power unit, Hun Yi, Hun Xiang and time-keeping of work unit

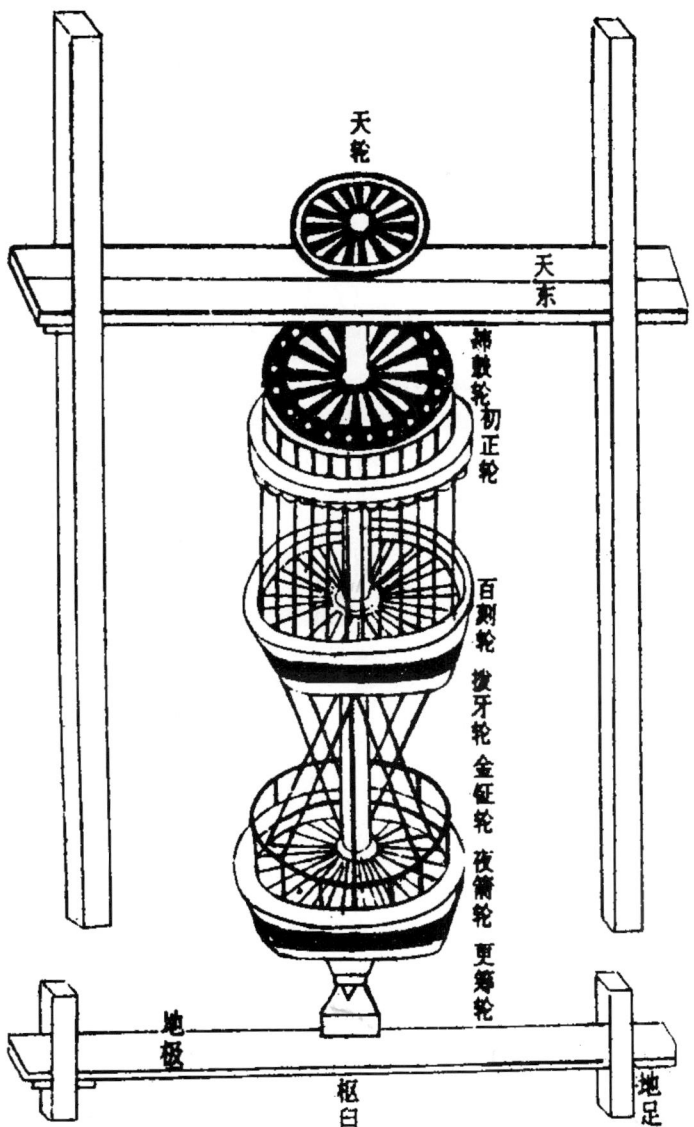

天轮

天东

枢数轮初正轮

百刻轮

拨牙轮 金钲轮 夜漏轮 更筹轮

地极

地足

枢臼

Fig 5. The main structure of time-keeping of the machine. They all come from the ancient book Xin Yi Xiang Fa Yao.

Fig.6 The integrated simulating model of main parts, units of the astronomical clock tower, we can use it to do our simulating experiment.

Then we know on our whole mechanism the driving ratio is 96:1, the liberation and collision can be ignored when they transmitting to the working unit. So we describe the motion of water-wheel with a periodic function getting from a program. In the experiment, we make use of the program Visual C++ and Matlab 6.1 to programming a periodic function and saving it to a file with the format *.dat, which can be identified by the ADAMS platform.[7] See fig 8. Then we can get a series of motion produced by the periodic function.

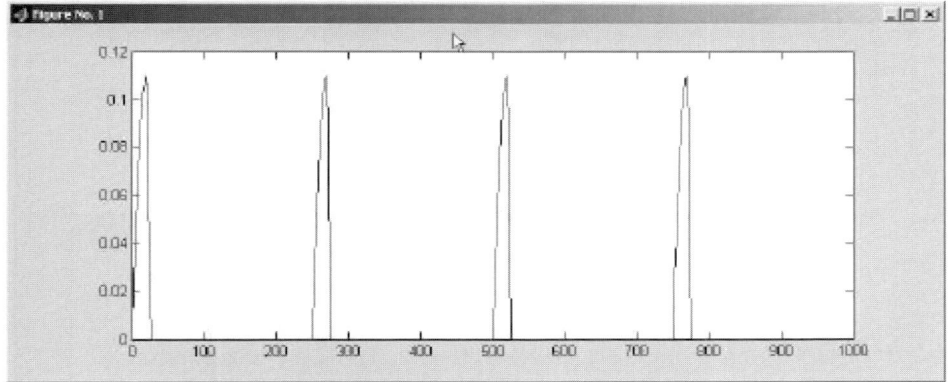

Fig.7 The periodic function graph

5. Process and results of the simulating experiment

We used the simulating model to do our athletics simulation in ADAMS platform. We can analysis the strength put in different parts with ADAMS, then input it to the software ANSYS to do our study. At last get our results. Later we will give a example to show this process.

Please let us think about the main moving origin of the astronomical clock tower, water-wheel.

The water-wheel is driven by a series of clepsydras. Even water from the last clepsydra falls to water jug in the wheel. When the water jug comes to a definite weight, it will open the control parts and let the wheel running a definite space, and go on. Main operations from control unit to the wheel are left and right up-brake on upside and the sustained switch on right side of the wheel. We study the water-wheel on the strength including its direction, numerical value, etc, like I show them in the Fig. 8.

Fig.9 is a part of strength analysis of the water-wheel. Using the analysis we can get the calculated result of strength and moving trend. We apply the result to simulating the motion of water-wheel, and put a moving to the rotation pair

$$\text{Velocity} = - \text{CUBSPL(time}, 0, \text{SPLINE\_2)}$$

CUPBPL is a synthesized function. It can form a curve of derivative continuity by linking all the points smoothly in SPLINE_2, see Fig. 11. It's a famous result of our simulating experiment.

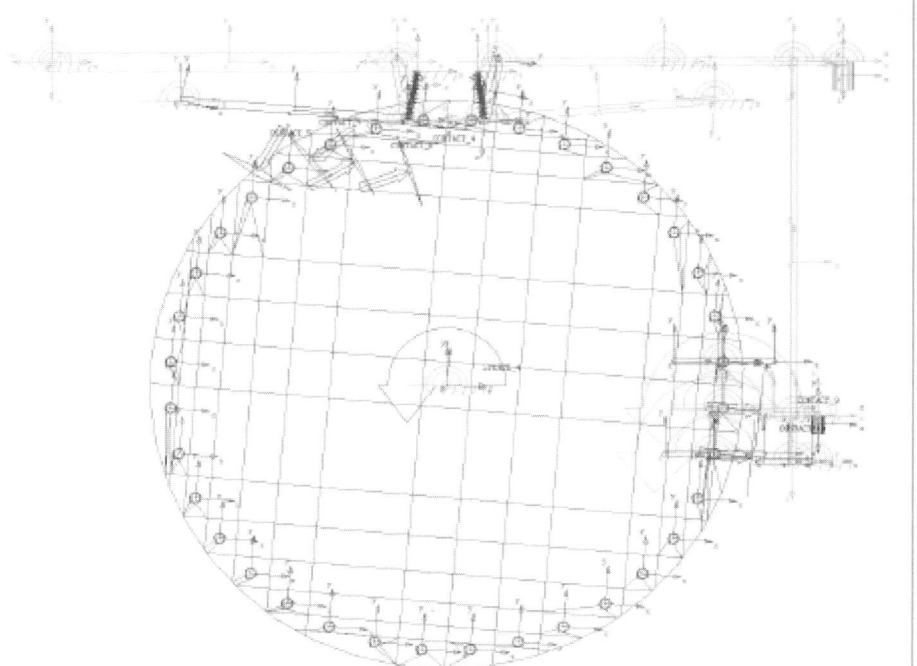

Fig.8 A analyzed graph on strength of water-wheel from the simulating model, we can use it to do our dynamical experiment and get helpful data for reconstruction of the astronomical clock tower in the future.

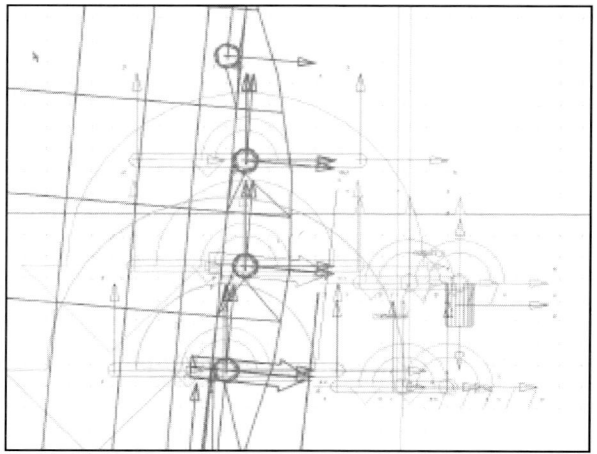

Fig.9 A part of strength analysis of the water-wheel, it show the direction and trend of strength between some water jugs and a sustained switch of control unit.

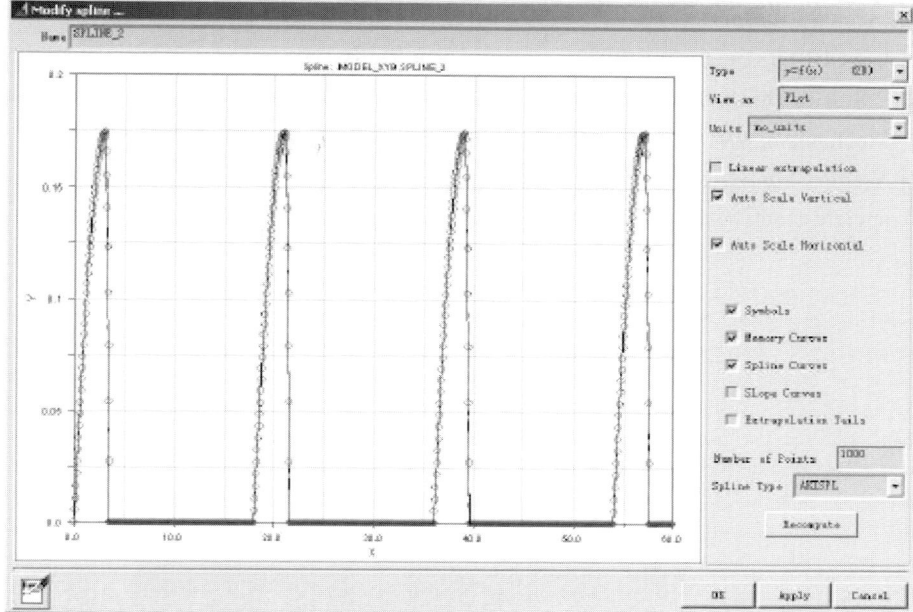

Fig.10 The curve synthesized by our simulating software.

Now we are analyzing the results of our simulating experiment. I hope that we will get more and important study results for reconstructed the great ancient machine.

Acknowledgement

The researches in this paper have been supported by the National Nature Science Foundation of China. (No. 59975047)

References:

[1] Editor in chief: Tuo Tuo (Yuan Dynasty), *History of Jin Dynasty*, p. 520, Zhong Hua Shu Ju Press, Beijing, 1975.

[2] Su Song, *Xin Yi Xiang Fa Yao*, Edition of Qing Dynasty, Published in 1852.

[3] Wang Zhenduo, *Collection of Discourses on Scientific and Technological Archeology*, pp. 132-168, Culture Relic Press, Beijing, 1989.

[4] Editor in chief: Lu Jiaxi, *History of Science and Technology in China*, Vol. On Measures, pp. 352-370, Science Press, Beijing, 2001.

[5] Lu Zhen, Gao Xuan, Research on the Records of *Xin Yi Xiang Fa Yao* about the

68

Mechanisms and Parts of the Astronomical Clock Tower, The Colloquia of 3rd ISACBRST, Tuebingen, 2003.

[6] Editor in chief: Jiang Chunfang, *The Chinese Encyclopedia*, Vol. On Machine and Mechanism, p.918, Chinese Encyclopedia Press, Beijing, Shanghai, 1987.

[7] Wang Chunjie, Ma Liangwen and Gao Xuan, Reconstruction & Research of an ancient clock based on ADAMS, Mechanical Design, Vol. 6, 2003.

Engineering Aspects of the Collapse of the Colossus of Rhodes Statue

Neville F. Rieger
STI Technologies, Inc.
1800 Brighton-Henrietta TL Road
Rochester, New York 14623
e-mail: nrieger@sti-tech.com

ABSTRACT-The collapse of the Colossus of Rhodes statue in 224 B.C. is usually ascribed to an earthquake. Although details of this statue are available from Philo of Byzantium [1] and from Pliny [2], there appears to have been no scientific assessment as to the reasonableness of the earthquake argument. The investigation described in this paper uses the descriptions of Philo and Pliny to propose a likely model for the statue, and employs the present-day technologies of finite element analysis with representative earthquake engineering data to investigate the effects of an earthquake on the Colossus. The results shed new light on the manner in which the collapse of the Colossus occurred.

KEYWORDS: Colossus of Rhodes, earthquake spectrum, ancient technology, bronze casting, finite element engineering

THE COLOSSUS OF RHODES

The Colossus of Rhodes was a huge bronze statue of Helios, the god of the sun. It stood about 110 feet tall, possibly mounted upon a pedestal of some form, and situated close to the harbor of the city of Rhodes in the Dodecanese islands. The statue was erected by the Rhodians to honor Helios for allowing the unsuccessful siege of the city by Demetrius Poliocetes to be lifted in 305 B.C. It was designed and built during the years 292-280 B.C. by the famous sculptor Chares of Lyndos. Tradition says that the Colossus was felled by a strong earthquake around 224 B.C. The shattered pieces of the statue lay where they had fallen until 653 A. D., when they were finally broken up and sold as scrap to a merchant from Edessa, Syria. No trace remains of the Colossus today.

HISTORICAL SOURCES USED

Brief commentaries with descriptions of the Colossus have been provided by Philo of Byzantium [1], by Strabo, and by Pliny [2]. Philo was a Greek mechanist who lived in the 2nd Century B.C., included the Colossus in his list of the Seven Wonders of the Ancient World. He apparently saw the wreck of the Colossus around 150 B.C., and details from his account of its construction will be used later in this text. Pliny (Plineus Caius Secundus) was a Roman naturalist who lived in the 1st Century A.D. He died during the eruption of Vesuvius in 79 A.D. Pliny commented on the Colossus as follows:

This statue was thrown down by an earthquake fifty-six years after it was erected. Few men can clasp the thumb in their arms, and its fingers are larger than most statues. When the limbs are broken asunder vast caverns are seen yawning in the interior. Within it, too, are to be seen large masses of rock by whose weight the artist steadied it while in process of erection. It is said that it was twelve years in the making, and that three hundred talents were spent upon it-a sum raised from the engines of war abandoned by Demetrius after his futile siege. **(Pliny ca. 50 A. D.)**

Medieval writers devised and perpetrated the picturesque, but erroneous idea (endorsed by Shakespeare), of a 'bestride' or legs-apart Colossus which stood across a narrow entrance to the inner harbor of Rhodes. This concept is still widely depicted today.[1] General historical details concerning the Colossus and of its period, the contemporaneous eastern Mediterranean, as used in this study were obtained from a variety of sources, including Durant [3], Romer [4] and other writers. Details of the development of metal working and craftsmanship were obtained from Singer, et.al. [5] (articles by various authors), from Bronowski [6], and from the Encyclopedia Britannica [7] (articles by several authors). One depiction of the Colossus is shown in Figures 1.

Figure 1 Colossus IV Standing Near the Harbor Entrance. The Posture Resembles the Apollo Belvedere. Wood Engraving by Sidney Barclay 1875. © Bettman Archive, used with permission

PURPOSE OF THIS WORK

This essay presents the results of personal research conducted over a period of years together with thoughts and calculations into conditions which may have led to the destruction of this great statue. The purpose here was first to develop a plausible theory for the form and structure of the statue, consistent with both the ancient writings and culture of the region, and to use present-day finite element and earthquake engineering technology to explain *why* the Colossus was destroyed by that earthquake.

---

[1] All postcards depicting the Colossus available in Rhodes today show only the 'astride' form of the statue.

STATEMENT OF THE PROBLEM

Apart from the comments of Philo and Pliny, we know little of the form and construction of the statue, and virtually nothing concerning its destruction. We are likewise unaware concerning the details of its manufacture and assembly, aside from the same comments. As there are no remains of the statue to inspect, we are also ignorant of the details of the materials used, and of the metallurgical condition of the broken components. From these we might have learned something concerning the metalworking techniques used during construction, the casting practices used to create the component sections or plates, the assembly procedures used, and of the quality of workmanship involved. We know nothing of the manner whereby the Colossus was mounted upon its base, and nothing concerning the occurrence of those storms, winds, and minor earthquakes which must have occurred during the fifty-six years when the statue stood erect in Rhodes. Lastly, there are of course no records of that fatal earthquake which ancient writers claim led to the collapse of the Colossus. We therefore begin by determining the contemporary state of technology and of the relevant skills of workmanship and construction which existed in classical Greece, and second, we will establish those aspects of style pertaining to bronze statuary which prevailed on Rhodes at the start of the third century B.C.

EVOLUTION OF METALS TECHNOLOGY

The Bronze Age began with the discovery of copper smelting, and continued with the use of copper for the creation of an increasing complexity of utensils, tools and weapons. By 3200 B.C. the principles of alloying had been 'worked out' not only for a variety of bronzes, but for gold, silver, lead, antimony, tin, and also for alloys of these metals; Forbes [5]. By the close of the fourth millennium, the working of copper and other metals was widespread, annealing was understood for the relief of brittleness, and the use of bronze as stronger alloy of copper and tin was known and growing. The first copper tools (hammer-head, chisel, drift-wedge) evolved between 3500 and 3000 B.C. throughout Mesopotamia, and were followed during the next millennium by development of the (bronze) saw, the file (flat and round), and the copper-head cudgel, or club. For details of the development of mining, copper working, bronze technology, and of tool development, see various articles in Singer et al. [5].

Carbon steel was first developed in Syria, and in Anatolia by the Hittites at a time when the Hittite empire was at its prime (1800 to 1200 B.C.) and rivaled the Egyptian empire in area and power. Between 900 and 700 B.C., the technology of iron and steel advanced further with the development of water quenching to maximize hardness. Tempering to relieve brittleness did not truly evolve until Roman times (200 B.C. to 100 A.D.). During the first millennium B.C., the making of hard but ductile iron components became a consummate art, and was perfected nowhere to a finer degree than in Damascus, Syria, around the time of the Colossus. Furnace technology for iron smelting became available around 1400 B.C.. The forming of wrought iron components by forging soon followed, together with the joining of iron bars and other components by swaging. Iron swords, repaired by swaging during Roman times and possibly earlier, have been found [5], and an iron dagger was found in Tutanhkamon's tomb (1325 B.C.).

DEVELOPMENTS IN FORMING TECHNOLOGY

An important development in manufacturing technology concerns the technique of riveting. A copper pot of riveted construction was discovered during excavations at Ur in southern Iraq, dating from the period around 2500 B.C. By 2500 B.C., thin sheets of soft malleable copper were being hammered and formed into component parts for many utensils. By this time these sheets could have been cut to size by hammering over a sharpened, hardened bronze edge. A suitable punch for copper could readily have been made at that time from an antimony or leaded bronze (bronze alloying principles were then well known), and bronze tools could be hardened by working or possibly by quenching (air or water). In this manner, bronze tools for forming and punching holes for rivets could have been prepared. Finally, the components of the pot could have been joined by the hammering of soft copper rivets.

The technique of punching may have been used by the artisans of the Colossus. While cold punching of thin sections could readily have been undertaken by 300 B.C., the punching of thicker sections would probably have required a furnace for local heating for hot punching. Long pliers could have been used routinely for holding the punch at a distance while making rivet holes along edges of the adjoining sections. These sections could then be assembled using a line of rivets. The long pliers removed the worker from both the heat of the plate and the danger of the punching hammer. A massive cast hammer-head with a long fitted wooden handle would have been commonplace in those times, wielded by a skilled hammerman (as on railroad ties today). Both hot and cold bronze could have been worked in this way, probably using a cast wrought iron anvil foundation containing suitable punching die holes. It is noted that the surface finish of such punched holes could easily have become frayed or even jagged due to these punching operations. It is also likely that no steel files (and certainly no reaming tools) would have existed at the time of the Colossus, and so any jagged surfaces around the punched holes would have to be smoothed by rubbing with stone files. Such jagged edges are still likely to have retained some pattern of small radiating cracks from the punching operations. To the extent that these cracks existed, such cracks would have been logical sites for crack growth as the Colossus swayed, driven by ground-and wind-induced motions. If the method of fabrication by punching and riveting was used in the construction of the Colossus, the finished statue could have contained many cracks, some developed, some latent.

Another technology which was needed for manufacture of the components of the Colossus is mold casting. Casting is said to have begun in Mesopotamia around 3500 B.C.[5] Capabilities for molding and casting bronze implements and figures grew steadily over the next 2000 years. A tomb painting has been found at a temple in Egupt, dating from around 1500 B.C. Each door probably weighed around one ton, and possibly more (a one-inch thick bronze slab of plan 8ft x 12ft weighs two tons). By 800 B.C., casting of exquisitely detailed animal shapes had been achieved without the need for further hand finishing of surface details apart from burnishing [3]. By 500 B.C., the *cire perdue* (lost wax) process for casting of wall thicknesses of 1/8 in. or less for hollow bronze figures (of moderate to life-size, or slightly greater) was being practiced in Greece, having been developed in Syria and Egypt somewhat earlier. Durant [3] reports that Theodorus of Samos and Rhoecus introduced the hollow bronze casting process into Greece.

FORM AND POSE OF THE COLOSSUS

With the above understanding of the construction technology of the time, it is now appropriate to consider the problem of the likely form of the Colossus structure; and secondly, its probable manner of construction.

Romer [4] mentions that colossi were fairly common throughout the Egyptian, Greek, and Roman world, and he states that there are numbers of smaller, contemporaneous statues and bronzework still in existence from the period of the Colossus. There also are descriptions of at least two other huge (but again smaller) statues of gods from a somewhat earlier period available to us - the Zeus of Pheidias at Olympus and the Athena Promachos by the same sculptor[1]. These works provide useful insight into possible manufacturing techniques and capabilities of that period, and are also of value for identifying the artistic concepts which must have influenced the sculptors who designed the Colossus.

The constructional technologies needed to build such a large structure in the third century B.C. included the alloying of bronze, iron working, metal casting, metal forming, riveting, and constructional procedures for large structures. The specific skills required are again finely conditioned by the details of form and shape which were concomitant to the form selected by the Rhodian artisans for the Colossus. Both form and shape were ultimately conditioned by the vast size of the statue. The assumption of a specific form for our purpose must be made with great care, as a wrong choice can be expected to lead to erroneous results from the studies that follow. For example, although the 'pedestal' Colossus (like the Statue of Liberty) shown in Figure 3 is more likely to be stable against ground motions than the 'astride' Colossus forms, this observation alone would be insufficient to accept the pedestal form as the true form. Some clues to the question of form may be obtained from knowledge of the shapes of certain statues that remain from that time, but the knowledge needed is so specific that only general guidance can be expected. Confirmation of metalworking practices used during the classical Greek era has also been obtained from workshops such as that of Pheidias at Olympus, which has been examined by archeologists. Here, the database is much more extensive. To address the question of why the Colossus fell we will therefore bear in mind the following thoughts; first, the statue was that of a male god; second, it was most almost certainly made from bronze; third, it was exceedingly tall in height; and fourth, it most likely stood on two legs. These fundamentals are in agreement with the words of Philo and Pliny, and so it seems most likely that from these witnesses the insight needed

---

[1] The statue of Zeus by Pheidias in 435 B.C. was 60 ft tall, was seated on a throne, and was made from gold with an ivory skin. Romer has discussed its manufacture in detail, including the workshops of Pheidias. Similarly, the bronze statue of Athena Promachos, also by Pheidias, with its pedestal rose to a height of 70 ft. This great statue had a standing draped female form, which made it much more stable and easier to construct. Helpful discussions of aesthetic style in bronze statuary during the late classical period of ancient Greece are given in the works of both Durant and Romer.

to solve the riddles of form, construction, and ultimately collapse of the Colossus can be derived.

Figure 2 Warrior "A" (detail of statue, 200 cm in height) Bronze with Naturalistic Decoration C 460-450 B.C. Found in the Sea of Riace Marina, Italy. Museo Nazionale Reggio Calabria. ©Flynn [9], used with permission

Speculations concerning the pose of the Colossus fall into three categories, and these poses are somewhat related to conjectures as to where the Colossus stood. The first of these is the figure astride the harbor entrance, which despite its widespread popularity has generally been dismissed by scholars for over a century. Such a pose is now regarded as; (a) too *undignified* for a god of such importance as Helios, (b) *impractical*, because the corresponding harbor entrance could then only have been about 30 feet wide, (too narrow and too dangerous for safe passage of laden vessels in and out of the harbor), (c) structurally *unsound* for practical construction and long-term survival of the statue and (d) *too inaccessible* during the construction phase due to the extensive surrounding area required under all construction hypotheses for ramp-works, etc. Objections (a) and (b) were also recognized by Romer. Such ramp-works around an astride Colossus during construction could have blocked the flow of traffic through and around the harbor entrance, possibly for years. The astride Colossus will not be pursued further for these reasons.

The second Colossus concept was the fully draped figure, mounted on an elevated masonry pedestal, and possibly standing within the grounds of a temple complex somewhere distant from the harbor. This concept is also somewhat fanciful, but the form is structurally plausible because the legs are shown together with a draped gown (again like *Liberty* though less extensive). Such a grouping if draped to the ground can be expected to receive valuable structural support and rigidity from this added bracing. Romer rejected the draped concept as being 'more Roman than Greek' in its style. Some form of fully draped Colossus would be both plausible and attractive from a *structural* viewpoint, but a statue of a god *draped in a toga-like garment* may not be acceptable from an aesthetic standpoint, given the reverence afforded the male form by the ancient Greeks. One other deterrent is that if the Colossus was required to serve as a harbor beacon, a statue situated closer to the water would seem more likely for this purpose than one in a temple complex removed from the harbor.

The third Colossus idea, Figure 1, is a semi-nude figure standing *near* the harbor entrance, possibly upon a masonry platform near the present site of the fort of St. Nicholas. This form of this concept is similar to that of the Apollo Belvedere which was of Greek craftsmanship dating from the same period as the Colossus. The form shown in Figure 1 appears to be relatively weak at the ankles and lower legs, and its torso and arms are relatively massive. This suggests an inherent structural weakness.

For the structural analysis given in this paper, a figure with a closer-legged pose mounted on a single platform was chosen. The selected pose permits straight-forward computer modeling, but retains all the essential cultural and aesthetic aspects which would affect the dynamics of the statue. After consideration, it seemed likely that the true pose was some closer-legged version of the pose shown in Figure 1, with an aesthetic form similar to that of the Apollo Belvedere, mounted on a platform..

ANCIENT DESCRIPTIONS OF THE STRUCTURAL DETAILS
Romer [4] cites in detail specific writings from Philo of Byzantium concerning the construction of the Colossus. For detailed examination, these comments are laid out below in numbered sequence to aid the development of an hypothesis concerning the most likely method of construction used. The following statements have been taken *verbatim* from Romer, who quotes Philo as saying:

1. It was secured firmly from the inside with iron frames and square blocks of stone.
2. The horizontal bars exhibit hammer-work.
3. The hidden part of the work is bigger (*more substantial?*) than the visible parts.
4. With what workforce was such a weight of poles (*rods?*) forged?
5. The molten image of the structure was the bronzework of the world.
6. A base of white marble was laid down, and on this …
7. The sculptor first set the feet of the Colossus up to the ankle-bones.
8. The soles of the feet on the base were already at a greater height than other statues.
9. It was impossible to lift up the rest and set it on top.
10. The ankles had to be cast on top, and as it happens in building houses
11. The whole work had to rise on top of itself.
12. Artists first make a mold, then divide it into parts, cast them, and finally put them all together and erect the statue.
13. It was not possible to move the metal parts.
14. When the casting had been done on the earlier worked parts, the intervals of the bars and the joints of the framework were taken care of.
15. The structure was held steady with stones that had been put inside.
16. An immense mound of earth was continually poured round the finished parts of the Colossus, hiding what had already been done underground.
17. The artist carried out the next stage of casting on the flat surface of what was underneath.

Durant [3] and Romer [4] also attribute the following descriptive points to Pliny:

1. Few people can get their arms around one of its thumbs.
2. Its fingers are bigger than most statues.
3. Where the limbs have broken off there are huge gaping cavities.

4.   Inside it one can see rocks of enormous size which Chares had used to stabilize it when he was building it.

POSSIBLE METHOD FOR FABRICATION AND CONSTRUCTION

The following guidelines were use: (a) the state of construction and fabrication technology which is outlined in the preceding sections, (b) the seventeen criteria of Philo, and (c) the four criteria of Pliny. The hypothesis, which follows is based on the above items and is here submitted as satisfying the above criteria. This hypothesis thus represents a workable concept for the fabrication and construction of the statue (C = criteria numbers with author, referred to the earlier criteria listings):

Base: A substanstial base platform of stone surfaced with marble at least 60 feet in diameter and 10 to 15 feet in height would first have been constructed. The form of this platform was possible circular or polygonal in shape. Philo, C6, 7 and 8.

Feet: The feet of the statue were built using massive blocks of carved stone. The feet were then covered with thin sheets of beaten bronze, riveted together. This stone block construction of the feet was carried up to the ankles. Philo, C1, 7 and 8.

Ankles: Eight or so major iron bars were each made by forging and swaging, and were set into and beneath the base structure in a horizontal, radiating pattern. The horizontal bars in the basework converged inwardly upon the ankle where they turned and became vertical. These vertical bars became smaller in diameter, and followed the shape of the statue up the feet and inside the leg. In this manner continuity, strength, and stability were imparted to the lower regions of the structure. These bars were further used to constrain the bronze outer surface plates into a continuous circular structure. Philo, C1, 2 and 7.

Legs: From the ankles up, the structure of the statue was formed from many curved, cast bronze plates joined together into circular rings by rivets, or T-clamps. These clamping devices are set into precisely- aligned, adjoining lugs or holes in the surface plates. Each plate was about 60 inches by 60 inches at the sides and about one inch thick in the lower leg, and successively less in thickness at the higher levels of the statue. These inch thick plates weigh from 600 to 1000 lbs depending on their outer detail. The higher plates were of thickness about ¼ in. or so, and they weighed correspondingly less. Philo, C7, 8, 9, 10 and 11. Pliny C3.

As the structure rose, the thickness of the plates could be reduced as the surface detail would allow, except where needed for strength (joints, loins, arm-shoulder region, etc.). Thus a 60x60x1/4 inch plate could weight 150 to 250 pounds with the same thickness of rim for strength of assembly and for the lugs which are needed for attachment. Such a surface thickness is well within the casting capabilities of that time. (No comment from Philo or Pliny).

It is here suggested that such plates may each have been cast to the particular local shape required for the statue by the artist, possibly with their vertical edges slightly angled inwards, so that when assembled the plates would readily form the circular or elliptical ring shape required by the surface shape. These plates were possibly clamped together using rivets fitted through carefully aligned holes formed during the casting process (and by punching with thinner sections), as shown in Figure 3. The metal casting technique proposed is practical and within all aspects of the technology of the time. The fine level of skill achieved during this period is shown in Figure 2. Moreover, the casting of plates could have been undertaken off-site, and the finished plates could have been transported to the site of the statue, and then assembled. The first rings of the legs could have been assembled into position and secured to the stone ankle bases. Subsequent rings could have been secured into position above the lower rings by the same riveting or T-clamping procedure. A possible casting procedure for the face of the Colosus is suggested in Figure 4.

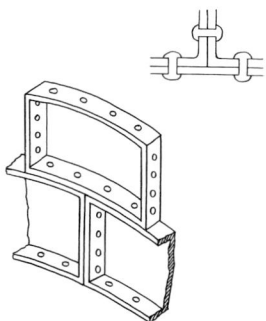

Figure 3 Details of Joining Plates by Riveting

If the structure rose on average by about 10 feet per year (two rings) up through the legs, torso, arms and head, this would require the manufacture and assembly of about 30 to 60 of such plates per year over the 12 years of construction. Philo, C11, 12, 13 and 17.

A circular mound of earth might have been built around the structure to provide access for transportation and construction. This mound could have been enlarged as the statue grew upward, but it would also grow radially outward. The size and stability of this mound could have been limited somewhat if it were strengthened by a massive wooden spiral retaining wall or fence, perhaps reinforced by forged chains or ropes. The mound might have been raised and extended outward as each successive ring of plates was completed. The legs of the structure could still have been stabilized by filling them with masses of stones, some large, some small, at least up to the knees for greatest effect. Philo C15 and 16. Pliny C4.

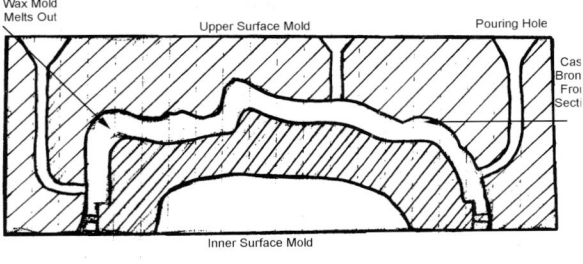

b) Mold for Casting Front of Head

Figure 4 Details of Construction

It seems possible that many successive rings of the massive structure might first have been created as full-size wax impressions each mounted upon a central hollow core in the artist's studio near his foundry. The outer surface of each ring could then have been worked on in great detail by artists for months to ultimately create the desired form of the god. The inner surface could also have been worked on by foundrymen to develop the flanges, holes and assembly features. Upon artistic completion each molded ring of the structure could have been subdivided for final forming into many separate vertical plates about 5ft x 5ft in size. The completed molds for each of these plates could then have been poured individually in the foundry using the *cire perdue* process, to create the required of interlocking bronze plates for the statue.

Philo in Comment 16 mentions the 'mound-of-earth' idea for the construction process, for which there are precedents in the ziggurats of Mesopotamia and likely with the pyramids of Egypt. But though it is a practical possibility, a stabilizing surrounding circular ramp could require stout retaining walls (perhaps of wood, bound by ropes or chains), plus an extensive adjacent area upon which the earth mound might grow outward. A clear circle of about 300 ft or more in diameter would be needed to reach the top of the Colossus, with a 30° angle of repose for the soil. And of course a *vast* amount of earth fill. Such an area is larger than the footprint of the present fort of St. Nicholas at the end of the sea wall of Mandraki harbor. Romer has mentioned the possibility that the great wooden towers left behind from the siege of Demetrius 15 years before may have been transported to the site of the Colossus, and might have been used in its construction. One tower is said to have been 100 ft tall, 200 ft long, and possibly 75 ft broad. Stripped of its armor and engines of war, such a huge platform would appear to constitute a very useful assist for the peacetime construction of another tall structure of similar height.

NATURE OF EARTHQUAKE LOADING

Earthquake loading is not pure-harmonic in nature. Typically it is spectral (multiple-frequencies) and transient in time. That the Colossus was felled by a strong earthquake is widely accepted, and it is also known that the terrain of the Aegean and of Asia Minor contains is generally rocky. One strong earthquake which is well-documented in its time-history, and which also occurred in rocky terrain is the El Centro, California earthquake of 1940. This earthquake is relevant because it took place in ground similar to that of the terrain of Rhodes. Acceleration, velocity, and displacement vs time traces of the surface horizontal motion from this earthquake are shown in Figure 5, from Newmark and Rosenbluth [10]. Nothing, of course, is known concerning the Rhodes earthquake of 224 B.C. Its strength, duration and directional properties remain as mysteries. The El Centro earthquake registered around 8.0 on the Richter (modified Mercalli) scale, i.e., of similar strength to the San Francisco (1906) quake which is to have measured Richter 8.1.

It was ultimately decided that the most representative earthquake excitation at the Colossus site could be best defined by using a median Newmark-Hall forcing spectrum anchored in a 0.30g zero period ground acceleration (ZPGA). This spectrum is the normalized average of many earthquakes in similar terrain. This excitation was applied with equal strength in two orthogonal horizontal directions, and the applied vertical peak acceleration was assumed to be 2/3 of the horizontal peak accelerations. The Newmark-

Hall spectrum is widely used as representative in a probabilistic sense for firm soil sites all over the world, including Asia and Europe. Having in mind the physical uncertainty related to the definition of the seismic hazard at the Colossus site, the acceptance of a median Newmark-Hall spectrum appears to be most reasonable for present purposes. Two Newmark-Hall analyses and two El Centro (1940) analyses also were performed, and as with the Newmark-Hall spectrum using 2 percent and 5 percent equivalent damping for the statue structure. Lastly, it was verified that the shape of the spectrum is consistent with other earthquakes, which have occurred in the Aegean area.

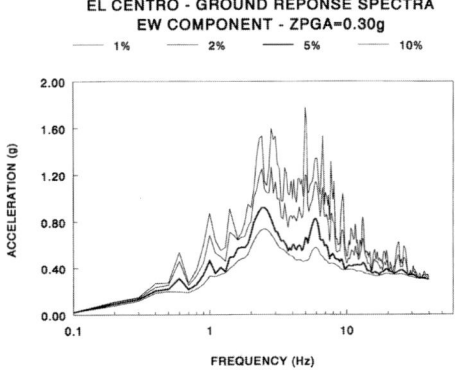

Figure 5 El Centro (1940) Acceleration Spectrum

FINITE ELEMENT MODEL

A representative computer model of the Colossus was created using the solid modeling capabilities of the ANSYS [11] code; see Figure 6. The structure was modeled as a surface continuum of plates, with identical boundary displacements and slopes throughout. The legs were modeled as a series of nine hollow circular cones per leg, as shown in Figure 8. The lower abdomen was represented by three hollow elliptical cones while the upper part of the abdomen was created using a series of four hollow elliptical cones. In order to model the chest, two hollow elliptical cones were used. The shoulders were next created with a series of five hollow partial-circular surfaces between the front and back portions of the chest. Next, the head was created (as a hollow sphere) together with the neck, as a hollow circular cylinder. The head and neck were then overlapped together with the shoulders using the Boolean capabilities of the ANSYS code. Finally, the arms were modeled using hollow circular truncated cone sections,

Figure 6 Finite Element Mesh

ending with an overlapping hollow sphere at each end of each arm, representing the hands. The arms were angled between defined spatial points at each elbow and wrist. After consideration, the system of iron rods was not included in the model.

Different values of plate thickness were assigned to various regions throughout the Colossus model as follows:

legs and feet-cones below knees	1.00 in
legs above the knee	0.75 in
lower abdomen	0.50 in
upper abdomen	0.50 in
shoulders	0.50 in
arms	0.50 in
neck and head	0.50 in

The geometry was converted into a finite element model using the ANSYS Automatic Mesh Generator capability. This technique automatically led to the model shown in Figure 6, which comprises 11004 shell elements and 6552 nodes. It was assumed that the statue was made of bronze for which the following mechanical properties were selected in the analyses, Marks [14]:

Young's modulus	15×10^6 psi
Poisson's ratio	0.32
Density	0.332 lb/in^3

It was assumed that below the knee the lower part of the leg contained rock fill. This condition was incorporated into the model by changing the density of the leg material below the knees. This was done to represent the mass of the rocks without adding stiffness to the model. Using a rock density of 150 lb/ft^3 and a fill/volume ratio of 0.75 for the leg and 0.9 for the foot, the density in the foot section was changed to 2.195 lb/in^3, while that of the leg it was altered to 1.882lb/in^3. The rocks in the leg were of great size (Pliny), which is likely to have made the fill ratio less than in the foot, where the stone was carved and consequently the fill was more complete there (Philo). Finally, the feet were represented by short conical sections, the soles of which were constrained against both displacement and rotation to simulate the effect of the rugged stone-and-iron foot and foundation structure, and of the foundation itself, on the motions of the Colossus.

FINITE ELEMENT ANALYSIS
Static Stresses
Static stresses in the model were first computed under gravity loading. Generally the largest stresses were found at the joints of the arms and legs. Away from the joints the stresses were always lower. The maximum stresses occurred in the arms and shoulder joint were due to gravity bending. Von Mises stresses of about 48 KSI were found in these regions. Static gravity stresses above the knees and in the legs and ankles were found to be in the order of 3.3 KSI.

Natural Frequencies and Mode Shapes

Natural frequencies and vibration mode shapes were calculated for the statue. The shapes of the lowest two modes are shown in Figures 20 and 21. Properties of the first ten natural modes are described in the following table with their corresponding frequencies.

Mode	Frequency Hz	Shape of Mode
1	0.6702	front/back oscillation
2	1.2747	up/down oscillation of left arm
3	1.4992	twisting of both arms and shoulders
4	2.0721	up/down oscillation of right arm
5	2.2418	front/back movements of both arms in-phase
6	2.8176	right arm side-ways oscillation
7	3.4378	left arm side-ways oscillation
8	4.4742	twisting motion of whole body
9	5.0951	right arm and upper body up/down motion
10	5.3605	body twist around waist, no arms movement

Spectral Dynamic Response Analysis

The model Colossus was subjected to seismic excitation applied at the base of the platform. The main seismic spectrum used was a median Newmark-Hall ground response spectrum having a peak ground acceleration of 0.3g. This excitation was applied in all three X, Y and Z directions, with relative strengths in the ratio of 1, 1 and 0.67. Two different spectra were applied to the model. These spectra represented the ground acceleration applied at damping ratios of 2 percent and 5 percent respectively. The spectrum results were determined as the Square-Root of the Sum of the Squares (SRSS) of all the contributing modal stress responses.

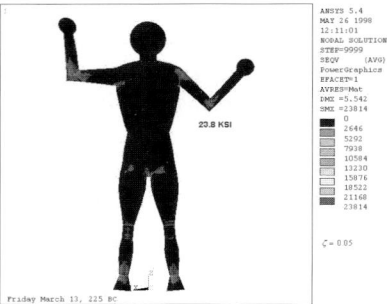

Figure 7 Maximum Dynamic Stress Response to Acceleration Spectrum. All Modes

The maximum spectral Von Mises (SRSS) stress found by this method was ±23.8 KSI. This stress occurred in the elbow joint, as shown in Figure 7. The corresponding (SRSS) stress in the ankle was ±13.4 KSI, while that above the knee was ± 9.0 KSI. These results are shown in Figure 8. Again, low spectral dynamic stresses were found throughout the body of the Colossus. This clearly indicates that without the presence of substantial reinforcing construction in the body joint regions, the first signs of damage from earthquake forcing can be expected to occur in the vicinity of the body joints. Hand calculation showed that such stresses could have been reduced by a maximum of about six percent by the proposed system of vertical iron rods.

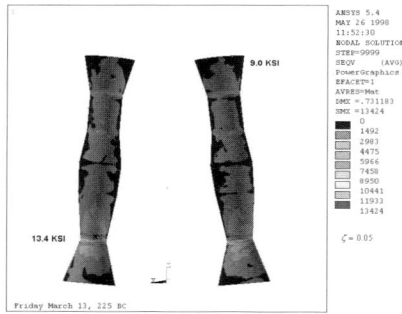

Figure 8 Acceleration Stresses in Legs

DISCUSSION OF STRESSES AND OF FAILURE POTENTIAL

1.   How failure by earthquake ground motions influenced form and pose of the statue.

A seated colossus form is not likely to have broken at the knees or ankles because the throne and feet would likely represent a more substantial and stable restraint. A standing, draped female form (like *Liberty*) would be more resilient for the same reasons. A male form draped to the ground is not likely to have been the form of the god. Statues of male gods in Hellenist Greece were characteristically nude, especially when standing.[1] A standing nude form was frequently braced at the leg to impart stability (the Apollo Belvedere), and though such bracing was *possibly* used with the Colossus, this form seems less likely, because if the statue were situated near the harbor to serve as a welcoming beacon, the sculptor's desire to portray the beauty of the god would have been encumbered by such attachments. Internal reinforcement therefore seems *much* more likely, and perhaps this is why the iron bars are mentioned by Philo. Thus it appears that the most likely form was also the one with potentially the least structural stability. The action of an earthquake on such a statue is unlikely to have been mitigated to any extent by leg bracing, because if such bracing were effective, it would have stiffened the frequency of the destructive lowest mode *directly into the greatest strength of the excitation spectrum*: see Figure 5. The strong excitation region lying between 1.0 Hz and 5.0 Hz is sufficiently broad that the stiffening of one leg is unlikely to have saved the structure. The arms would still have failed, and probably the unstiffened leg also, thus bringing down the remainder. Similarly, removing mass from the upper levels of the statue by using thinner plates would also have increased the lowest natural frequency into the strongest excitation region. Height and chosen form were thus the major aesthetic factors which influenced the collapse of the Colossus.

2.   The internal fastening procedure must be both constructable and practical for the times.

Given the proposed construction using cast plates, and accepting that these might have been crafted for assembly by grinding the mating surfaces with stones, only a limited number of assembly techniques are available. Assuming consistent skilled craftsmanship (but not necessarily precision), rivets appear to be the most likely selection for fasteners between the surface plates. They could be manufactured in great quantities off-site by mold casting - one head, and a shank for peening. There were of course no bolts; the metal screw thread dates from as recently as Leonardo da Vinci's era. Another possibility was the clamp, which had a two-headed cross section (like a peened rivet), but needed to be inserted from the outside into an open slot, cut into the two mated surfaces. Clamps obtain their clamping action from shrinkage after being inserted while hot.[2] This shrinkage typically gives rise to high residual tensile stresses. Clamps are considered less likely for the Colossus because they would have been less convenient to manufacture and install, and required greater precision. The rivet therefore emerges as the most likely fastening used in the Colossus, for reasons of convenience, of manufacture, and of installation. A simple installation procedure would have been to heat the rivet tip in a brazier at the site, then install it in the mating holes using a pair of tongs, and then to peen down the second head, using a metal anvil or a large rock to absorb the transmitted peening blows. When the rivet cooled, the joint would have clamped tight from the residual tensile force developed by the shrinkage.

---

[1] Such was the level of respect for athletic male forms and for the greater deities among Greek sculptors that a standing god offered the greatest scope a sculptor to demonstrate his skill in depicting the sublime beauty of the divine form. As the architect of the Colossus was an eminent sculptor of the day, it seems certain that veracity of form in the statue must have powerfully influenced Chares' choice for Helios's form.

[2] Clamps were widely used throughout ancient cultures. Mayans and Incas used them to bond stones together, by pouring molten metal into matching carved slots in the two mated components [13].

3.   What was the net effect of all likely loads acting on the knee joint rivets.

The loads acting on the cross-sections of the knee joint rivets consist of combinations of the following forces:

Static Loads

Gravity　　　　　～ 2 KSI on flanges
Horizontal Wind ～0.5 KSI from 30 mph wind
Rivet Shrinkage　+15.0 KSI from 500°F shrinkage

Dynamic Loads

Horizontal Wind ～ ± 0.5 KSI from 30 mph wind
Earthquake　　　± 45.0 KSI average 0.3g acceleration

The most significant static load evidently results from rivet shrinkage. This load is expected to impose local strains beyond the yield strength of a well formed tight-fitting rivet. An even greater region of yielding can be expected to develop at the shank/head fillet radius. In some locations, the compressive load of gravity may slightly relieve some of the rivet tension..

The most important rivet dynamic load is the earthquake load S_{fem}, which was found to lie somewhere between ±30 KSI and ±100 KSI (elastic), depending on rivet diameter and number. An average elastic dynamic stress of ±45 KSI was used for making rivet strain estimates. This dynamic strain alone would be sufficient to yield the outermost rivet to failure (cracking) at the rivet head fillet in very few load cycles. Repeated cycling could be expected to propagate such a crack to failure in a short period of time, noting that upon cracking (or even substantial yielding) the load is expected redistribute to other rivets, until they too cracked. Combining a high static stress with the high earthquake load would further hasten the development of this cracking mechanism.

SUMMARY

Ample sources of high stress loading combinations sufficient to eventually cause rivet failure existed in the day-to-day conditions surrounding the Colossus. Such dynamic plus static load combinations existed during the fatal earthquake, and these must have been well able to cause significant rivet yielding. High static residual stresses may also have existed in the copper rivets following manufacture, especially in the rivet head fillets. Load shedding due to yielding would have spread the high cyclic loads to other rivets, and likely caused these rivets to crack also.

When the extreme fiber rivets failed, the rivet failures could thus have cascaded until the structure broke up at its joints (arms, legs, knees) and toppled to the ground. The Colossus likely shattered at several locations during the quake, as noted from the multiple sites where high dynamic stresses existed. The fundamental breakup most likely occurred while the statue was standing and whole, though further destruction undoubtedly occurred when the pieces hit the ground.

CONCLUSIONS

1.   An El Centro (1940) strength earthquake could have brought down a Colossus of traditional size and form, built with the constructional details described in the paper.

2.   The most probable primary cause of the collapse of the statue was the inherent weakness of wrought copper as a joint material, and possibly the manufactured quality of the cast rivets and holes.

3.   The weakest points of the Colossus under earthquake conditions occurred where the joints were located. Stresses in the body and limbs away from joints were low.

4.   The arm-body joints were the highest stress locations, from both gravity load and from the earthquake dynamic load. It appears likely that an arm, or arms, would have failed first.

5.   The ankle joint was also a critical location. The survival of this joint would appear to have depended on the constructional details of the ankle joint, the quantity of iron used there, and the quality of detailed workmanship employed. Copper rivets at this location seem certain to have yielded under earthquake loading.

6.   The knee joint is less vulnerable than the ankle, but the actual distribution of stresses between these locations would depend greatly upon the internal strengthening details employed.

7.   All major technologies required to build a metal Colossus were available in 290 B.C. The required constructional procedures and techniques were also available, and had been successfully demonstrated on other statues and structures prior to this time.

8.   The construction method proposed herein is plausible for the times and is consistent with the writings of Philo and Phiny.

REFERENCES

1. Philo of Byzantium, "On the Seven Wonder," Written ca. 225 B.C. in Alexandria, Egypt. From a Ninth-Century Byzantine Manuscript, University of Heidelburg, Germany, Trans. H. Johnstone, from J&E Romer, ref. 4, 1995.

2. Pliny, "Natural History," Vol. 34, Art. 18, pp. 41-42, English Transl., E. W. Bostock 1855.

3. Durant, William, "The Story of Civilization: Part 2 'The Life of Greece'," Simon & Shuster Publishers, New York 1939, pp. 622, 143.

4. Romer, John and Elizabeth, "The Seven Wonders of the World," 1994.

5. Singer, C. J., Holmyard, E. J., Hall, A. R., "A History of Technology," Vol. 1, Oxford University Press, 1954.

6. Bronowski, J., "The Ascent of Man," Little, Brown & Co., Boston, 1973.

7. Encyclopedia Britannica, Various Volumes, University of Chicago Press, 1966.

8. Konstantinopoulos, G. G., "The Archeological Museum of Rhodes," Adam Editions, 275 Mesogion Ave., Athens, Greece, 1987.

9. Flynn, T., "The Body in Sculpture," Everyman Art Library, Orion Publishing Group, Upper St. Martin's Lane, London, WC2H 9EA, 1998.

10. Newmark, N. M., Rosenbluth, E., "Fundamentals of Earthquake Engineering," Chapter 7, Prentice-Hall Publishers Inc., Englewood Cliffs, N.J., 1971.

11. ANSYS User's Manual, ANSYS Inc., 275 Technology Drive, Canonsburg, P.A.

12. Bannantine, J. A., Comer, J. J., Hardrock, J. L., "Fundamentals of Metal Fatigue Analysis," Prentice-Hall Publishers Inc., Englewood Cliffs, N.J., 1990.

13. White, P. T., "The Temples of Ankor., Ancient Glory in Stone," National Geographic Magazine, Vol. 161, No. 8, p. 579, May 1982.

14. Marks, L., Baumeister, "Mechanical Engineers Handbook," McGraw-Hill Book Co., Inc., New York, N.Y., 1978.

15. Walkosz, L., "Stars Web-Site Home Page," April 25, 1998 On-Line Internet (6/16/98), Available www.http://www.mmmpcc.

"DEUS-EX-MACHINA" RECONSTRUCTION AND DYNAMICS

Thomas G. Chondros,
University of Patras, Department of Mechanical Eng & Aeronautics
265 00 Patras, Greece, E-mail: chondros@mech.upatras.gr

ABSTRACT-In some ancient Greek drama, an apparently insoluble crisis was solved by the intervention of a god often brought on stage by an elaborate piece of equipment. This "god from the machine" was literally a *Deus Ex Machina*. Archaeological evidences and descriptions of mechanisms used in the ancient Greek theatre were investigated in an attempt to reconstruct the *Deus Ex Machina*. None of these machines, made of perishable materials is extant. However, from the numerous references to such machines in extant tragedies or comedies and vase paintings, information about its design and operation is available. Static and dynamic analysis and simulation of the mechanism kinematics were performed. The reconstructed mechanism is a spatial three or four bar linkage designed for path generation.

KEYWORDS: Deus Ex Machina, Kinematics, Path Generation, Reconstruction.

INTRODUCTION

The ancient Greek theater was born in the 6th Century BC out of a form of a single actor dance drama that flourished in the Dorian parts of Greece. It reached a period of maturity by the 5th Century BC, following the victorious wars against the Persian Empire. Aeschylos, who was himself a war hero, won his first literary victory in the dramatic festival of 484 in Athens. In 472 he presented the Persians, where he introduced the second actor besides himself, and thus created the multi actor theatrical play. Sophocles, Euripedes and Aristophanes made great contributions to the Greek theater of the 5th Century.

The word *Mechanism* is a derivative of the Greek word *mechane* (which meant machine, more precisely machine element meaning an assemblage of machines). While it was used for the first time by Homer in the Iliad to describe the political manipulation, it was used with its modern meaning first in Aeschylus times to describe the stage machine used to bring the gods or the heroes of the tragedy on stage, known with the Latin term *Deus ex machina*. *Mechanema- Mechanism*, in turn, means an *assemblage of machines* and was used by Aristophanes. At the same time, the word *mechanopoios*, meaning the machine maker or engineer, was introduced for the man who designed, built and operated the *mechane*.

Dimarogonas [1] (1991&1992) refers that *mechanes* are large mechanisms consisting of

---

[1] This paper is based on the work that the author has developed with Professor Andrew Dimarogonas, in an attempt to reconstruct the mechanism and operate it in an ancient Greek theater. Andrew D. Dimarogonas (1938-2000) W. Palm Professor of Mechanical Design in the School of Engineering and Applied Science at Washington University, St. Louis, Missouri was widely recognized as a distinguished authority in various specialties of mechanical engineering. As an engineer-historian he made important contributions to the mechanical design and vibrations and was member of the PC for History of MMS.

88

beams, wheels and ropes which could raise weights up to one ton and, in some cases, move them back and forth violently to depict space travel, when the play demanded it. They were well balanced and the finger of the engineer could operate them, with some exaggeration perhaps. The vertical dimensions were over 4m while the horizontal travel could be more than 8 m. There is indirect information about the timing of these mechanisms. During the loading and the motion there were specific lines of the chorus, from which we can infer the duration of the respective operation. The fact that the designer himself operates the machine could only point to a complex mechanical device.

None of these machines, made of perishable materials is extant. However, there are numerous references to such machines in extant tragedies or comedies and vase paintings from which they can be reconstructed.

A computational algorithm for the design of various *mechanes* configurations was prepared. Static and dynamic analysis as well as simulation of the mechanism kinematics is performed. The reconstructed mechanism is a spatial three or four bar linkage designed for path generation. Furthermore, a detailed design of the various parts of the mechanism is presented. In this way the construction and operation of the mechanism can be evaluated and thus, reconstruction and operation could be feasible.

ARCHAEOLOGICAL EVIDENCE

A substantial number of ancient Greek theaters have survived the time to the extent that we can reconstruct the theater architecture with some precision. We shall make use of the architecture of the Athens theater of Dionysus Eleuthereus on the southeast slope of Acropolis, which later was modified under the orator Lykourgos, in the times of Alexander the Great. A reconstruction of the Athens theater by Bulle (1928) is shown in Figure 1 while the top view of the theatre in the 5th Century by Fiechter (1930) is shown in Figure 2. About 30,000 spectators could be accommodated according to the ancient authors (Platon, Symposion). Performances were given throughout the day and lasted for several days.

Fig. 1: Reconstruction of the Athens theater by Bulle (1928)

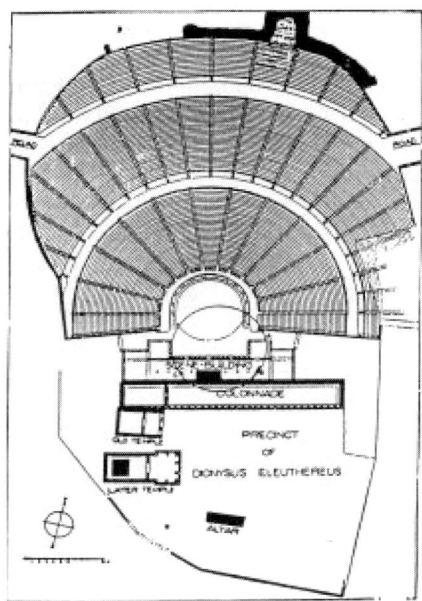

Fig. 2: Plan of the Athens theater in the 5th Century B.C., Fiechter (1930)

In earlier times, the *skene* building of the Athens theater was about 4 m high and 27 m wide. At later times the *skene* was a two-story building, according to some authors.

The *skene* building had an almost flat roof on which the actors performed in some plays. Otherwise the play was performed in the circular area called Orchestra, the original meaning of which was dance-area. Aeschylos introduced the idea of a god, which would emerge suddenly from behind the *skene* building, over the roof and (perhaps) into the Orchestra area to conclude the drama. The preparation and pick-up of the actors (and in some cases of horses and chariots too) was done from behind the *skene* building. The instrument for this operation was called *mechane* equivalent to the Latin term *machina*. From the dimensions of the *skene* building we can figure out the spatial envelope in which the *mechane* would operate.

Other pieces of stage machinery were the *periaktos* and the *ekkyklema*. *Periaktos* was a revolving piece of scenery, which was used as todays revolving stage for quick change of scenery. It had three faces, and was supported on a vertical shaft, which was itself pivoted on the floor in a stone bearing. The *ekkyklema* was a wheeled low-profile vehicle used to carry to and from the stage heavy scenery (such as a throne) and also dead bodies of the play since murders and suicides were not allowed to occur on the stage of the ancient theater.

Periaktos and *ekkyklema* evolved from the pottery wheel and the wheeled vehicle, respectively, and they do not have substantial engineering significance.

The *mechane* does not appear to have any close predecessor and its development is of great engineering significance since it appears that it was designed to meet very close requirements of the play and it was not arrived at by long evolution.

Although there are numerous illustrations of plays involving the *mechane* in vase paintings there is a very little information about the mechane itself, which seems to be invisible. In an Apulian vase in the Metropolitan Museum of Art in New York we can see the body of Sharpedon, son of Europa, carried through the air by Sleep (Ypnos) and Death (Thanatos) most probably from Aeschylos' Europa (Figure 3). There is an important piece of information in this, however, since it indicates that the *mechane* was not very visible. This also agrees with aesthetic considerations, since a highly visible mechanism would not be in tune with the scenery, costumes and masks they were using.

Fig. 3 Sharpedon, son of Europa, carried through the air by Sleep and Death , Aeschylos' Europa.Apulian vase.

Fig. 4 Medea in a snake drawn chariot of the Sun, Euripides' Medea. Early Lucanian hydria, ca 400 B.C.

Another vase painting illustrates a scene from Euripides' *Medea* (Figure 4). Medea is ready to drive off to Athens in a snake drawn chariot of the Sun, which would have been on the *mechane* while below there are her dead children, on the left is Jason and on the right the teacher and, probably Aphrodite. Again the *mechane* seems invisible.

Another vase painting shows a scene from Euripides' *Stheneboia* (Figure 5), Bellerophon already on the Pegasos both on the (not visible) *mechane*.

Fig. 5 Bellerophon on the Pegasos,Euripides' Stheneboia. Attic red-figured krater, 400-375 B.C.

Fig. 6 Zeus and Heracles threatening Apollon, Rhinthons' Heracles.Phlyakes vase, Ermitage, St Petrograd.

From a Phlyakes vase (Figure 6) we see Zeus and Heracles threatening Apollon with the thunderbold and the club, probably in the comedy *Heracles* of Rhinthon. It is interesting that Apollon is seated on a scaffold, which seems out of place and, most probably, is a caricature of the *mechane*. If this is the case, this is an indirect illustration of the mechane.

EVIDENCE FROM THE CLASSICAL TEXTS

Although none of the *mechane* itself or its illustrations are extant, there are numerous references in extant plays, which will yield information about its design and operation.

Detailed information on stage scenery and machinery in the ancient theater are given by Vitruvius (1st Century AD) and Pollux (3rd Century AD). It is described by Pollux as follows "The mechane shows gods or heroes -Bellerophons or Perseuses- in the air. It is situated by the left parodos at a height above the *skene*" Parodos (side walk) is the narrow pass to the Orchestra on each side. This defines the position of the main body of the *mechane*, though it is not known at which side. It defines also the range of operation since "Deus ex Machina" appeared at about the middle of the *skene* building.

According to Dimarogonas (1992) the use of the *mechane* is always preceded by a notice in the text, several lines of covering dialogue to give time for the *mechane* to swing into place. This can be used to time the respective operations. It should be noted, however, that many scholars argue that not always the entrance of Deus ex Machina was preceded by a proper dialogue.

Most authorities agree that the *mechane* was first used in Aeschylos times. In *Eumenides* (403) Athena flies in on a chariot, which means that a load of about 4000 N was transported. Wilamowitz (1907) however, suggested that the entrance of Athena was on foot and that the respective line was added later when the *mechane* was available.

There is further evidence that the *mechane* was used by Aeschylos: in *Prometheus Bound* the entrance of the chorus of the Oceanids by a flying chariot is introduced by Prometheus by thirteen lines, while in Euripedes plays the respective introductions are 6 to 7 lines. One explanation is the substantial load to be transported: Since the number of the chorus members at this time was fifteen, it can be estimated that the load transported would have been about 10,000 N. The door is open, however, for a different interpretation: A double entry using two *mechanes* as suggested by the scholiast Lucianus (2nd Century AD) one on each side. This reduces the load on each to 5,000 N still a formidable one for that time. Pollux describes the *mechane* as a crane, which comes from above and grabs the body of Memnon in the Aeschylos tragedy.

Aristophanes used the *mechane* too. In *Peace* Trygaios flies to heaven on a dung beetle and appeals to the *mechanician* not to let him fall (line 173) "...engineer be careful, for if you are not, I will become food for the dung beetle..." and he further complains that he is dizzy from the swinging motion. He recites eighteen lines (154-172) during which time he is swung violently enough to convey a sense of space travel and then lowered back to the point where he had started. In *Daedalus* Aristophanes mentioned that the *mechanopoios* - machine maker operates the wheel. The fact that the designer himself operates the machine could only point to a complex mechanical device. Further, in *Gerytades* he moves the crane fast using the word *periagon*, which means general rotation about a pivot. In *Clouds* he describes Socrates is coming from the air in a net hanging from the *mechane*. In *Birds* there are references to flight.

Several conclusions can be drawn from these:

a) The machine could provide vertical motion in addition to horizontal.

b) The operator could control the vertical position of the load.

c) A wheel was used by the operator for some control function.

d) A pivoted beam was probably the main party of the *mechane*.

e) In some cases the load should move fast.

It is important to note that Prometheus Bound, where the most powerful *mechane* on record appears to have been used, is a hymn to democracy and technology. Product of Aeschylos age of maturity, it was produced in 458, shortly before his death in 456, in Cicily. This tragedy has a powerful symbolism: Prometheus taught the people the science, the technology and the use of the fire. With them, they could tame the natural powers and make their life better.

This brings Prometheus, the Democracy, in a collision course with Zeus, the oligarchy or Tyranny. Zeus orders the state to arrest him and nail him on a rock on the mount Caucasus. The parallelism with Jesus crucifixion in the Christian tradition is apparent.

Of course *Prometheus Bound* and the Attic Drama in the 5th Century were no isolated events. In the preceding century, the Ionian School of Natural Philosophy and the Pythagoreans have founded natural science (Dimarogonas 1990) and introduced the idea that "nature can be explained" (Schroendiger 1954). Technology is not only the work of unknown craftsmen but learned people like Thales, Pythagoras, Democritos and Aristoteles are developing the theoretical foundations for it, causing sometimes bitter comments by other philosophers like Platon. Technology now is an accepted occupation of free men and the mechane development together with Aeschylus' Ode to the science, technology and Democracy in *Prometheus Bound*, are not irrelevant events.

The *mechane* becomes an established stage machine with Euripides. In *Medea*, she flies on the mechane in the snake-driven chariot of the Sun. In *Bellerophon*, and probably in *Stheneboia* Bellerophon rides the flying Pegasus. The horse and the rider must have been moved through the air in a platform and the load should have been about 4000 N. Other gods or heroes of Euripides used the *mechane*, such as Lyssa and Iris in *Heracles*, Dioskouroi in *Electra* and *Helen*, Perseus in *Andromeda*, Apollon and Helen in *Orestes*, Medea, Athena in *Ion*. In *Phoenisae* he makes reference to the actor hanging from a harness.

Antiphanes, making fun of the tragedy poets, said that "...they move the *mechane* as they raise their finger, when they have nothing to say...". This indicates that mechanical advantage and load balancing was used. Platon in *Cratylos* has a similar reference to the tragic poets.

Pollux wrote that the mechane was placed towards the left parodos but it is not clear in what respect he meant that.

According to Pollux and Ploutarchos the gods were hanging from the *mechane* by way of straps and ropes.

This places less load on the *mechane* but this method could be used only for individual actors.

MECHANE RECONSTRUCTION

From the archaeological and historical evidence discussed above, the following specifications for the *mechane* seem rational:

1. It was situated by the left parodos.
2. It operated at a height above the *skene*.
3. The actors were carried to the proskenion at about the middle of the *skene* building.
4. The load carried was at least 4000 N but it could have been as high as 10000 N.
5. It could swing violently.
6. The main element was a beam having general motion about a pivot.
7. The machine could provide vertical motion in addition to horizontal.
8. The operator could control the vertical motion of the load.
9. A wheel was used in some control function by the operator.
10. In some cases the mechane should move fast.
11. It would have substantial mechanical advantage.
12. The load should be, at least, partly balanced.
13. The actors were supported, depending on the needs of the play by a harness or a trapeze or in a chariot which itself was hanging from the mechane.
14. The spectators, even during the day, could see very little of the mechane.

Two main models have been avocated by classicists: Bieber (1939) suggested a sliding device on the roof of the *skene* on which the actor(s) would be carried forward. Though it could have been used in some occasion, this does not agree with the most of the features observed above.

Mastronarde (1990) suggested a pivoted beam, which, in the horizontal position, would be slightly above the *skene* roof and would have rotations about the pivot both about the vertical pole and about a horizontal axis fixed on the vertical pole. This suggestion incorporates most of the above observations but it has some operation and technical problems.

  a) It cannot unload the actor to the orchestra floor.

  b) It is very visible since the size of such beam should have been substantial.

  c) It would be impossible for the operator to have any control of the motion.

Dimarogonas (1991) suggests a pivoted beam with ropes and pulleys.

In attempting a rational reconstruction effort, we will follow the above suggestion by locating the support of the *mechane* behind the *skene* and near the left parodos of the theater of Athens (Figure 7). The length of the beam between the pivot and the hook would be about 10m. To attach the balance weights, one needs an additional length of about 4m. This yields a total length of 14 m, about the maximum length of a cypress tree found in Greece that could be used for the beam. The diameter for a tree like this at the pivot point is about 0.4 m. The maximum end load that this beam could carry at the end, assuming strength of 20 MPa, would be about 12,500 N which seems to be the limit for the single tree design of the beam. It seems that the 10,000 N load at the Prometheus Bound was quite feasible.

It is difficult to hide in the daytime where the performances were given, a 14 m long, 0.4 diameter beam. Therefore, the pivot point ought to be behind the *skene* building and below the level of the roof.

Therefore, in the horizontal position the mechane is completely hidden. It is natural to assume that the horizontal (or nearly horizontal) position of the beam was the resting position for loading.

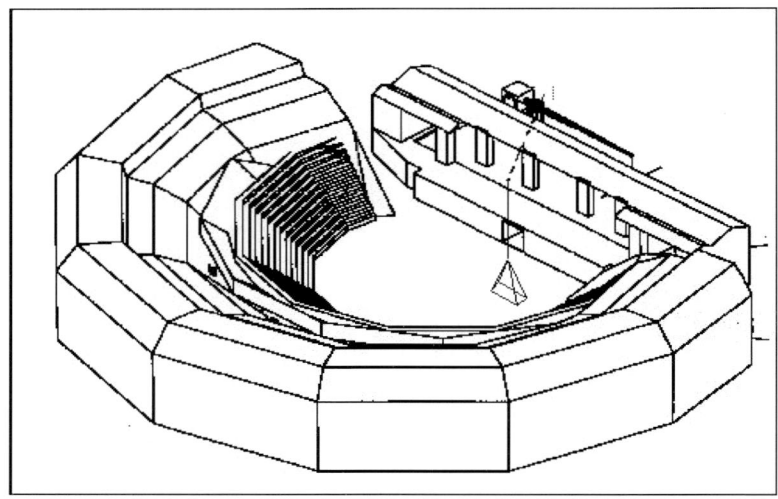

Fig. 7: 3D sketch of the ancient theater of Athens with the mechane

To reach the Orchestra over the top of the *skene* the beam has to tilt to about 30° so that the bottom of the carriage is clear of the top of the *skene*, and then rotated around the vertical pole by about 30° (Figure 7).

To lower the actors to the Orchestra level, one or more pulleys must have been used. To lower a 10,000 N load, a mechanical advantage of 20 must have been used. A pulley with the hook and a 10/1 crank to wheel diameter were considered. The beam had two supports, one at the pivot point for the main joint pin and the other at about 1 m far from the beam's end for the rope drum. The drum's rotating wheel resembles a lever of about 0.5 m length.

Moreover, since the load had to be lowered only with the winch, the rope friction at the drum and the beam end would help the operator during unloading.

It is impossible to time exactly the motion of the *mechane*. We can safely assume, however, in view of the above observations and contemporary theatrical interpretations that it would be in the order of a few seconds. If 5 seconds were assumed for a 60° angle, the angular acceleration would be about 0.04 rad/sec^2, the mass moment of inertia 1.2E5 kgm^2, the required moment 5000 Nm and the force at 4 m from the pivot would be 1250 N. One man can hardly apply this force but one can assume that for very heavy loads two operators were required, since there are only two known occasions that this would be necessary. For the usual 2,000 to 4,000 N loads one operator would suffice. With such forces required, however, the operator could not have precise control of the motion. This makes the path control essential. From what we know about load handling machines in antiquity (Dimarogonas 1986) ropes were used to provide circular path of a point of the beam. In the *mechane*, the most probable design would be a link between the beam pivot and the hook, giving circular motion to this point and a curved motion of the hook. This would result in a four-bar spatial mechanism for generation of a curved path of the end of the beam coupler. Figure 8 shows the components of the mechanism

There remains a question whether the vertical pole was rotating or stationary. From the circular holes found on the ground of many theaters it appears that such poles were in wide use for the periaktos and probably for the *mechane*. Since bearings and lubrication were well-known at that time (Dowson 1979, Dimarogonas 1992) it seems that the vertical pole was rotating.

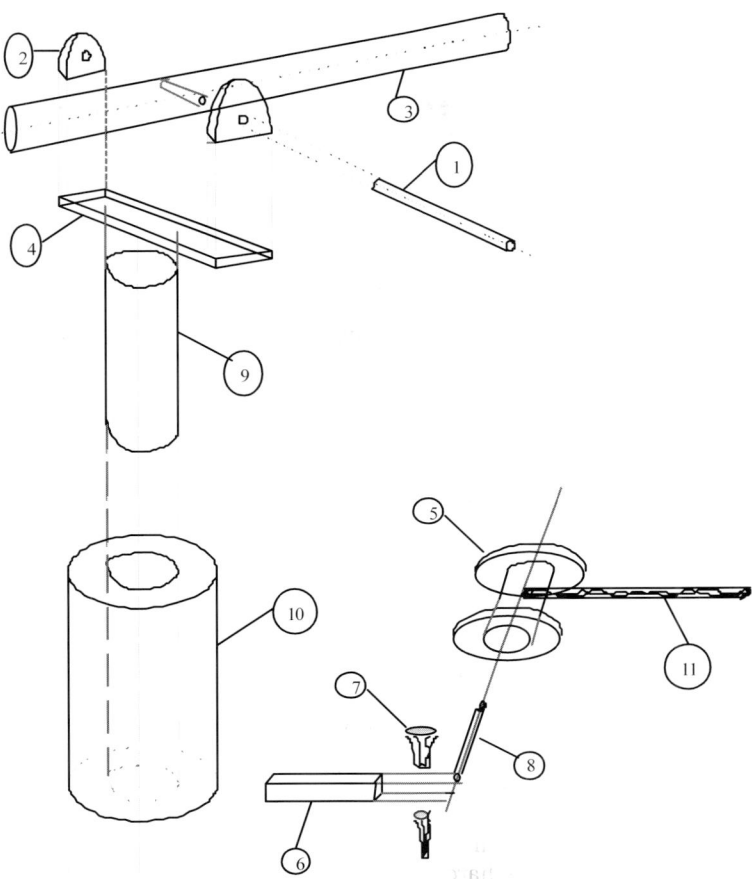

Fig. 8: The mechanism's components 1. Pivot pin 2. Pivot pin bearing 3. Main beam 4. Pivot bearings arm 5. Rotating drum for the rope 6. Drum rotating lever 7. Pin connecting lever and drum 8. Pin connecting drum with the main beam 9. Vertical pole 10. Stone lower bearing 11. Rope for load rise

DYNAMIC ANALYSIS

Figure 9 shows the horizontal and vertical motion models of the *mechane*. In this Figure, O is the pivot point, S is the mass center of the beam, A_c is the *mechane* mass center acceleration, e the angular acceleration on point O, P_a is the inertia force due to the angular acceleration of the *mechane*, M_a is the moment due to the inertia forces, T the friction moment due to the vertical pole rotation, R_o'' is the reaction of the pivot pin in the direction of the main beam, R_o' is the reaction of the pivot pin in the direction perpendicular to the main beam, N is the force applied by the operator, For the vertical motion of the *mechane* the inertial force P_a due to its acceleration is given by:

$$P_a = -(m+M+q+\gamma)A_0 \tag{1}$$

where m is the beam mass, M the mass of the table, q is the counterbalance mass, γ the mass of the actors and A_0 the acceleration of the mechanism center of mass.

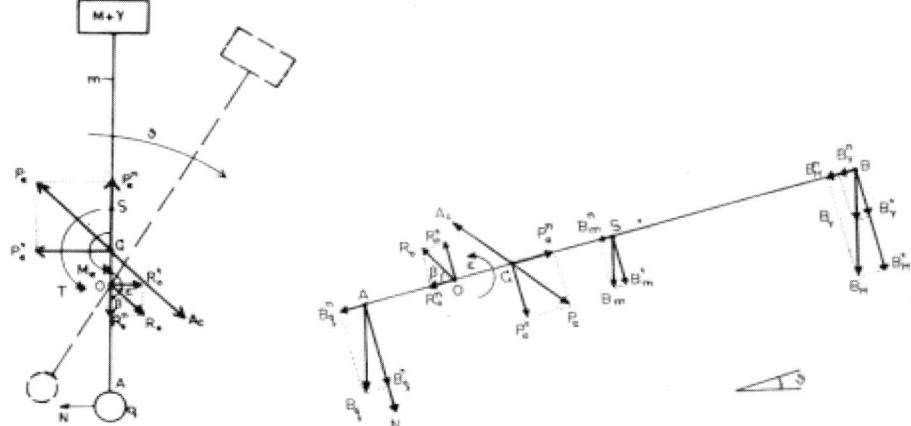

Fig. 9 Free body diagrams for the horizontal and the vertical motion of the mechane

The moment M_a required to rotate the mechanism around the pivot point when the beam is forced to accelerate towards the upper position is:

$$M_a = I_o \, \varepsilon \tag{2}$$

where I_o is the mass moment of inertia and ε is the angular acceleration.
From the moments equilibrium at pivot O the rquired force N by the operator is:

$$N = (M_a + T + B^t \, OS + W^t \, OB - Q^t \, OA)/OA \tag{3}$$

where T is the friction moment of the pivot Q, B, W the weights of the balance, the beam and the load respectively.

Besides, from the equilibrium of the forces components towards the beam axis and the direction perpendicular to it the components of the force R_O acting on the pivot pin, R_O' is the reaction to the pin perependicular to the main beam axis direction and R_O'' is the reaction to the pin in the main beam axis direction yield as:

$$R_O' = N + (M + m + b)g \cos \theta$$
$$R_O'' = (M + m + b)g \sin \theta \tag{4}$$

and

$$R_O = \sqrt{(R_0{}^t)^2 + (R_0{}^n)^2} \tag{5}$$

The angle β shown in Figure 9 is calculated as

$$\beta = \arctan(R_O' / R_O'') \tag{6}$$

Assuming that:

$$\omega(0) = \omega_0 = 0,$$
$$\theta(0) = \theta_0 = 0$$

for $t = 0$ and a constant angular acceleration $\varepsilon(t) = 0.04 \ rad / \sec^2$ the operators' force N and the reactions R_O', R_O'' and R_O acting at the pivot pin, can be calculated for different time steps.

Following a similar path the dynamic characteristics for the horizontal motion of the *mechane* can be estimated.

FRICTION CONSIDERATION

In mechanisms, even when lubrication is sufficient, dry friction appears at contact points. Figure 10 shows a representation of the vertical pole bearing as well as the horizontal bearing frictional model. According to Figure 10 the friction torque T_f for the vertical pole bearing is calculated as (Dimarogonas 2001):

$$T_f = \frac{2\pi \cdot r^3 \cdot l \cdot \mu \cdot \omega}{c} \tag{7}$$

where: ω is the vertical pole angular velocity, μ the lubricant's dynamic viscosity, r the pole's radius, c the radial clearance and l the length of the bearing. Similar calculations apply for the horizontal bearing of the main beam.

The frictional torque for both the vertical and the horizontal motion of the *mechane* were found not to contribute significantly to the whole reaction felt by the operator.

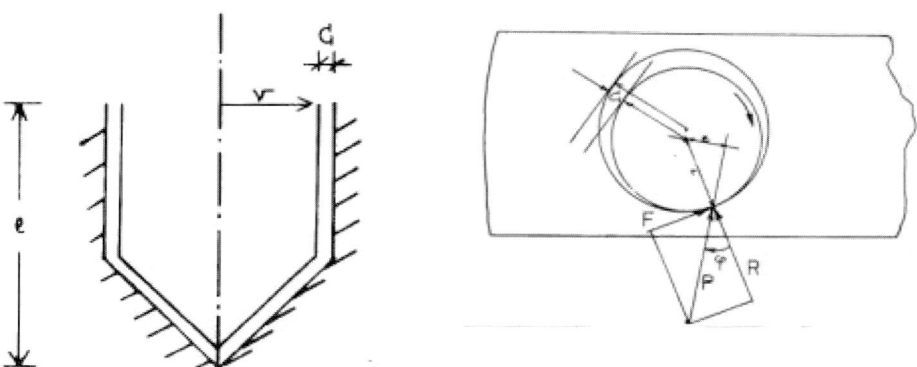

Fig. 10. The vertical pole bearing and the horizontal bearing friction model

DESIGN EVALUATION

A computer algorithm for the design of various *mechanes* configurations was developed. The inputs are the geometrical data of the *mechane*, i.e. the total length of the beam, the main beam's mass center, the pivot point location on the main beam, the angular acceleration and the operators' handle distance from the pivot point. A preliminary kinematic study of the *mechane* provides data for the beam angular velocity, and the end point velocity for specified time intervals. Furthermore, the mass of the table and the actors are set as inputs and the required counterweight mass is defined. The main beam mass and the *mechane* mass moment of inertia are calculated from the algorithm and a static and dynamic analysis is performed (Vitzileos 1993, Vaitsis 1996).

The vertical and horizontal reactions on the vertical pole and the friction forces on the rotating pole are calculated and furthermore, the lubrication requirements are estimated. Then, the force required by the operator(s), stresses and deflections of the beam for various tilt angles of the *mechane* are calculated too. Thus, an optimization procedure for the *mechane* design can be performed easily.

A 14 meters long, with 400 mm diameter in the pivot point cypress tree was selected as the main beam element. For the estimation of the force required to accelerate the *mechane* from rest to the upper position and then rotate clockwise for about 60 degrees the free-body diagrams shown in Figure 9 shall be considered. Let M the actors mass and γ the table mass both varying between 100 and 1000 kg, the beam's mass $m = 800$ kg and the counter-balance mass $q \cong 3000$ kg. Angular acceleration for the *mechane* motion is defined as $\varepsilon = 0.04$ rad/sec^2. Under this acceleration the mechane turns around one rad (57,3 degrees) in 7 seconds. The mechanism's mass moment of inertia around the pivot joint is calculated as $I_o = 205000$ Nm and the bending moment at the same point is 125000 Nm. The axial and shearing forces on the pivot joint are calculated for each time step as well as the resultant of those forces and its direction (angle β in Figure 9).

Table I shows rotational angle θ, angular velocity ω, axial and shearing forces in the pivot pin, the resultant force, the operator's force and the friction torque during the first five seconds of the *mechane* operation for the horizontal motion.

t sec	Angle θ degrees	Angular velocity rad/sec	R_O^I N	$R_O^{\prime\prime}$ N	R_O N	Operators Force N	Friction Torque Nm
0	0.0	0.00	48812.6	0.0	48812.6	1661.2	0.0
1	1.15	0.04	48803.2	943.0	48812.3	1661.3	2.4
2	4.58	0.08	48661.8	3768.1	48807.5	1661.3	4.8
3	10.31	0.12	48050.9	8441.5	48786.7	1661.3	7.2
4	18.33	0.16	46419.1	14832.2	48731.2	1661.4	9.6
5	28.65	0.20	43040.6	22605.6	48615.9	1661.4	12.0

TABLE I *Horizontal motion of the mechane*

Then, the force required by the operator is calculated as shown in Figure 11 for 10, 20, and 30 degrees tilt angle of the main beam. Figure 12 shows deformations of the *mechane's* main beam for a maximum loading of 10000 N as suggested in *Prometheus Bound*

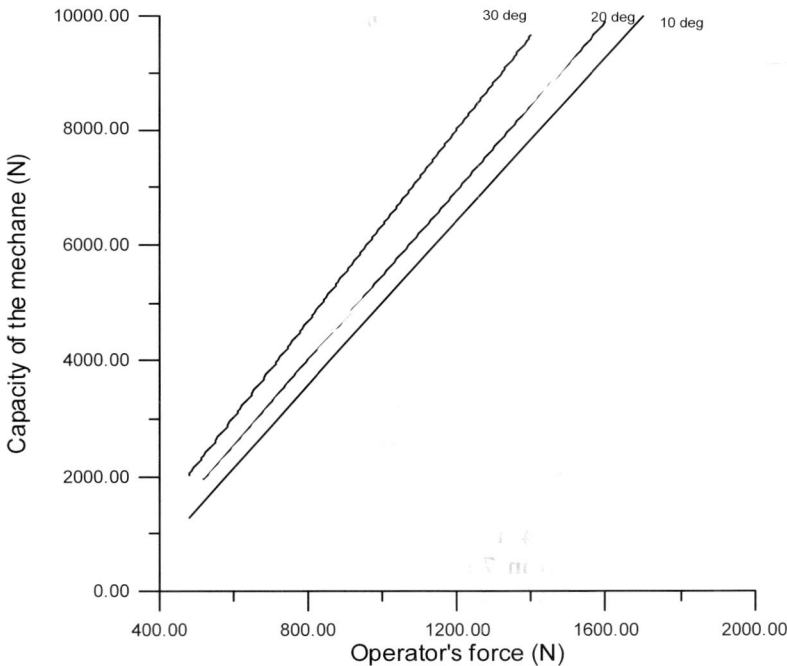

Fig. 11 Force required by the operator for the horizontal motion of the mechane for 10, 20, and 30 degrees tilt angle

Table II shows angle β as shown in Figure 9, angular velocity ω, the operator's force N, axial and shearing forces in the pivot point, the resultant force and the friction torque during the first five seconds of the *mechane* operation for the vertical motion under angular acceleration $\varepsilon = 0.04$ rad/sec^2. Angles of vertical rotation of the beam are identical with those in Table I. Again, the force required by the operator to rotate the main beam upwards for a maximum loading of 10000 N is available by the use of the algorithm.

t sec	Angle β degrees	Angular velocity rad/sec	$R_O^{\prime}$ N	$R_O^{\prime\prime}$ N	R_O N	Operators Force N	Friction Torque Nm
0	90	0.00	49760	0.0	49760	1700	0.0
1	89	0.04	49720	960	49760	1700	0.3
2	86	0.08	49610	3840	49760	1700	0.6
3	80	0.12	48990	8600	49760	1700	0.9
4	72	0.16	47327	15120	49760	1700	1.2
5	62	0.20	43830	23060	49760	1700	2.0

TABLE II *Vertical motion of the mechane*

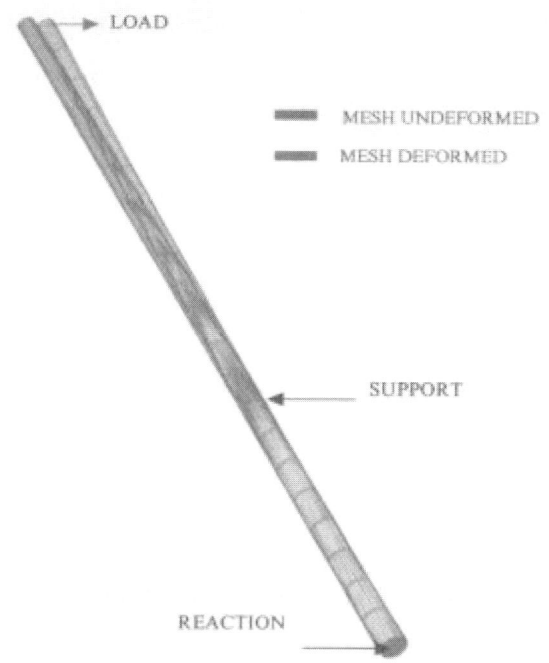

Fig. 12: Schematic representation of deformations for the mechane's main beam with a finite element analysis algorithm.

Static and Dynamic analysis were performed without considering the presence of the rope causing the path generation of a point of the beam. The rope's presence might change some of the characteristics investigated, but the preceeding analysis holds for all the extreme possible situations that could arise during the mechanism's functioning.

VIBRATION ANALYSIS

A finite elements dynamic analysis program was used to compute the harmonics for the beam's vibration (Dimarogonas 1996). The eigenvalues computed for the first three harmonics are 1.63 Hz, 2.02 Hz and 14.20 Hz.

The first two vibration frequencies are 1.6 and 2 Hz. Those frequencies are in the area around 2 Hz, which is well affordable by humans. The modes of vibration of these first harmonics are shown in Figure 13, from which it may be concluded that they were well sensed by the actors, while in the operator's side the vibration amplitude was too low.

Static and Dynamic analysis was performed without considering the presence of the rope causing the path generation of a point of the beam. The rope's presence might change some of the characteristics investigated, but the preceding analysis holds for all the extreme possible situations that could arise during the *mechane's* functioning.

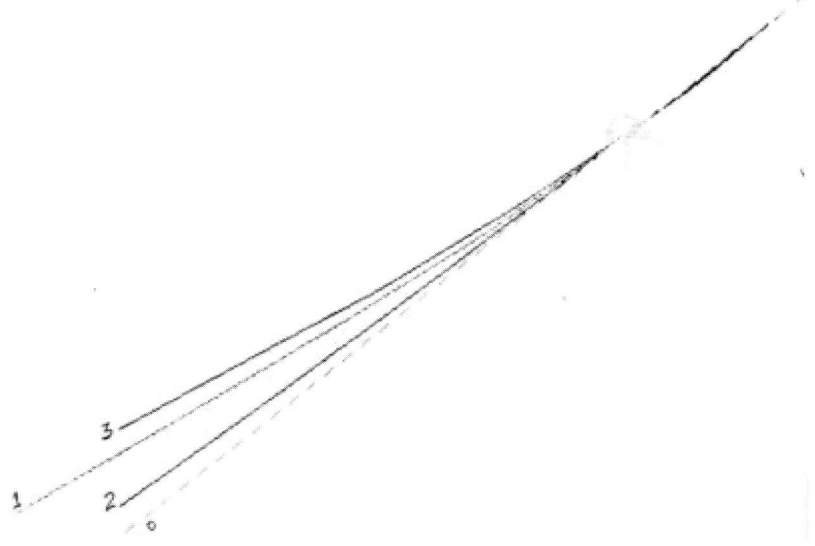

Fig. 13: Modes of vibration for the first three harmonics

ANTHROPOMETRICAL STANDARDS

In order to validate the ability of the operator to handle the mechane, some evidences about people's anthropometrical standards in antiquity must be brought. Those evidences were found out of the archaeologists' and anthropologists' research.

Skeletons found in graves after the excavations in Mycenae Palace, shows that the ancient man was not much different than the contemporary one in height and body dimensions. A mean height of 1.70 m was calculated for the bodies found in those graves. Besides, anthropologists imply that the muscle strength of the man in antiquity was equivalent, if not greater, than the respective one of contemporary man. This theory is supported by two evidences of great significance: first, it is inferred by the thickness of the muscle roots remaining on bones found in graves, a fact that shows the way the muscles were tightened on these bones also proving that men had substantial strength in muscles, and second by the fact that young boys in antiquity started training themselves by the age of seven. Archaeologists' and anthropologists' conclusion is that ancient man had greater muscle strength than the average man of today.

It may be assumed that the *mechane* operator was able to push or pull with a force equal to his weight about 750 N. Under this assumption, two operators might be needed to handle the beam's horizontal movement in the extreme case of the 10000 N load at the beam's end, while, out of the calculations made for the horizontal movement, another operator was necessary when the load was more than 3220 N at the beam's end.

For the vertical movement of the main beam the presence of the rope with the appropriate counterbalance greatly facilitates the effort of the operator in the up and down direction. The 10:1 mechanical advantage of the winch asked for a 1000 N force either to push or pull the lever for the heavier loading conditions. Since we accept 750 N to be the operator's capability, it is inferred that a second operator was necessary for loads greater than 7500 N for lifting or lowering the load.

For the same loading a 1661 N force was required for the movement of the mechane in the horizontal direction with the load lifted upwards.

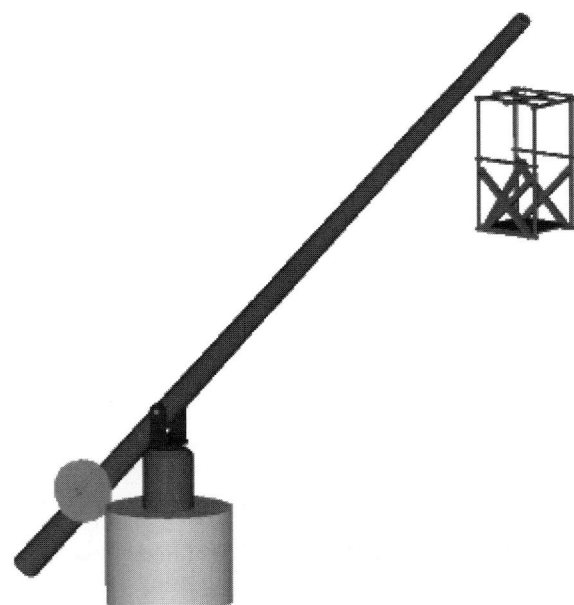

Fig. 14. The Mechane reconstruction

CONCLUSION

The *mechane* together with other mechanical devices such as the *periaktos* and the *ekkyklema*, used as stage machinery in the ancient Greek theater, are the very early heritage of mechanical engineering. Mechanical devices were made and used before, but arrived at by craft (evolution) or by invention (revolution). *Mechane* was the achievement of engineering (intelligence) in response to specifications imposed by the needs of the stage production. Mechanical engineering (a name with redundancy since mechanical and engineering are derived from the respective roots in Greek and Latin of the same thing) has its roots in the *mechanopoioi* of Aristophanes, the pioneers who designed, built and operated the *mechane*. Static and Dynamic analysis make possible the actual reconstruction of the mechane. It is expected that a mechanism will be erected at a Greek ancient theater in the future to investigate its operation and characteristics.

REFERENCES

Aeschylos, 525-456 BC.
    Europa
    Eumenides
    Persians
    Prometheus Bound
Euripides, 480-407 BC .
    Andromeda
    Bellerophon
    Electra
    Helen
    Herakles
    Orestes
    Ion.
    Phoenisae
Aristophanes, ca 445-385 BC.
    Peace
    Daedalus
    Gerytades
    Clouds
    Thesmophoriazousai
Platon 429-347 BC. *Republic*
Homer, 850-750 (?), *Odyssey.*
Dimarogonas, A.D., 1991. *The Origins of the Theory of Machines and Mechanisms. Proceedings 40 Years of Modern Kinematics: A Tribute to Ferdinand Freudenstein* Conference. Minneapolis, Minn.
Dimarogonas, ?. D., 1992, *Mechanisms of the Ancient Greek Theater*, ASME Design Conference, Phoenix Arizona.
Bulle, H., 1928. *Untersuchungen an Griecheschen Theatern*. Abhandlungen, Bayrischen Akademie der Wissenschaften, 33.
Fiechter, E. 1930-1937, *Antike Griecheschen Theaterbauten*
Vitruvius M. P., 1st Century A.D, *De Architectura* V. 7.
Pollux Julius, 3rd Century AD, *Onomastikon*, edited by Bethe, Pollucis Onomastikon, Stuttgart 1890.
Wilamowitz–Moellendorf, U. 1907, *Einleitung in der GriechischenTragoedie*, Weidmann Buchhandlung, Berlin.
Lucianus Sophista (2nd Century AD). *Philopseudes*. Ed. C. Jacobitz, Leipzig 1906.
Bieber, M. 1939. *The History of Greek and Roman Theater*. Princeton Univ. Press, Princeton NJ.
Mastronarde, D. J. 1990. *Actors at High: The Scene Roof, the Crane and the Gods in Attic Drama,* Classical antiquity, Univ. of California - Berkeley Press.
Dimarogonas, A.D., 1986. *History of Technology*, Symmetry publ. Athens.
Dowson 1979, *History of Tribology.* Longman Group Ltd, London.

Vitzileos G. 1993, *Deus Ex Machina Kinematics,* Graduate Thesis, University of Patras, Mechanical Engineering Department.

Vaitsis J. 1996, *Deus Ex Machina Design,* Graduate Thesis, University of Patras, Mechanical Engineering Department.

Dimarogonas, A.D., 2001, Machine Design A CAD Approach, John Wiley and Sons, Inc. N.Y.

Dimarogonas, A.D., 1996, *Vibration for Engineers*, 2nd Edition, Prentice-Hall Int.

3. Past Ideas for New Designs

Borisov A., Golovin A., Ermakova A.: *Some Examples from the History of Machinery in Teaching TMM*

Chen I.-M., Tay R., Xing S., Yeo S.H.: *Marionette: From Traditional Manipulation to Robotic Manipulation*

SOME EXAMPLES FROM THE HISTORY OF MACHINERY IN TEACHING TMM

1) Alexander Borisov,
postgraduate student,
Material Plastic Treatment Technologies and Equipment Department,
Moscow State Technical University named after N.E. Bauman,
Ave. Enthusiastov 100-1-24, 111531 Moscow, Russia
e-mail: borisanet@mtu-net.ru

2) Alexander Golovin,
professor,
Theory of Machines and Mechanisms Department,
Moscow State Technical University named after N.E. Bauman,
Str. Koroleva 4-1-379, 129515 Moscow, Russia
e-mail: aalgol@mail.ru

3) Anastasia Ermakova,
postgraduate student,
Material Plastic Treatment Technologies and Equipment Department,
Moscow State Technical University named after N.E. Bauman,
Str. Brateevskaya 25-1-142, 115612 Moscow, Russia
e-mail: eav_nastya@list.ru

ABSTRACT - The course of "History of Machinery" or "History of Machines and Mechanisms Sciences" does not exist in the current syllabus of Moscow State Technical University named after N. E. Bauman. However, when describing engineering sciences, for example TMM, exposing students to some fragments of the history of machinery is not only interesting but also useful. With these examples, one can show general trends of development of machinery and the succession of core ideas of design. Using the exhibits from the cabinet of mechanisms of the TMM department of BMSTU helps to explain the specifics of operation of machine elements based on geometrical analogies.
The paper addresses two examples of using of historical fragments in the education of students.
The first example: The concept of machine design from the idea to the implementation and the relationship between technology and design are set forth. The brief history of creation of steam engine from Newcomen to Watt is expounded. The concurrence between main stages of machine design is shown: idea – ambiguity of engineering solutions – capabilities of technology – design – operation. The comparison with the process of design of a new machine is also given.

108

The second example: In the cabinet of mechanisms of TMM department of BMSTU, there are fragments of gearings of a motor car of a local electrical train (the time of production – the end of 20's of the 20th century). The value of these exhibits is in the great wear of the tooth profile, which is not allowable in modern machines. That helps to explain quite obviously the mechanism of the possible wear of the higher pair on the basis of geometrical and kinematic analogues of force and energy characteristics.

KEYWORDS: TMM education, history, design, steam engine, wear and tear of a higher pair

INTRODUCTION

The course of "History of Machinery" or "History of Machines and Mechanisms Sciences" does not exist in the current syllabus of Moscow State Technical University named after N. E. Bauman. However, the history of machinery, in particular the history of mechanisms and machines sciences, contains a great number of examples that are useful in the educational process in technical universities [1-9 etc.].

The syllabus of the discipline contains the following sections:

1. The Stages of Machine Design – System Approach
2. Classification of Typical Problems of Machine Design
3. The Type of Mechanisms, Classification of Mechanisms by Type and Purpose
4. Mechanisms with Lower Kinematic Pairs Including Mechanisms With Close and Open Looped Kinematic Chains.
5. Higher Pair, Cams and Gearings
6. Dynamics of Mechanisms, Balancing

We may illustrate each of these sections with spectacular examples from the history of machinery. For example in the section 2 of syllabus, we may show the evolution of concepts of type and classification of mechanisms from the time of Hachette, Lanz and Betancourt to nowadays. In this paper, we will take note of sections 1 and 5. Why? In the process of design, there can be no firm division between science branches. However, in the syllabus we have to distinguish disciplines: ..., TMM, Machine Elements, Strength of Materials... Given this approach, we loose some connection between disciplines. In our opinion, the most suitable place for expounding the concepts of design is the course of Theory of Mechanisms. This could be done by posing it in the first section – "The Stages of Machine Design – System Approach". The history of creation of the steam engine serves as a great illustration of principles of a machine design. Not going beyond the period from Newcomen to Watt, we can explain all basic principles: the connection between design, technology, and operation, the ambiguity of engineering solutions, the development of original ideas and stimulation of new ones. All these elements are present in the modern machine-building but they are "compressed" in time.

Expounding the section 5 – "Higher Pair, Cams and Gearings", we may base on geometrical images the idea of force and energy characteristics of a higher pair and give a probable conception of some possible wear mechanisms. But the confidence in the concept must be proven by real examples though. In the cabinet of mechanisms of the TMM department of BMSTU, there are remarkably worn fragments of gearings of a motor car of a local electrical train (the time of production – the end of twenties of the twentieth century). The value of these exhibits is in the great wear of the tooth profile, which is not allowable in modern machines. The complex character of wear, the influence of construction, different types of wear mechanism and influence of geometry of a pair can be shown on these gearings.

THE CONCEPT OF MACHINE DESIGN AND THE HISTORY OF CREATION OF A STEAM ENGINE FROM NEWCOMEN TO WATT

At the heart of the concept of machine creation lies next theses: the need for its creation, the capabilities of existing technologies, "price-quality" ratio, past experience of operating the existing machines of a given category. This can be outlined in the form presented in the Fig. 1 [10]. For the course of TMM, the third block "Design" is of great interest.

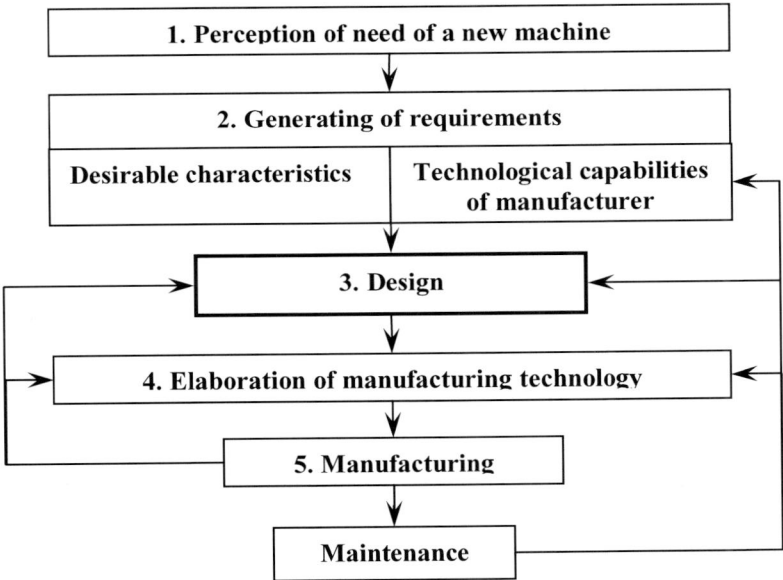

Fig. 1: General approach to machine design

At the base of the concept of machine design lies multivariance of solutions of technical problems and two levels of their comparison: the comparison of possible variants of schemes with the purpose of separation of less promising (weak, "probable" criteria), and comparison of design solutions with the purpose of selection of the most promising one (strong, "impartial" criteria). This can be illustrated by Fig. 2 [10].

110

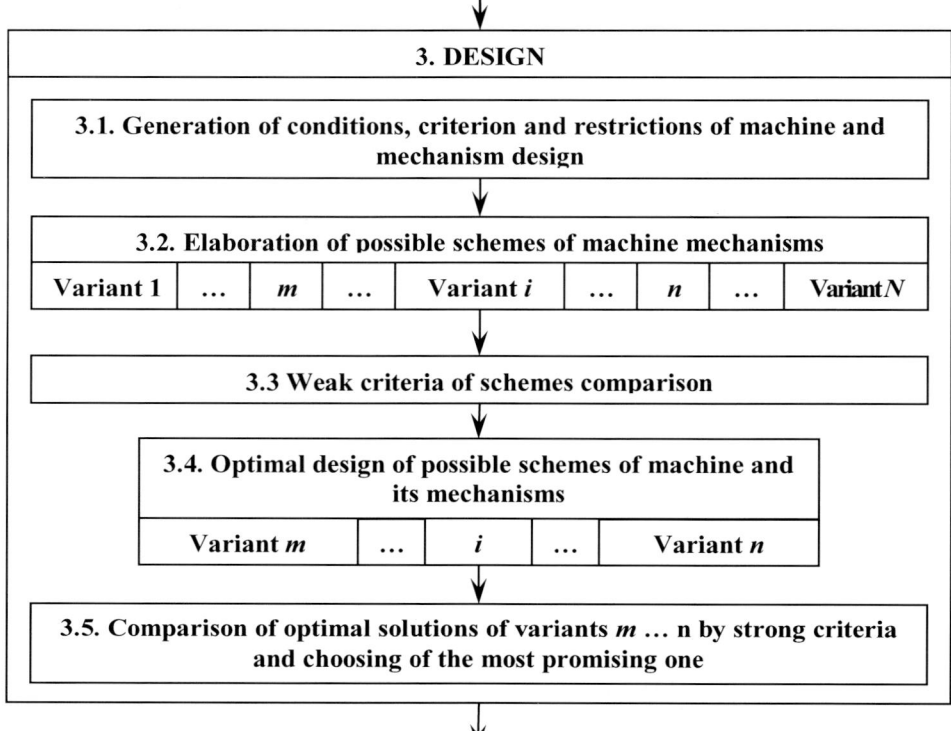

Fig. 2: Order of designing of a new machine

A good illustration of this natural concept of machine design is a story of creation of the steam engine from times of Newcomen to Watt [1], [2]. One can manage to show the connection between the technology, the design of machine and its characteristics (Fig. 3), the development of idea (Fig. 4), multivariant solutions of one technical problem (Fig. 5).

One can consider this question ab ovo, for example from Ctesibios [2] or from Segner's wheel approach but it is hardly expediently. It is sufficient for us to turn to the history of 18th century. The steam engine is maybe the greatest achievement of that century. We may distinguish three steps – three stages of its creation: 1712 – Newcomen's machine (Fig. 3a), 1727 – Leupold's machine (Fig. 3b) and finally the first Watt's machine (Fig. 3c) – 1769. Here we can see direct dependence of quality of a machine (machine efficiency, specific speed, control system), of its design on achievements of a technology.

In Newcomen's machine (Fig. 3a), a single-acting cylinder was used, i.e. with one working chamber. The working stroke is carried out by means of atmospheric pressure: the atmosphere, acting on the open top of the piston in the steam cylinder, caused the engine end of the beam to be pulled down when the steam beneath the piston was condensed, and the backstroke – by means of counter weight (gravity).

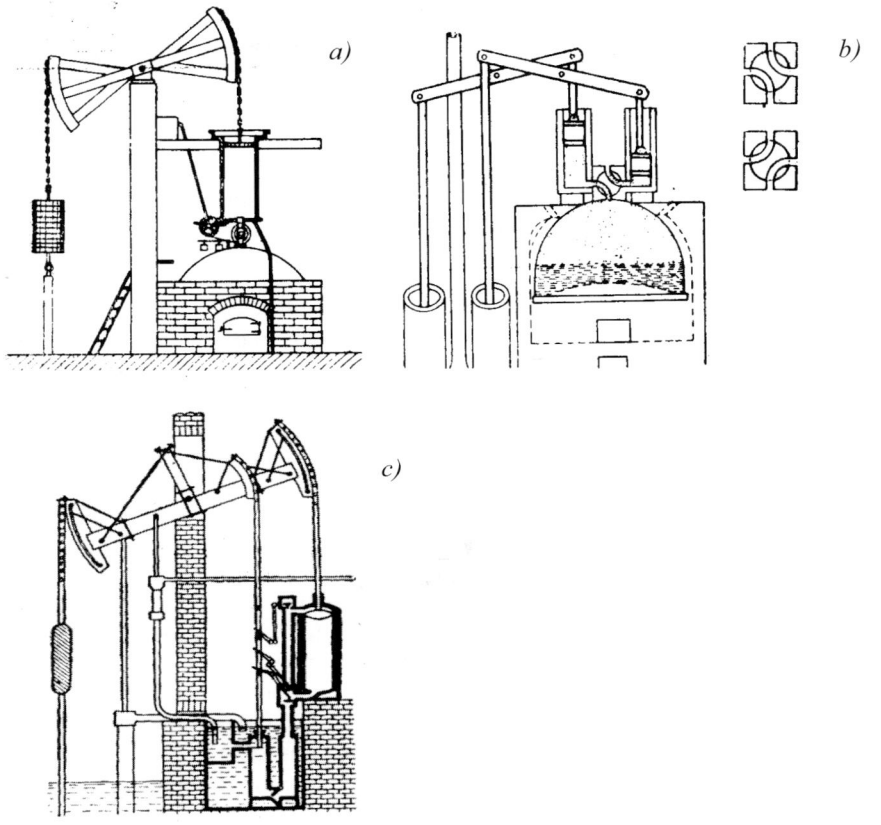

a)

b)

c)

Fig. 3: Newcomen's, Leupold's and Watt's machines

Characteristics of the machine:
- Imperfection of the technology of cylinder manufacturing, therefore
- The lack of a good gaskets in kinematic pairs of the cylinder, therefore
- It was impossible to use a rigid constrain piston-beam pair, which was replaced in Newcomen's mechanism with a chain, therefore
- Extremely low efficiency (0.63%), therefore
- Low specific speed (the upper estimation can be calculated), therefore
- There was no need in automatic control system – the manual control was enough.

As for Leupold's machine, the same technological disadvantages were common. However, in order to improve the specific speed Leupold used two single-acting cylinders (fig. 3b). The innovative idea contrary to the Newcomen "atmospheric" engine was the use of the steam pressure instead of vacuum for a higher efficiency. For the

112

purpose of synchronization of cylinders work, it required to apply a system for switching over the pressure from one cylinder to another. However, specific speed of the machine was still low and the control system was operated manually.

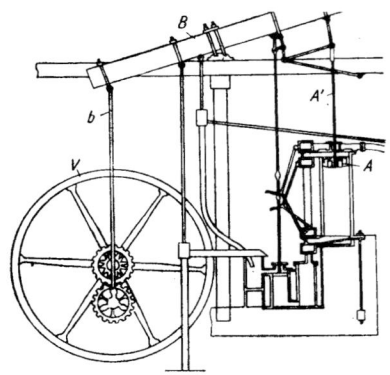

Fig. 4: Watt's machine of 1784

In 1765-1766 yrs James Watt created, and in 1769 patented his first steam engine (fig 3c). In his first construction, Watt added a simple control system and a separate condenser that resulted in significant growth of efficiency.

In 1775 in Birmingham Matthew Boulton, at the Soho Works, wrote to his partner James Watt and commented upon receiving the cast iron steam engine cylinder that had been finished in John Wilkinson's boring mill:

"... It seems tolerably true, but is an inch thick and weighs about 10 cwt. Its diameter is about as much above 18 inches as the tin one was under, and therefore it is become necessary to add a brass hoop to the piston, which is made almost two inches broad." [1]

This cylinder indeed marked the turning point in the discouragingly long development of the Watt steam engine. Although there were many trials ahead for the firm of Boulton and Watt in further developing and perfecting the steam engine, the crucial problem of leakage of steam past the piston in the cylinder had now been solved by Wilkinson's new boring mill, which was the first large machine tool capable of boring a cylinder both round and straight.

In 1784, Watt patented his brand new engine (fig. 4) for transformation of the steam energy into the continuous rotary motion of a flywheel (a crank). From the first, the rotative engines were made double-acting – that is, work was done by steam alternately in each end of the cylinder. The double-acting engine, unlike the single-acting pumping engine, required a piston rod that would push as well as pull. Therefore, in the patent Watt decided to replace a chain with a rod. As the result of these conditions – good gaskets, double-acting cylinders – the efficiency of the machine and its speed grew significantly, and so appeared a need for a good control system and Watt invented a valve – a closed-loop control. Watt solved the problem of transforming of a straight-line motion of a piston rod (a slider) into a rotary motion of a rocker by means of a connecting rod – so called Watt's straight-line linkage.

Pafnutii L'vovich Chebyshev later developed this area with the creation of the theory of the best approximation. Nowadays it is solved by a slider-crank or crank-and-rocker mechanisms.

Further Watt's work was related to development of that machine: transforming of translational motion of a piston rod into a rotary motion of an output link, Watt's regulator, a device for recording of an indicator diagram etc.

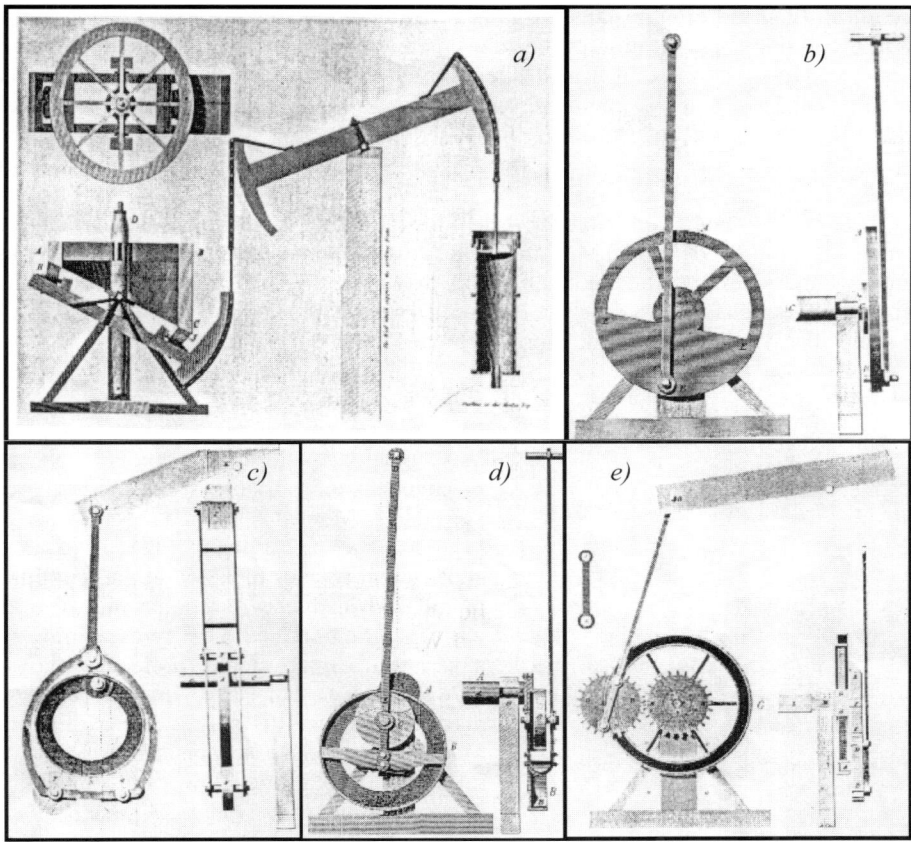

Fig. 5: Watt's variants of output links

Watt worked over several designs of output links – the existence of the patent for a simple crank (Wasbrough, Pickard) forced Watt to search for a substitute. In his patent Watt suggested the following mechanisms: an "inclined wheel" – a mechanism with a swash wheel (fig. 5a), one that was easily recognizable as a crank (fig. 5b), two "eccentric wheels" (fig. 5c & 5d), and sun-and-planet gearing (fig. 5e). Watt and Boulton tried first device but discarded it. Watt adopted the sun-and-planet arrangement, with gears of equal size, for nearly all the rotative engines that he built during the term of the "crank patents". This arrangement had the advantage of turning the flywheel through two revolutions during a single cycle of operation of a piston, thus requiring a flywheel only one-fourth the size of a flywheel needed if a simple crank were used.

After the demonstration by Boulton and Watt, that large mechanisms could be wrought with sufficient precision to be useful, the English tool builders Maudslay, Roberts, Clement, Nasmyth, and Whitworth developed machine tools of increasing size and truth. The design of other machinery kept pace with – sometimes just behind,

114

sometimes just ahead of – the capacity and capability of machine tools. In general, there was an increasing sophistication of mechanisms that could only be accounted for by an increase of information with which the individual designer could start. Reuleaux pointed out in 1875 that the "almost feverish progress made in the regions of technical work" was "not a consequence of any increased capacity for intellectual action in the race, but only the perfecting and extending of the tools with which the intellect works." [1]

This material can be accompanied by demonstration of models from the room of mechanisms of the chair of TMM in BMSTU and theme may be followed by an example from modern practice of creation of a new machine. For example, creation of a complicated radial-forging machine allowed substituting over 30 technological operations that were carried out on more simple machinery.

HIGHER PAIR

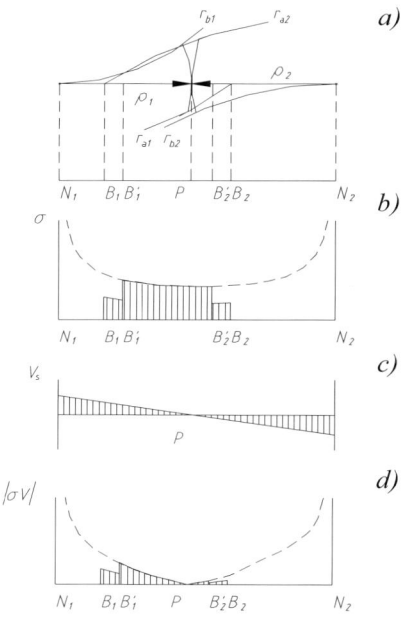

Fig. 6: Diagrams of contact stress, slip velocity, and their multiplication

Expounding the section "Higher pair", we may give an illustrative explanation of mechanisms of possible wear of a higher pair. N. I. Mertsalov [12], L. N. Reshetov [9], and G. G. Baranov [8] used that approach in courses of TMM in the first third of the 20th century. A reliable picture of a wear may be obtained only by examining of a wear on a microscopic (molecular) level. However, we can take into account a macroscopic level as a first approximation. Macroscopic wear mechanism is based on a relatively rough modeling and generally obeys simple but reliable models. An approach of simplified interpretation and orientation towards applying fundamental origins is of interest for an engineer.

Interpretation of the processes of wear by methods of TMM may be carried out by means of creating a geometrical image of a source or a result of a wear. Among these images, the form of mating surfaces and every possible epures and diagrams (pressure profile, velocity profile, stress diagram etc.) may be placed. The Fig. 6b shows a diagram of an adjusted radius of curvature – analogue of contact stress σ, which is calculated by Hertz formula. This analogue may be used to explain the origin of bearing-spalling mechanism of wear. A diagram of analogue of energy dissipation in the point of contact $\sigma \cdot V_s$ (V_s – slip velocity) is shown in the Fig. 6d. This analogue may be used to explain the origins of galling. Naturally, real wear bears a complex character and supposed analogues are only dominating factors of wear of a higher pair. The Fig. 7 explains wear of gearing with different mechanical characteristics of gears operating in conditions of insufficient lubricating. Confidence for this explanation must be proved by real examples though. The Figs. 8, 9, 10 show fragments of gearings of a motorcar of a local electrical train (the time production – the end of 20's of the 20th century). These exhibits are the property of the room of mechanisms of the chair of TMM in BMSTU. In Fig.8 (gear set), we can see that the dominating factor of wear of the pinion could be the contact stress and dominating factor of wear of the driven wheel can be galling. In Figs. 9, 10 (helical gear), the traces of yielding in the area of the pitch point, bearing of surface in the area of tooth point (a burr), traces of spalling and galling can be seen. Therefore, the suggest explanation does not interfere with a real situation.

CONCLUSION

Offered material is only a brief of those themes. Excursus into the history of machine and mechanisms made in the article is not unique. However, this material must be carefully "dosed" because of the limit of study hours.

It is well-known that it is difficult to keep students attention for a long time that's why a lecturer ought to find time even during a very serious lecture for a joke or/and a little interesting essay. Therefore, methodically these examples are well justified.

The experience of teaching shows that these examples help for better understanding of basic ideas of those sections.

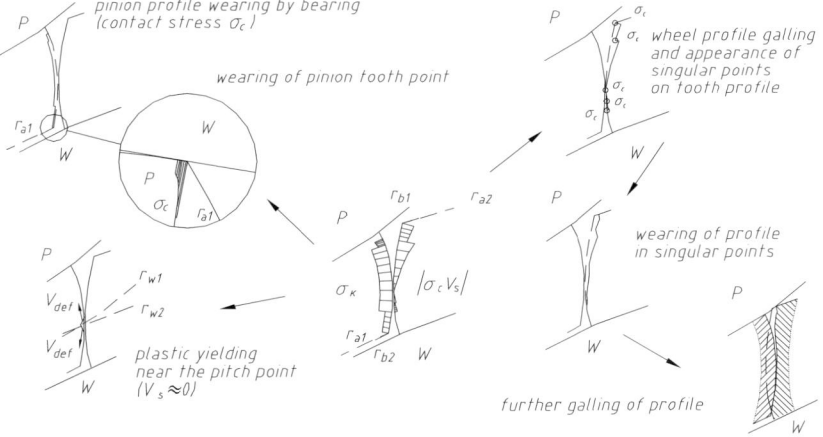

Fig. 7: Possible explanation of wearing

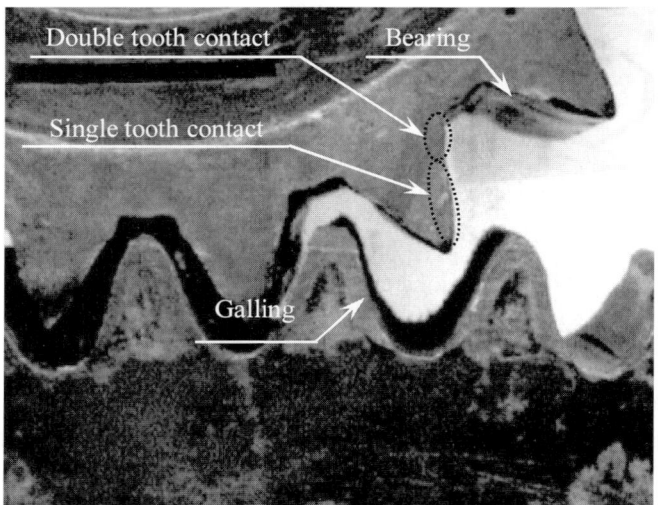

Fig. 8: Fragments of gearings of a motorcar

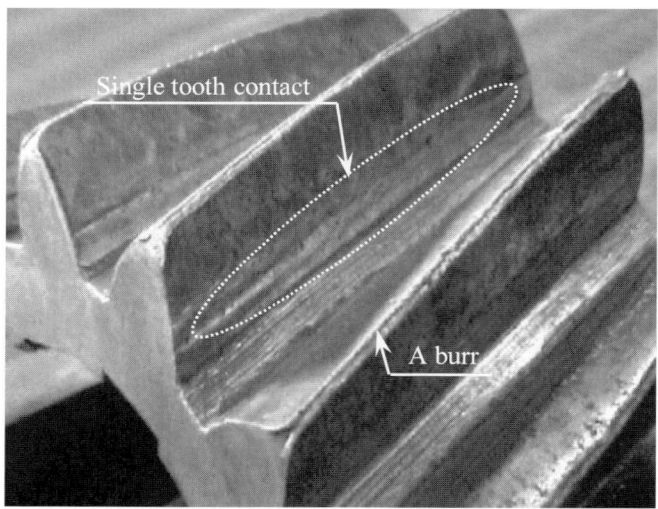

Fig. 9: Fragments of gearings of a motorcar

Fig. 10: Fragments of gearings of a motorcar

REFERENCE

1. E. S. Ferguson, Kinematics of mechanisms from the time of Watt, US National Museum, Bul.228, Washington, 1962
2. F. Dudi !, D. Diaconescu, G. Gogu, Mecanisme Articulate/ Inventica Şi cinematica 'n abordare filogenetic!, Editura Technic!, Bucureşti, 1989. - 424 pp. (in Romanian)
3. L. Zhen, G. Xuan, The development of water-powered machines of China in 10-14th Century / Proceeding of International Symposium on History of Machines and Mechanisms, Edited by Marco Ceccarelli, Kluwer Academic Publishers, 2000.
4. Y. Kerle, M. Helm, Animal kinematics – A review of study of Franz Realeaux about restrained animal motions / Proceeding of International Symposium on History of Machines and Mechanisms, Edited by Marco Ceccarelli, Kluwer Academic Publishers, 2000.
5. A. Rovetta, A. Emanueli, I. Nasry, A. Helmi, Ancient Egiptian chariots – Design and functional aspects/Proceeding of International Symposium on History of Machines and Mechanisms, Edited by Marco Ceccarelli, Kluwer Academic Publishers, 2000.
6. H. S. Yan, T. Y. Lin, Comparison between the escapement regulators of Su Song's clock-tower and modern mechanical clocks / /Proceeding of International Symposium on History of Machines and Mechanisms, Edited by Marco Ceccarelli, Kluwer Academic Publishers, 2000.
7. B. Roth, The search for the fundamental principles of mechanism design / Proceeding of International Symposium on History of Machines and Mechanisms, Edited by Marco Ceccarelli, Kluwer Academic Publishers, 2000.
8. Baranov G. G. Kinematics and Dynamics of Mechanisms, part 1. Moscow – Leningrad, 1932. (in Russian)
9. Reshetov L. N. Involute gearings correction. Moscow – Leningrad, 1935. (in Russian)
10. Golovin A. A. Design of multibody linkages schemes. Edition of BMSTU, 1995. - 76 pp. (in Russian)
11. Golovin A. A. Theory of Mechanisms from Gaspard Monge to the present: A Science and an Academic Discipline // Russian Science: the way of life. Collection of the best popular articles written for the contest, organized by the Russian Foundation for Basic Research. Editor V. Skulachev. "Octpus" Publishing, 2002. - 416 pp. (in Russian)
12. Mertsalov N. I. Dynamics of mechanisms / Lithographic course of lectures, read in Imperial Technical School. Processed and published by M.I.Phelinskii. – M., 1914. – 610 pp., il.

MARIONETTE: FROM TRADITIONAL MANIPULATION TO ROBOTIC MANIPULATION

I-Ming Chen   Raymond Tay   Shusong Xing   Song Huat Yeo
School of Mechanical and Production Engineering
Nanyang Technological University
Singapore 639798
E-mail: michen@ntu.edu.sg

ABSTRACT- Marionettes are string-operated puppets. It is an ancient and universal form of performing art which still evolves slowly today. From the engineering perspective, the marionette is a wire- (or string-) driven multi-limbed under-actuated mechanism under gravity influence that exhibits rich kinematic and dynamic behaviours. This article introduces the evolution and the engineering aspect of traditional marionette design and manipulation skills. Based on the mechanics of marionette, a novel robotic marionette system that manipulates the puppet through mechatronic means is developed. The marionette may create life-like movements according to programmed motion commands issued from the computer and a motor-driven puppeteer mechanism. A simple demonstration of the robotic marionette performance is shown as well.

KEYWORDS: Puppet. Marionette history. Robotic marionette. Puppeteer mechanism.

I. INTRODUCTION

Puppetry and Puppet Theater are popular art forms having a long and fascinating heritage in many cultures. The survival of this art form is due to man's fascination with the inanimate object animated in a dramatic manner [1], and human's curiosity to "reproduce" an exact artificial copy of him (or her). Recent advances in the humanoid robots speak for this quest of curiosity. Most types of puppets in use today fall into four broad categories: *hand* (or *glove*) *puppets*, *rod puppets*, *marionettes*, and *shadow puppets* [1]. The glove puppet is used like a glove on the operator's hand. The rod puppet is held and moved by rods, usually from below but sometimes from above. The marionette is a puppet on strings, suspended from a control mechanism held by the puppeteer. Shadow puppets are usually flat cut-out figures held against a translucent, illuminated screen. There is also a variety of combinations among the four categories: hand-rod, glove-rod, rod-hand, and rod-marionette puppets. Though people are mostly interested in the theatrical and artistic content of puppetry, basic puppet fabrication and manipulation techniques follow the physical laws and engineering principles. From the engineering point of view, the puppet, the puppet control device and the puppeteer form an interesting complex mechanical system that embodies life-like movement of artificially made entities. However, it is rare to see engineering study of the puppet-

119

120

making and manipulation skills. Though there are quite a few literatures dedicated to the practice of puppetry, specific puppet making and manipulation skills are usually learned from the puppet masters through apprenticeship, and are usually not publicly available. This article first introduces the evolution of traditional marionette design and manipulation skills from the engineering perspective. Then, a novel robotic marionette system (ROMS – *RObotic Marionette Systems*) developed in Nanyang Technological University that manipulates the puppet through mechatronic means based on the mechanics of marionette instead of human operation, is described. The marionette may create life-like movements according to programmed motion commands issued from the computer and a motor-driven puppeteer mechanism. The marionette is chosen as the object of study for three reasons: 1) mechanically, it is equivalent to a cable-operated multiple rigid body system that exhibits rich kinematic and dynamic behaviours; 2) theatrically, the marionette performances are graceful, charming, and sometimes mystic because of the "invisible" string control; 3) the marionette is versatile and can be simple as well as sophisticate in both construction and control. Through this novel system we hope to compliment the trade of traditional puppetry in a mechatronic approach.

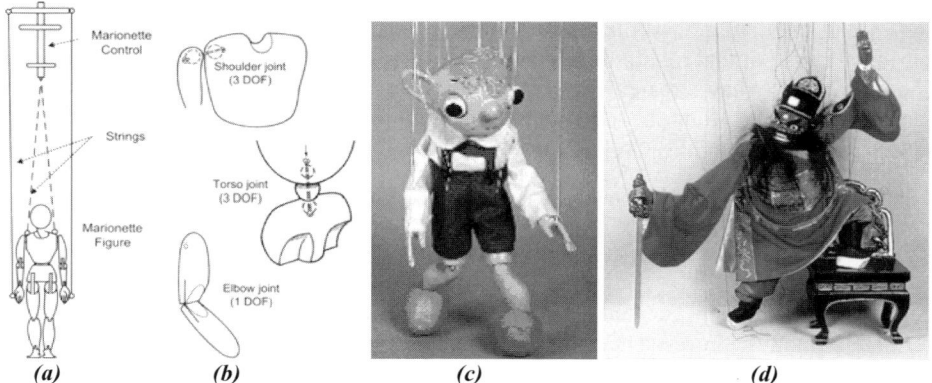

(a)   *(b)*   *(c)*   *(d)*

Fig.1: (a) Anatomy of a marionette. (b) Types of joints [1]. (c) Czech marionette (Hurvinek by Sotak). (d) Chinese marionette (Zhong Kui) [12]

II. ANATOMY OF MARIONETTES

The term "marionette" was first associated with string puppets in 16th century Europe. The origin of the word may be traceable to the Virgin Mary, often the principal character of puppet plays during the 1500s, either as a diminutive of "Maria," or in its literal translation "little Marys," from the French reference to the Virgin [2]. In China, it is called "Xuan Si Mu Ou" (懸絲木偶) or "Ti Xian Mu Ou" (提線木偶), which is a direct interpretation of the operation of the puppet. In the early history of Europe, the marionettes were used to entertain people – nobles or walk of all lives, whereas in China, the marionettes were mainly used for ceremonial purpose. The marionette theatres were used to entertain gods instead of ordinary people [3]. Hence, the appearance and fabrication techniques of marinettes have strong cultural influences. Regardless of cultural differences, the basic structure and anatomy of the marionettes are similar. In fact, this structure also changed very little through the history. A marionette usually consists of three essential elements: a puppet figure, a puppet control device, and a bundle of strings tied to the control device and various locations on the puppet figure as shown in Fig. 1(a). Together they form a string-operated multi-limbed mechanism.

2.1 Puppet Figure

The puppet figure is usually duplication or simplification of a human, an animal, or any kind of living creatures. The design of the figure follows three major steps: 1) artistic or theatrical feature development, 2) puppet movement or trick design and development, and 3) mechanical design of the puppet components. Mechanical design of the puppet covers the task of determining the number and shapes of limbs for the figure, types of joints connecting the limbs, and additional features on the puppet, such as blinking eyes, moveable mouthpieces, articulated palms for human figures, etc. The dexterity of puppet movement increases with more segments of limbs added to the figure, such as the inclusion of the torso segment in a human figure for the turning and bowing motions. The design of puppet joints is versatile. Joints with one degree of freedom (DOF) revolute motion, 2- and 3-DOF spherical motions are frequently used in the marionette. The DOFs in the joints are realized using different kind of connecting materials, such as leather, cord, or screw-eyes. Practical construction methods of puppet figures and joints can be found in [1, 4-11]. As shown in Fig. 1(b), the shoulder, torso, and limbs of a human puppet are all connected with each other through joints of different DOFs. A famous modern Czech marionette and a traditional Chinese marionette – Zhong Kui (鍾 魁), a ghost-fighting god [12], are shown in Fig. 1(c) and 1(d) respectively. The size of marionette figures performed in live is usually between 45 to 75cm in height with consideration of the visual effect and convenience for the puppeteer operations.

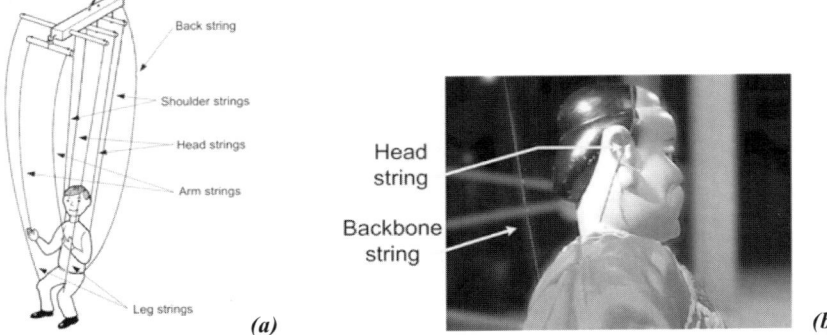

Fig. 2: (a) Functions of strings. (b) Head & backbone strings on a traditional Chinese marionette.

2.2 Strings

In robotics terms, the strings for the marionette allow the puppeteer to "remotely operate" the puppet from a control device. By manipulating the control device and plucking the strings, the puppet will produce life-like motions. The strings act like "actuators" to the puppet. Hence, using proper type of strings for puppetry becomes an important issue. From our study [13], it has been noticed that western puppetry did not impose a standard type of strings used in the marionette, i.e., the control strings can be of any material. However, in the traditional Chinese puppetry, embroidery strings are frequently used. The embroidery strings are multi-threaded with little elasticity. Therefore, the strings will remain the same length under prolonged or large tension.

Functions of strings. The strings in a marionette can be classified into three categories: strings for *support/reference* , strings for *motion control*, and strings for *special effect*. The supporting strings hold most of the weight of the puppet figure and are usually kept stationary during the performance. In western human marionettes, the two shoulder strings usually serve this purpose (Fig. 2a). The traditional Chinese marionette includes a single reference string which acts as a *backbone* to the marionette (Fig. 2b). This is to ensure that the marionette is in an upright position. All remaining strings will take reference from this string. The strings attached to the arms, legs and the head of the figure are mainly for motion control purpose. The manipulation range of these strings attached to the limbs may be very large depending on the gesture of the figure. With the inclusion of head strings, the marionette may shake or nod his head. For a very basic human marionette, at least 8 strings are needed for full body motion: two for the shoulders (support), two for the arms, legs and the head each (control). Usually one back string is added to the western marionette having a torso joint (Fig. 2a). With this additional back string, the puppet figure may perform bowing motion.

Number of Strings. Using 8 to 9 strings in a generic marionette is only a rule of thumb. The actual number of strings used in a marionette depends on the desired performance of the figure. Western puppeteers tend to keep the number of strings to a minimum by using a counter-balanced design which allows gravity to hold a puppet in a "normal" pose when at rest. A typical western marionette with human figure uses 8 to 12 strings. Traditional Chinese marionettes normally use 16 to 24 strings, sometimes up to 50 strings, for a fully detailed human-like manipulation. These detail controls usually focus on the facial expressions, arm, and palm movements. In a Chinese marionette play called "Drunken Zhong Kui" (鍾魁醉酒) by the puppet master Mr. Huang Yique (黃奕缺) of Quanzhou (泉州), China, the figure of Zhong Kui (Fig. 1d) comprises 50 over strings. In the play, the marionette can perform very sophisticated movements like picking up wine cups, drinking the wine, pulling out and holding back the sword, etc.

2.3 Control Device

The control device for marionettes, or the "controller" here, is a simple mechanism held by the puppeteer to operate the puppet through a bundle of strings. The design and geometry of the marionette controllers vary in different cultures but evolve slowly with time. The Chinese use a simple platform type of control, termed "Gou Pai" (鈎牌 or hooked plate) which has passed down from Song dynasty for about a thousand years [14] whereas the Europeans generally adopt a cross platform control, sometimes called "airplane control" which took shape in the late 18th century [15]. Such difference may probably due to the design and stringing of the marionettes and the philosophy of marionette performance. However, they do have a common feature of symmetric layout, which is also the fundamental feature of human body and most of the living beings.

2.3.1 Western marionette controllers

Unlike modern marionettes controlled by strings hung vertically using the balance of the tensions in the strings and the gravity, early in the history string-operated puppets were controlled by the strings laid horizontally. Figure 3(a) illustrates that puppets on the desktop controlled by horizontal strings were performed in the street in medieval France [16]. An early type of European marionette, known as "A La Planchette" in French or "fantoccini" in Italian was also controlled by horizontal strings as shown in an early 19th century painting (Fig. 3b) [2]. It consisted of a string, secured to a post at one end, passing through the body of a jointed puppet or a group of puppets. Manipulation of the free end of string tied to the leg of a minstrel, caused movement and action thereby bringing the puppet to life.

(a)   *(b)*

Fig. 3: (a) Puppets controlled by horizontal strings. (b) A La Planchette

Marionettes with vertical strings came in two forms in the 19th century Europe: *dramatic marionettes* that were designed to perform in the dramatic repertoire and *variety marionettes* that were mainly for tricks and variety. Dramatic marionettes adopted a hybrid rod-string puppet design. In early days, the puppet figure was usually carved out of wood. Most puppets measured between 70 and 90 cm and, and when fully dressed and armed, their weight could be well over 10kg [15]. Using a rod attached to the head to support the weight is necessary. Control rods attached to the arms also allowed for strong clear expressive movement. However, as puppets adopted more dexterous design to give more vivid and lively performance for realism, the rod control became awkward for handling due to its rigid structure and number of rods used. In the late 19th century, family-run puppet theaters in Europe gradually replaced control rods by strings on the puppets. The head rod was replaced with support strings after 1880 in English and German theaters. Nevertheless, some French and Italian theaters continued using the head rod until the companies ceased to function in 1930s [15]. Some rod-string marionette controllers used by European puppet theaters are illustrated in Fig.4.

Variety marionettes were those puppet figures that can dismember, give birth to other figures, expand or contract their bodies. Usually they were fully strung figures to maintain a sense of wonder because of their "invisible controls" [15]. From the late 18th century, fully strung trick marionettes were usually controlled by one or more horizontal bars to which the strings were tied (Fig. 4). One bar took the hand and the head or support strings and the other, the ones for the legs. Later, British showmen mounted two or more of these horizontal bars onto a horizontal central bar making a single composite control which may have been one of the features that contributed to the success of the

124

so-called "English marionettes". The vertical control for the all-string marionettes appeared in the end of 19th century which was probably developed out of the rod-marionette grip. The control consisted of a vertical grip, with a horizontal bar at the bottom to take head strings, hand strings, and other special ones, and a rocking bar at the top for operation of the legs. The vertical control became standard in Germany and Britain, where it superseded the horizontal bars [9,15]. Some all-string marionette controllers used by European puppet theaters are illustrated in Fig. 4.

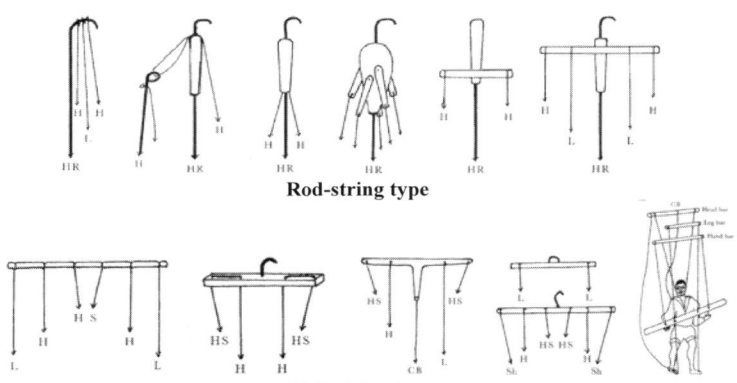

Rod-string type

Full-string type

Fig.4: Various marionette controllers [adopted from 15]

The horizontal controller later proliferated in American in early 20th century through European immigrants. Hence, we see a mixture of vertical and horizontal marionette controllers in modern puppet troupes. Figure 5(a) and 5(b) demonstrate typical vertical and horizontal controllers. They are also called airplane controls for the marionettes because their shapes resemble that of aircrafts. A variation of the airplane controller is the *Angle controller* invented by F. H. Bross [6,9]. (Fig. 5c) This controller has a central rod to be held by the puppeteer in tilted position. One major horizontal bar is pivoted to the central rod to produce rocking movement. Head strings are joined at additional horizontal bar on top and shoulder strings may join at the tail which is pivoted to the central rod as shown in the figure to produce swinging motion back and forth.

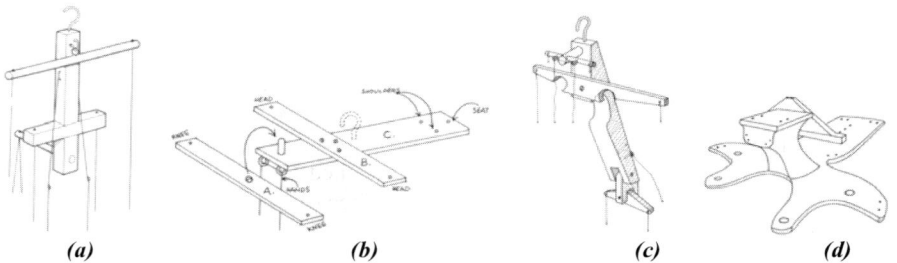

| (a) | (b) | (c) | (d) |

Fig. 5: (a): Vertical controller [1]. (b) Horizontal controller [5]. (c) Angle controller [6]. (d) Paddle controller [10].

A third type of marionette controllers is the paddle controller developed by W. A. Dwiggins [17,18] as shown in Fig. 5(d). This tiny (a little over 6 inches) controller is shaped like a sweep-wing aircraft and can make the subtlest movements in 12" puppets. On the wing are strings for the head, forearm, and hands. On the main bar are strings for the forehead, shoulders, and back. The legs are controlled by a small piece of wood, vertical to the main bar and fixed on a pivot so the puppeteer can control the legs simply by rocking the bar with a finger [2]. For more variations of the vertical, horizontal, angle and paddle controllers, please refer to [9].

 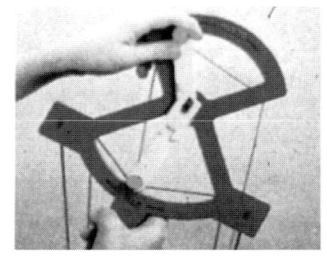
(a)   *(b)*

Fig.6: Toy type marionette controllers

The advancement of marionette controllers was fueled by commercialization. Until the late 19th century, marionettes were not readily available to the public. It was not until the 1920s that the first commercially made marionette toys were available in the US. which were designed by Tony Sarg [19]. Because of the use of new materials, like plastics, for the puppets, the weight of the marionette can be further reduced and new types of marionette controllers can be developed. Simple bar control with a central ring allows children to use only one hand to handle the marionette [2] (Fig. 6a). An "action frame control" made of plastics was designed by Madison Ltd. in 1977 (Fig. 6b). The action frame control consists of two 90° arc frame that the two control bars can slide through the slots in the respective arcs. The strings are attached to the central bars so that the puppeteer can manipulate the marionette by simply turning the central bars.

2.3.2 Chinese marionette controller
Chinese marionette theaters had been popular for more than a thousand years. Most of the repertoires of the marionette theater were adopted from those performed by the traditional Chinese theaters like Peking Opera. Hence, the marionette figures had strong connections to their human counterparts and the design of marionette figures was for realism. Therefore, the number of strings used by Chinese marionettes is usually more than the European marionettes. With so many strings to control, Chinese puppeteers made use of a controller known as the "*Gou Pai*", which is a single piece of plate on which all strings are nicely tied in systematical and symmetrical sequence (Fig. 7b and 7c) [20]. The plate is made of wood or halved bamboo segment in rectangular shape about 25 cm long and 15 cm wide. The plate is attached to a long handle bar to be held by the puppeteer. A hook is nailed to the opposite face of the plate so that the controller can be hooked to a beam on the stage for the ease of manipulation. Because the number of strings is large and most of the puppet movements are produced by manually plucking the strings, using a single plate controller design is a practical and simple solution. The puppeteer's attention can be focus on the string manipulations rather than doing controller maneuver and string plucking at the same time. This marionette

126

controller design has been passed down from Song Dynasty as illustrated by a Chinese painting in Song period (AD 960-1279) in Fig. 7(d) [21].

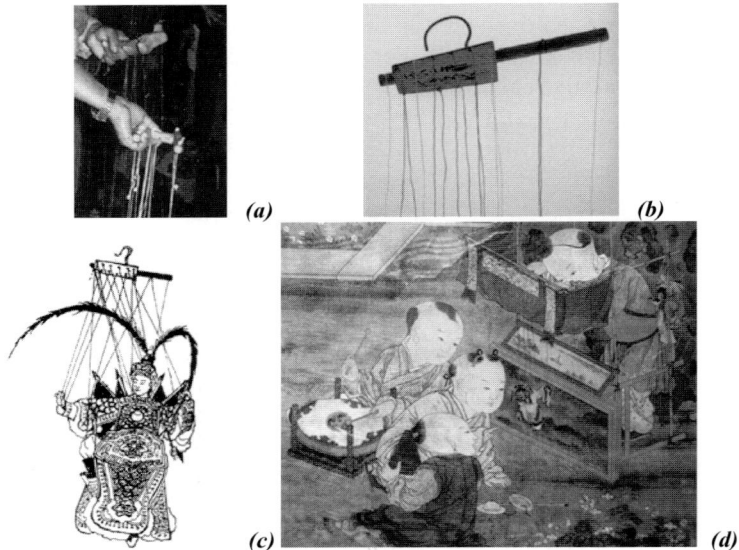

(a)   *(b)*

(c)   *(d)*

Fig. 7: (a) Burmese marionette controller. (b) Chinese Gou pai and (c) its operation [20]. (d) Puppet painting from Song Dynasty.

2.3.3 Control devices from other cultures

Figure 7(a) illustrates the controller of *Burmese* marionettes [22]. It is a palm-sized cross type control tightened with eight to ten strings. Because of its size, the puppeteer can manipulate the marionette with only one hand. For a sophisticate marionette with more than 10 strings, two such palm-sized controllers are employed. Instead of using any assistive device as marionette controllers, some puppeteers in *India* used their hands as the controller for the marionette [4]. Strings are tied to the fingers and the puppet figure. Moving the fingers and the hand will create puppet movements.

III. CONVENTIONAL MARIONETTE MANIPULATION

Unlike the humanoid robots where all limb joints are fully actuated and controlled, the marionettes are only "actuated" and controlled by a handful of strings attached to the limbs. Usually the number of control strings is less than the total degrees of freedom of the puppet figure. Assume that the puppet figure consists of n rigid segments (or rigid bodies) and j joints, and the DOF of each joint is f_j. According to Kutzbach criteria for computing the DOFs of spatial mechanisms [23], the DOF of the puppet figure without any strings connection, M, can be obtained by

$$M = 6(n - j) - \sum_{i=1}^{j} f_i. \qquad (1)$$

From the perspective of mechanisms, each string connection between the puppet figure and the controller can be considered as a "1-DOF" linear actuator that gives only the distance between the two connecting points. Therefore, it is necessary to have M strings to fully control and manipulation of the puppet movements. However, in practice, the number of strings is always less than M, and there are uncontrollable DOFs in the marionette. If the number of strings is k, where $k < M$, the uncontrollable DOFs will be $M - k$. The marionette, then, can be described as an under-actuated string-operated multi-limbed mechanism. The configuration of these uncontrollable DOFs will be determined by the gravity or other external factors. As the marionette is under-actuated, manipulation and control of the marionette movement become a very tricky technique.

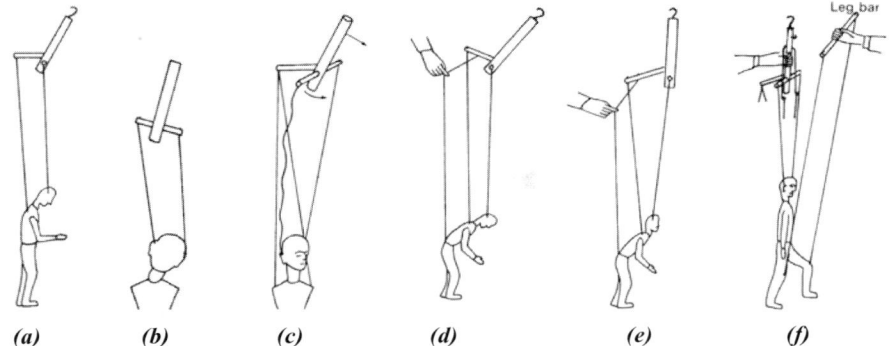

(a)   *(b)*   *(c)*   *(d)*   *(e)*   *(f)*

Fig. 8: Manipulating the controller to (a) nod the head; (b) incline the head; (c) turn the head; (d) bow the body; (e) keep the head upright; (f) walk the puppet.

Fig. 9: Walking a 6-string puppet using self-made Gou Pai

Manipulation of the marionette through the controller usually is achieved in three manners: pure manoeuvring of the controller, plucking the strings by hands, and the combination of both. With a properly designed controller, a wide range of puppet movement can be achieved by simply tilting or turning the controller. For example, Figures 8(a), 8(b) and 8(c) show how to nod, incline, and turn the head respectively by maneuvering the controller [8]. Such kind of manoeuvring usually produces fine puppet movement because string lengths are unchanged. Figures 8(d), 8(e), 8(f) illustrate how to use the combined hand and controller motion to make the puppet bow [8], lift his head, and walk. To make large and prominent gestures, such as walking, plucking the strings or pulling the detachable horizontal bars is necessary (Fig. 8f).

Usually one puppeteer can only operate one marionette from the controller unless the design of the controller can cater for multiple puppet figures. The puppeteer uses one hand to hold and maneuver the controller and the other to pluck or pull the strings and the detachable bars. With an ergonomically designed paddle controller (Fig. 5d), manoeuvring the controller and plucking the strings can be achieved using one hand simultaneously [10].

Manipulating the marionette using a Chinese Gou Pai is similar. Figure 9 shows a sequence of walking movement using a simple Gou Pai made in our laboratory with only six control strings. Because of the special layout, walking the puppet is achieved by rolling the Gou Pai clockwise and counter clockwise alternatively. Generally sophisticate movements can only be obtained by plucking various strings on the Gou Pai. Due to the large number of strings used in Chinese marionettes, maneuvering the Gou Pai becomes less obvious [20].

IV. ROBOTIC MARIONETTE SYSTEM

The basic robotic marionette system consists of a puppet figure, a puppeteer mechanism hosting an array of inexpensive RC servo motors, pulleys and strings, a motor control network, and a Pentium II 200Hz PC running Linux operating system (Fig. 10) [13]. The puppet can produce life-like movements according to programmed motion commands issued from the computer through the puppeteer mechanism. Besides the standard human gestures and motions, the robotic marionette can "defy" the gravity and "fly" in the air to perform various stunts. Our focus on this system is on the development of the robotic puppeteer mechanism. We intend to make it portable and universal. Portability means that the system can be used as a standalone unit or can perform puppet shows side by side with puppeteers using manually operated marionettes. Universality means that the system adopts a generic and modular design so that different puppet figures can be mounted under the same puppeteer mechanism to perform different shows by simply changing the computer programs. Certain puppet figures with strong cultural, artisitc, or historical values can be mounted under the puppeteer mechanism to perform dynamic demonstration instead of static exhibition.

Three versions of robotic marionette systems, ROMS-I, II, and III [26] (Fig. 10) are developed. ROMS-I was the first prototype used to explore the pulley-motor concept for puppeteering tasks and the basic structure of the mechatronic system. The puppet figure of ROMS-I was modified from a small wooden human dummy and fishing wires were used as the control strings. ROMS-II and ROMS-III were improved versions of ROMS-I with feedback from a professional puppet master and were developed simultaneously. The puppet figures of ROMS-II and III were life-like figures modified from toy soldiers with full clothes on for showing expressive behaviors. Standard embriodery threads were used as the control strings. The difference of the later two versions from ROMS-I is the design of the puppeteer mechanism. The pulleys were laid horizontally in ROMS-I whereas they are in the vertical position in ROMS-II and III. The portability and modularity of the puppeteer mechanism were also taken into consideration. Basic specifications of the three ROMS are listed in Table 1.

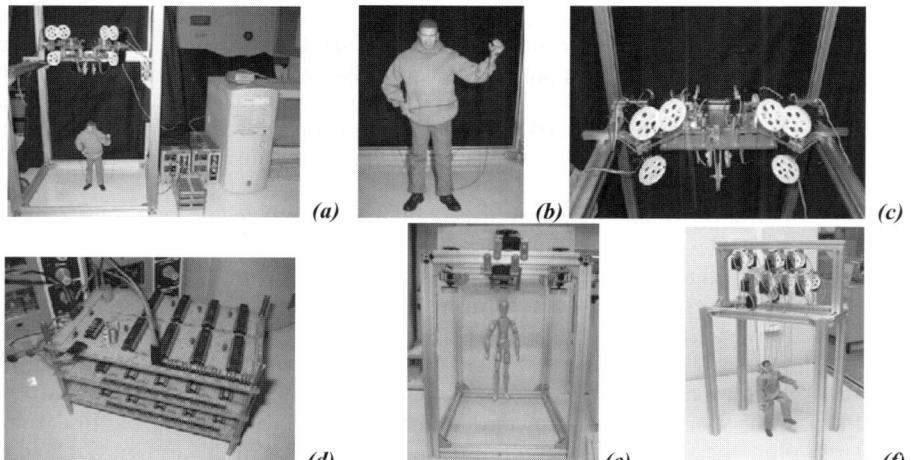

Fig. 10: (a) Overall marionette system (ROMS-II). (b) Puppet figure. (c) Robotic puppeteer Mechanism. (d) Motor control network. (e) ROMS-I (f) ROMS-III

	ROMS-I [24]	**ROMS-II** [13,25]	**ROMS-III** [26]
Motors	8	16	16
Strings	8	14*	16
DOF of puppet	30	35	23
Links of puppet	9	11	10
Joint of puppet	8	10	9
Type of joint	8 s-joint (3-dof)	9 s-joint (3-dof) 1 u-joint (2-dof)	4 s-joint (3-dof) 5 r-joint (1-dof)
Height (cm)	31.2	30.5	29.5
Weight (g)	314	214	259

Table 1: Specifications of ROMS-I, II, and III

4.1 Puppeteer mechanism

Here we focus on the puppeteer mechanism of ROMS-II and III, which becomes our reference design for further development. As we intend to develop the puppeteer mechanism as a generic platform for both Chinese and Western puppet performance, the maximum number of strings is set to be around 24 to 30, and each string is to be individually controlled.

String-retracting device. The strokes of strings can be large due to the limb motion of the puppet. Various concepts had been proposed for the string-retracting device including the linear actuator, a linkage mechanism assisted pulley system, and a simple pulley-motor assembly [13]. Considering the stroke of the string, interference of the devices in motion, and the space required, the simple pulley-motor unit is chosen as the string-retracting device. Each string is controlled by one pulley-motor unit so that the length of the strings can be lengthened or shortened by simply rotating the pulley on the motor clockwise or counter clockwise. The pulley diameter is determined by the speed and resolution of puppet motion which require a balance between large motion range and fine smooth motion. The space constraint needs to be addressed as well. Based on the size of the puppet figure, the pulley diameter for ROMS-II and III is set to be 50mm.

The selection of the motor has similar considerations. The motor chosen for the pulley-motor assembly is Parallax 900 continuous rotation servo motors. These motors accept speed control through continuous pulse-width-modulation (PWM) signals at 50Hz.

Mounting platform with extra DOF. A mounting platform combining the feature of Chinese and western marionette controllers is designed to house all pulley-motor units. As pulley-motor units can handle large as well as delicate string motions, motor units can remain stationary on the platform just like the all strings being tied to the Gou Pai. It is not necessary to mimic the manoeuvring of the controller as puppet masters do for this automated system. However, as strings drop vertically under gravity, performing puppet movements along the vertical plane, such as walking, nodding the head or bowing is much easier than sideway movements, such as arm opening and side-kicking, etc. Therefore, the two swinging armatures that can rotate horizontally are added to the mounting platform of ROMS-II to enable limbs to move sideways (Fig.10c). The movement of the two armatures are controlled by the servo motors. Three pulley-motor units controlling the arm and leg strings are mounted on each of the armature. The novel and extra degrees of freedom introduced on the mounting platform enable the marionette to produce more expressive behaviours. (Fig. 11a)

Motor/string functions. The assignment of limb movement to the corresponding pulley-motor unit is similar to the traditional marionette control. Among the total 16 motors installed on ROMS-II, there are two motors on each arm to control the elbow and hand respectively, three motors on each leg to control the knee, lower leg, and the ankle respectively, two for the head, one for the backbone string, one for the trick string for clapping hands, and two for the swinging armatures. The motor assignment on ROMS-III is similar to that of ROMS-II except that the two motors controlling the swinging armatures now control the shoulder of the puppet figure.

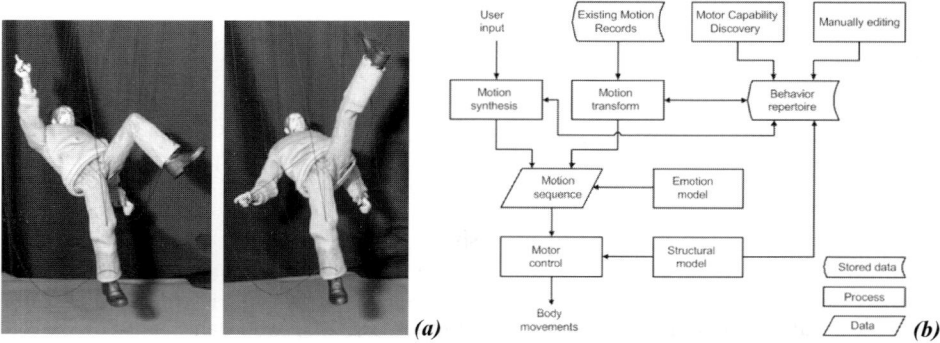

Fig. 11: (a) Puppet poses – Movie "Matrix"-like stunts. (b) Structure of Puppeteer software.

Motor layout issue. The design of the pulley-motor unit is modular. It is possible to reconfigure the pulley-motor arrangement for marionette with different number of strings and different characters. Due to the space and weight constraint, we adopt a *layered* motor mounting method to save the footprint of the puppeteer mechanism by stacking the pulley-motor units on top of each other. ROMS-II has two motor mounting layers on and below a single platform whereas ROMS-III has three motor mounting layers using extended fixtures.

4.2 Motor control network and interfacing

Motor control network. To cope with the large number (up to 30) of pulley-motor units, we develop a networked servo motor controller from scratch instead of off-the-shelf products [24,25]. The networked controller allows the RC servo motor to be added or removed from the controller easily when the configuration of the marionette changes. The networked motor controller consists of 16 PICmicro 16F876 microcontrollers with each controlling one servo motor for position and velocity control. The microcontrollers communicate through a common I2C bus network. One idle microcontroller with no motor attached serves as the bridge between the PC and motor controller on the I2C bus through RS-232 serial port. The networked motor controller can incorporate heterogeneous devices compatible with the I2C bus protocol, such as various sensors.

Interface programming. Low level servo motor control based on PWM is coded and implemented on the PIC microcontroller using the proprietary assembly language for this class of chips. The firmware development tools include the MPLAB IDE integrated development environment software and the PICSTART Plus programmer from Microchip Corporation.

4.3 Puppeteer software

As the marionette is mechanically controlled by an array of pulley-motor units, coordination of pulley-motor actions to create marionette movements is achieved through a high level marionette motion generation program on the PC, called *Puppeteer* [25]. The Puppeteer software developed on the Linux PC for the robotic marionette provides an interface to accept motion control instructions from the user. The user has three ways to input and edit the robotic marionette movements: *primitives*, *motion synthesis*, and *motion transformation*. Different motion generation methods give the user some selections of varied workload in order to make the motion input process comfortable and effective [25]. The major modules of the puppeteer software are shown in Fig. 11(b). Except for the motor control module, most parts of the software were developed in C language and with GCC 2.2 under Redhat 6.1 Linux OS.

4.4 Demonstration of robotic marionette behaviours

The visual results of the robotic puppet engaging behavioural expression are captured in a short motion sequence. The postures shown in Fig. 12 are directly obtained by combining existing posture primitives. These postures illustrate the diversity of behaviours. Feedback from the collaborating puppet master is that pros and cons of ROMS ideally compliment traditional puppeteers. ROMS can control delicate marionette motion that a human may find very difficult to achieve manually. As human hands have limited capacity, the robotic marionette may produce some movement that cannot be achieved using the manual marionette controller. For example, the stunt of "motion freeze" in the mid air shown in the movie – Matrix was coded into ROMS-II by one of our students (Fig. 11b) to demonstrate that our marionette can "fly". However, ROMS does have drawbacks, mainly in its dynamic behaviour. When the string motion becomes very drastic, the marionette may swing in undesirable directions frequently in addition to its controlled movement [27]. This is fundamentally similar to the pendulum effect that needs to be compensated with more advanced control algorithms.

Fig. 12: ROMS-II Puppet postures

V. CONCLUSION

Marionette, a disappearing art form in Singapore and also in other countries, is a universal cultural theme. The puppeteer techniques involved become a cultural or national heritage; the making of marionette figures becomes fine folk arts. In the modern elementary education, puppetry is also a very effective tool for shaping children's personality, creativity, and thinking. This work examined the basic marionette techniques (design, anatomy, and manipulation) in engineering perspective. Through the development of robotic marionette systems, ROMS-I, II, and III, we would like to explore the possibility of infusing modern robotic and mechatronic technology into the traditional art form of marionette, evoke and stimulate public interests in this art form, and thus provide the puppetry and puppet theatre new element and new features. As addressed by the response from the puppeteer master, the range and features of robotic manipulation compliment the traditional marionette due to its system characteristics, eventually this kind of technology infusion may stay side by side with the traditional puppetry. From the mechanics point of view, multi-cable driven marionettes may provide us insight in developing cable-driven multiple rigid body systems for engineering applications as well.

Acknowledgement: The authors appreciate invaluable comments and suggestions on the puppeteer techniques provided by Ms. Beng Tian Tan of *The Finger Players* (www.fingerplayers.com), Singapore, and encouragement from Prof. M. Ceccarelli. Efforts made by other members in this project, S. K. Tan, Ronald Yeu, Stefan Künzler, Wei Ji, Wesley Chia, are also appreciated.

REFERENCES

[1] D. Currell. *Puppets and Puppet Theatre*, Crowood Press, Wiltshire, UK, 1999.
[2] D. E. Hodges. *Marionettes and String Puppets Collector's Reference Guide*, Antique Trader Books, VA, USA, 1998.
[3] K. D. Sun. Origin of Chinese Puppetry (傀儡戲考原). Shanghai Publishing Co., Shanghai, China, 1952. (Chinese)
[4] C. Flower, A. Fortney. *Puppets: Methods and Materials*, Davis Publications, MA, USA, 1983.
[5] G. Latshaw. *The Complete Book of Puppetry*, Dover Publications, NY, USA, 1978.
[6] O. Batek. *Il Teatro Delle Marionette*, Ottaviano, Milan, Italy, 1981.
[7] H. Fling. *Marionettes: How to Make and Work Them*, Dover Publications, NY, USA, 1973
[8] D. Currell. *The Complete Book of Puppetry*, Pitman Publishing, 1974.
[9] L. Coad. Marionette Sourcebook, Charlemagne Press, Vancouver, Canada, 1993.
[10] B. Frascone, D. Frascone. *The Art and Technique of Marionette Making*, Vol., France, 2002.
[11] M. Steven. *Steven's Course in Puppetry*. Charlemagne Press, Vancouver, Canada, 1997.
[12] Q. Liu, S. L. Jiang. Chinese Puppetry Art (中國木偶藝術). China World Language Publishing (中國世界語文出版社), Beijing, China, 1993. (Chinese)
[13] B. K. Tay. *Development of a Robotic Marionette System*. Final Project Report, School of Mechanical and Production Engineering, Nanyang Technological University, 2003.
[14] Q. Y. Yang. *History of Puppet Theater in Hong Kong*. Cosmos Books, Ltd., Hong Kong, 2001. (Chinese)
[15] J. McCormick, B. Pratasik. *Popular Puppet Theatre in Europe, 1800-1914*, Cambridge University Press, UK,1998.
[16] M. Violette, History of French Puppetry, *Proc. Int. Puppet Theatre Conf.*, Taipei, Taiwan, 1999, pp234-261. (ISBN 957-02-4287-6)
[17] W. A. Dwiggins. *Marionette in Motion*, Puppetry Imprint, Detroit, USA, 1939.
[18] D. Abbe. *The Dwiggins Marionettes*, Abrams Inc., NY, USA, 1970.
[19] J. Bell. *Strings, Hands, Shadows: A Modern Puppet History*, Detroit Institute of Arts, USA, 2000.
[20] P. Tso. *Puppet Theatre in Hong Kong and Their Origins*, Hong Kong Municipal Government, 1987.
[21] Anon. *Children playing with string puppets*, Song Dynasty painting, National Palace Musuem, Taiwan.
[22] M. Thanegi. *The Illusion of Life*. White Orchid Press, Bangkok, Thailand, 1995.
[23] J. E. Shigley, J. J. Uickers. *Theory of Machines and Mechanisms,* 2nd ed., McGraw-Hill, NY, USA, 1995.
[24] S. S. Xing, I.-M. Chen. Design Expressive Behaviors for Robot Puppets, *Proc. 7th Int'l Conf. Control, Automation, Robotics, Vision*, Singapore, pp378-383, 2002.
[25] S. S. Xing. *Behavior-Based Physical Agents as Information Display Devices*, PhD Dissertation, School of Mechanical and Production Engineering, Nanyang Technological University, 2003.
[26] S. Künzler. *Development of Programmable Puppeteer Mechanism for Robotic Marionette Theatre*, Diploma Thesis, School of Mechanical and Production Engineering, Nanyang Technological University and Hochschule Rapperswil, Switzerland, 2003.
[27] K. Yamane, J K. Hodgins, H. B. Brown. Controlling a Marionette with Human Motion Capture Data. *Proc. IEEE Int. Conf. Robotics Automation*, Taipei, Taiwan, 2003, pp3834-3841.

4. People in MMS and Their Works

Golovin A.: *Bauman University TMM Department in the Former Half of the XX-th Century (Professor N. Mertsalov & Professor L Smirnov)*

Berkof R. S.: *History of Machines and Mechanisms by the Stevens Family and at Stevens Institute of Technology*

Zaharov I., Jatsun S., Jatsun S.: *Russian Scientist – Mechanics Ufimtsev A.G.*

Husty M.: *The Kinematic Work of W. Wunderlich*

BAUMAN UNIVERSITY TMM DEPARTMENT IN THE FORMER HALF OF THE XXTH CENTURY (PROFESSOR N. MERTSALOV & PROFESSOR L. SMIRNOV)

Alexander Golovin,
professor,
Theory of Machines and Mechanisms Department,
Moscow State Technical University named after N. E. Bauman
str. Koroleva 4-1-379, 129515 Moscow, Russia
Tel: +7-095-2168918; E-mail: irina@mx.bmstu.ru; aalgol@mail.ru

ABSTRACT

History of the "Theory of Machines and Mechanisms" Department of the Bauman Moscow Technical University in the first half of XX century is closely connected with Profs. N. I. Mertsalov, and L. P. Smirnov. Both of them graduated from two high schools: Moscow State University and Imperial Moscow Technical School (Bauman Moscow Technical University nowadays). Besides their study of the theory of mechanisms and machines (TMM), they were involved in other scientific fields: thermodynamics (N. I. Mertsalov), and steam machines (L. P. Smirnov).

In this paper the pedagogical activity of both the scientists before and after the October Revolution is described. Author's point of view concerns cases of different understanding of the TMM by these great scientists. In author's opinion, this can be explained by peculiarities of a typical pre-revolution and post-revolution student. Comparison of Prof. Mertsalov's book "Dynamics of Mechanisms" (1909-1912), and Prof. Smirnov's book "Kinetics of Mechanisms and Machines" (1926), is performed.

Keywords : History, Mechanism, Dynamics, Education

INTRODUCTION

The study of Mertsalov's and Smirnov's work proved that in general they had the same view on the subject of Theory of Mechanisms. However, contents of the course and methods of their approach to it were different in principal. What was the reason for these differences between two scientists who got basically the same education and were brought up in the same environment? In author's opinion, which is disputable, the answer for this question is in political and social aspects.

Russia made the change in political system in 1917. Before 1917 only rich enough people who graduated from classical or Applied Sciences gymnasiums had a chance to enter higher educational establishments. Most of the Russian population – workers and peasants – did not have that opportunity. The gymnasium graduates were well trained in Mathematics, Physics, Chemistry, and humanitarian sciences. The process of teaching was developed according to their educational level. The courses were taught on a high theoretical level without going into details concerning practical implementation. Often material was given as ideas. After 1917 the broad base of workers and peasants was given an opportunity to get a higher education. However, their school education obviously was insufficient: only few years in school and poor natural and humanitarian sciences background. "Work faculties" ("rabfacks" in Russian) were organized to prepare workers for studies in the institutes of higher education. The most talented but grown-up people who had to combine work and studies were sent to these faculties. Professors had to adjust their courses to their educational level. Basically, professors had to switch to a different method of teaching with specific technical calculations, in other words to create algorithms for all calculations.

Author wants to consider work of BMSTU professors N. I. Mertsalov and L. P. Smirnov from these positions. However, it's not possible to give a full survey of their pedagogical and scientific work in a small article. That is why analysis and comparison of their studies were accomplished with the examples of their works: N. Mertsalov "The Dynamics of Mechanisms" (1914) and L. Smirnov "The Kinematics of Machines and Mechanisms" (1926). For more complete understanding of these outstanding scientists their biography references are given below.

THE BIOGRAPHY REFERENCE

Professor N. I. Mertsalov – a student of a prominent Russian mechanical engineer Zhukovsky – graduated from a Mathematics faculty of Moscow State University in 1888. He worked at one of the German plants and attended lectures at Dresden Higher Technical School. He returned to Russia in 1892. The exams for Master's degree were passed at Imperial Moscow Technical School (IMTS – Bauman Moscow State Technical University nowadays) and got the rank of mechanical engineer in 1894. In 1895 with Zhukovsky's invitation Mertsalov became the Private-Docent at IMTS and "the keeper" of Applied Mechanics cabinet. At the same time (1895 – 1897) he taught Applied Mechanics at Peter's Agricultural Academy. Since 1897 N. I. Mertsalov was an adjunct professor of Applied Mechanics Department. Since 1921 N. I. Mertsalov combined the work at EMTS with one at the Applied Mechanics Department of Timiriazev (former Peter's) Agricultural Academy. He left BMSTU in 1929. In the end of the 30's N. I. Mertsalov founded the Engineering Mechanics seminar at the Institute of Machines of USSR Academy Sciences and directed it. I. I. Artobolevsky became the head of this institute afterwards. N. I. Mertsalov's main works concern the general theory of mechanisms, machines, theory of three-dimensional mechanisms and thermodynamics.

Professor L. P. Smirnov (1877-1954), a student of Zhukovsky, just like Mertsalov, graduated from EMTS in 1897 and with Mertsalov's recommendation stayed at the Applied Mechanics Department. From that time he was closely connected with this university. From 1929 to 1949 he was a head of the Applied Mechanics Department (later "Theory of Machines and Mechanisms" Department). In the 30's he founded and also directed the Department of "Steam engines and power plants". At the beginning of his scientific career he worked under Zhukovsky's supervision on the problems of Theory of Regulation and Theory of Machines. It was his assignment to Reshetov (later a professor, the head of a TMM Department) to begin the research concerning gearings. This research was completed with the creation of the unique gearing (for example the pinion with one tooth). In the 40's this research was continued by V. A. Gavrilenko who later became a professor and a head of TMM Department, and who led the science school of gearings at BMSTU. In the 40's under Smirnov's supervision the works concerning the balance of rotors began at TMM Department, these works were the base for establishing of another scientific school at BMSTU led by Professor Petrov. During these years, besides textbooks on kinematics and kinetics of machines and mechanisms, graph-analytic assignments on TMM, reference lists and teaching aides for these assignments were elaborated under the L. P. Smirnov's editorship.

Graphic procedures developed by L. P. Smirnov, including unpublished works, became the foundation for a number of TMM courses at other technical universities'. His influence on teaching TMM course at BMSTU still exists.

PROFESSOR MERTSALOV'S "DYNAMICS OF MECHANISMS"

Mertsalov taught the course of "Dynamics of Mechanisms" since 1909. In 1914 the lectures were organized and published by his disciple M. I. Felinski, later a professor. This course was taught after the course of "The Kinematics of Mechanisms" and consisted of three parts: the friction in machines, dynamic research of mechanisms, and the stability of machines during operation.

Four primary methods of exposing the material can be highlighted: exposing from simple to complex, combination of analytic method with graphic interpretation, extension of analytic method (with the elements of numerical methods) and its results for more complex cases and their engineering interpretation. It has to be mentioned that the book can serve as a guide but do not give the strict sequence of solving concrete problems. That is what Smirnov did, as a great admirer of graphic methods. His love for graph-analytic methods is objectively explained.

At that time a fast solution of a technical problem with accuracy sufficient for a practical use could be found only with graphic or graph-analytic methods. The most of the problems were reduced to numerous Athenian transformations and scale calculations with two arithmetic operations – multiplication and division. For grown-up students, who graduated "work faculties", that was easier than going deeper into the physical essence of a process. We also have to distinguish one more factor – different life experience of students before and after the Revolution. Before the Revolution students (Mertsalov's group) had good school fundamentals, but basically no practical experience. It was easier for them to follow the scientific basis, but they had no idea of actual machines. That is why Mertsalov always completed his theoretic computations with basic constructional representations of this computations and, what is very important, gave a substantial analysis of results and a prognosis of the system behavior after changing problem's conditions.

Fig.1

Students who entered technical universities after the Revolution (Smirnov's group) had poor school background but good practical experience. They were metalworkers, blacksmiths, locomotive engine-drivers, mechanics and so forth. They were familiar with the construction of machines, their operation and its result. That is why the substantial interpretation of computations was not as important as the strict technique of approaching the given problem. Of course, the class of problems being solved had to be decreased and the technique of graph-analytic approach had to reach its perfection. It would have been possible to return to the former "subject" method in a while but in 1941 the War started. Afterwards students mostly consisted of the War veterans. The history repeated ("the Revolution" situation happened once more). The author was a student at that time and remembers those people: the outstanding diligence, honesty, the faith in the future and not a very good educational background. Later the graph-analytic methods became a habit, routine exercise, and people with a insufficient mathematical background started to work on Theory of Machines and Mechanisms. When the informational revolution started, TMM at the technical universities wasn't ready it.

At the foundation of the part 1 (friction in machines) lies the ideas of Poncelet and the Coulomb's law. The following topics were considered:

- The motion of a solid body on a surface, including sloping, under the influence of positional forces
- The motion by the action of a spring, including auto-oscillation operation
- The friction on axis of rocking pendulum
- The friction on screw, wedge, higher pair, belt drive
- The friction in self braking mechanisms and the efficiency of mechanism
- The rolling friction and the conditions of slipping

To illustrate the method of teaching the subject of this part of the book we'll consider the fragment of the problem exposition of the friction in a higher pair. The dry friction is considered. In the basis of the reasoning about the nature of friction there is a number of factors: the force of interaction of profiles Q, their shapes (curvature radiuses of contacting profiles – r_1, r_2), kinematics of a transmission (p – the distance from the point of contact to pole of a gearing, thus giving in the non-obvious form one of the parameters of the speed of sliding), the way of interaction between rubbing surfaces (coefficient of friction – f). On the basis of the fig. 2a the analytic derivation of the elementary work of friction in the involute transmission is given:

$$dA_{fr} = fQ^n p\left(1/r_1 + 1/r_2\right) ds \tag{1}$$

Basing on this general case – non-involute transmission – is considered (fig. 2b):

$$dA = fQ^n ds \tag{2}$$

In my lectures, I present this as the multiplication of contact stress on the speed of sliding ($\sigma \cdot V_s$).

At the end of this fragment there is reasoning about the possible forms of wear and the statement of the problem of even wear of profile and the selection of contacting profile's shapes ensuring the minimal wear.

In the second part (the dynamic research of mechanisms) the mechanical system with forces, which depend only on the position, is considered (the positional system).To solve this problem the equation of motion is used in the energy form. Four fundamental requirements for the machine, operating at the periodic regime, were offered:

- strength of machine's elements
- sufficient stability on the foundation
- evenness of main shaft's movement
- the greatest efficiency in the given conditions

On the basis of these requirements the statements of problems of dynamic research are given. Nowadays these problems are formulated in the following way:

- 1st Problem: Given the forces and the initial velocity of any points of mechanism (at this part of the course N. I. Mertsalov considered only mechanisms with DOF $w = 1$) determine the law of mechanism motion and the reactions in kinematic pairs. This problem includes the determination of the law of the motion during acceleration-braking of machine. Basically, the results of this problem give the answers to the first two requirements.

- 2nd Problem: Given the forces determine the conditions that would provide the cyclic rotation of crank in the given limits of angular velocity variations. That would be the answer to the third requirement.

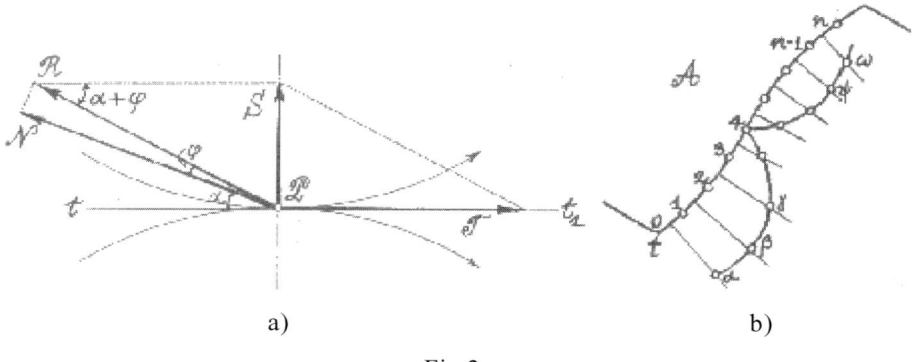

a)   b)

Fig.2

The answer to the fourth requirement – the determination of mechanism's efficiency – is formulated in the couple of lines, which help to understand that Mertsalov offered to evaluate friction losses with the method of the sequential approximation: at first determine the reactions at the kinematic pairs without friction, then – the losses by friction on the basis of the first part's material.

N. I. Mertsalov used D'Alamber's principle (force analysis), Lagrange's equation of second degree (design of dynamic model) and the movement's equation in energy form for solution of these problems. The terms: "the reduced moment of forces" and "reduced moment of inertia" did not exist. Wittenbauer suggested these terms much later, and L. P. Smirnov used them in his "The Kinetics...".

Since Mertsalov considered only positional systems, he used the movement's equation in the energy form for the solution of initial problem of dynamics (1st problem). 2nd problem represents a set of two boundary problems: the problem of mechanism's cyclic motion maintenance and the problem of slider's angular velocity variations in the given limits. It has to be mentioned that it was very hard for me to read lectures on this part of the course. The reason is extensive verbose explanations (nowadays they seem excessive, however, I think they were necessary at the time) and divergence with modern terms and symbols. At present, these problems are written down in a compact form, and something that was described in a number of pages could be put down in a number of lines. Unlike Wittenbauer's method Mertsalov's one could be presented as the first approximation of the solution of the problem for the general case of representation of the system of the forces, i. e. the forces that depend on the position, velocity and time.

In the third chapter (the stability of machines during operation) the following questions are considered:

- balancing of linkages
- mechanisms with elastic links
- the problem of the critical velocity of machines on the foundation
- balancing of flexible rotors

The problem of balancing the linkage is reduced to the consideration of the slider-crank mechanisms. The balancing, both with counterweight and crank-slider mechanism, is examined. The problem was brought into the consideration due to machine's anchor to the foundation and the influence of force of tightness of bolts, connecting machine with foundation, on the machine's stability. The question about the movement of mechanism with elastic links is examined on the example of slider-crank mechanism, the slider of which is connected with a coupler by cylindrical spring (fig. 3).

Initially, the simplified problem of critical velocity of crank's rotation, where the movement of point B is assumed harmonic and not dependent on the coupler's length, is examined. The dependence of B position on crank's position is represented in the form of harmonic series. It was shown that the accuracy of computation's results is provided with the allowance of only first two harmonics. Further the influence of other harmonics and the issues of friction were discussed. The paragraph was concluded with the recommendation on calculation of springs and the exposure of a number of methods of going through critical velocity. The issue of critical velocities on the foundation was considered as the continuance of previous issue and is summarized with the recommendation for designing the foundations for low- and high-speed machines.

In the conclusive paragraph the balancing of flexible shafts is examined. The mathematical apparatus for this problem's solution is simpler than of the previous one. The essential part is the discussion of practical methods of balancing: Mertsalov described the idea of the advantage of balancing mechanical system over balancing each element of it

.

PROFESSOR SMIRNOV'S "THE KINETICS OF MACHINES AND MECHANISMS"

This book is a natural sequel part to the Mertsalov's book "The Kinematics of Machines and Mechanisms".

It is appropriate to quote from the introduction to this book that gives us the image of Smirnov's credo: "...basing on his 17-year pedagogical experience, the author considered that it is necessary to put into all the formulas and graphic construction so called scales starting from the first pages...". The second quote from the book's introduction is: "...it had to be given a title "Mechanism's Dynamics" but author decided to name it "Kinetics of Mechanisms and Machines", meaning by this title to emphasize the dominating role that the study of capacity of kinetic energy of the system, obtained as the result of external forces work plays in the course...".

The book consists of 10 chapters. Topics considered in this book are closely connected with "The Kinematics…". There are changes in the sequence of topics, the issues of machines stability during movement are missing, but the terms of "the reduced moment of forces" and "the reduced moment of inertia" are included. The difference and the fundamental one is in the subject-matter of these topics.

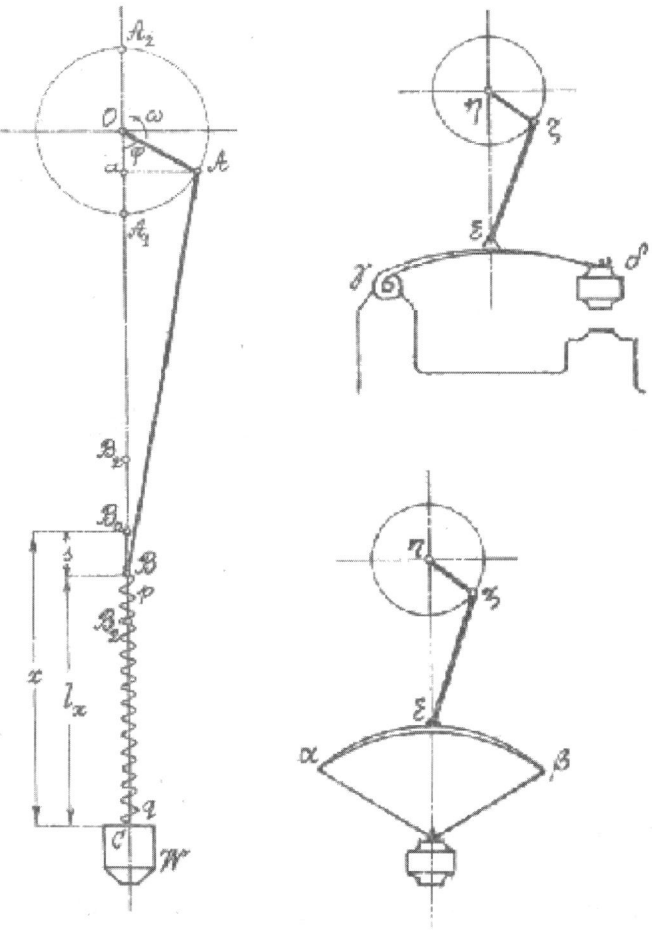

Fig.3

In the first and the second chapters the graphic approaches and experimental methods of simple movement's research are outlined. For example, in the fig. 5 experimental installation for studying of movement of a solid body with mass M under the action of elastic forces developed by springs is shown.

In Chapter 3 and 5 the static and dynamic balancing of rotors is considered. The graphic and experimental methods of determining moment of inertia of a part of arbitrary shape are shown in the Chapter 4. In the fig. 6 graphical method of moment of inertia determination is shown. This method is reduced to a number of geometrical transformations, that help to reduce the calculation of moment of inertia to summing up of line segments, which represent moment of inertia of elementary rectangles in scale.

 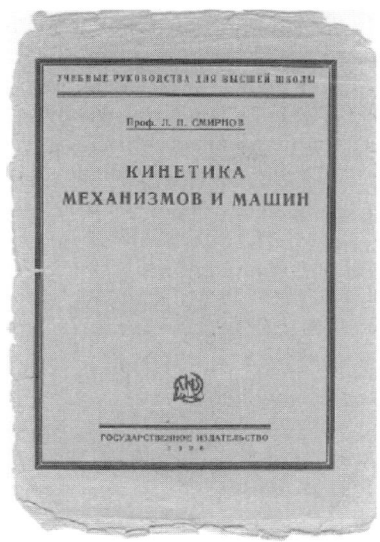

Fig.4

In the last four chapters the movement of mechanism is studied. Unlike Mertsalov, L. P. Smirnov used Wittenbauer's terms of "reducing" the forces and masses of mechanisms. Just like Mertsalov L. P .Smirnov worked on the dynamics of conservative systems and therefore used the equation of motion in the energy form. He used Wittenbauer's method for determination of flywheel's moment of inertia. It proved one more time his passion for graphic methods. It could be said that graphic methods for Mertsalov is the method of explanation while it's the method of calculation for Smirnov.

Introducing the term "reduced force" into the course allowed Smirnov to show the method of determination of friction losses in linkages – in fact, it was the explanation of sequential approximations method for efficiency determination used by Mertsalov (fig. 7). Friction losses were determined approximately basing on results of force computation assuming no friction in linkages.

Comparing to the method of "friction circles" from later published books, which is suitable only for simple mechanisms, this new approach is much more up-to-date.

CONCLUSION

It seems to me that the comparison of approaches to the teaching of "Theory of Machines and Mechanisms" by Mertsalov and Smirnov at the background of country's

146

social and political life does much of the explanation about the further way of the development of this branch of science. The convenience of using graph-analytic methods instead of numerical and analytical methods turned into an obsession. The substantial explanations of phenomenon gradually disappeared from the majority of books, only the solutions of narrow class of problems remained. The connection with "The Strength of Materials" and "The Elements of Machines" courses slowly disappeared. The subject began to be perceived as some kind of theoretical mechanics continuation. The subject wasn't ready for Era of computers. Nevertheless, it is not that sadly. In Russia the engineer training is accomplished by specialization, therefore the problem-oriented subject development is promising, like "TMM for robot's specialists", "TMM for textile machinery specialists", etc.

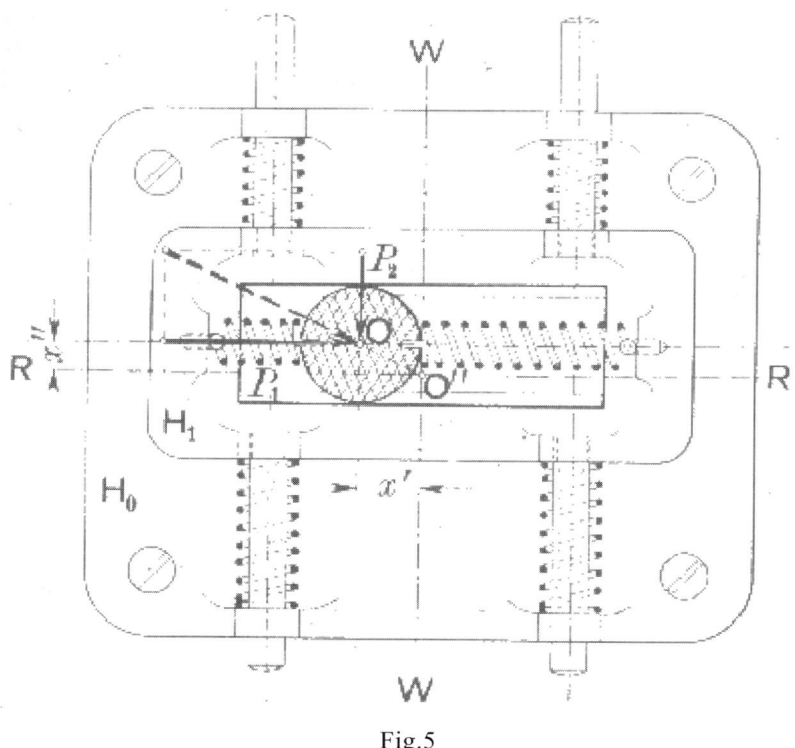

Fig.5

ACKNOWLEDGMENTS

I would like to give my acknowledgments to the authors' colleagues at Moscow State Technical University named after Bauman: Ass.-Prof. V. A. Nikonorov for taking part in discussions, giving the material and information, and for advice, Ass.-Prof. A. V. Ivanov, and to my students: Anna-Maria Eroshenko, Irina Petuhova and Evgenia Palevskaya for designing the article.

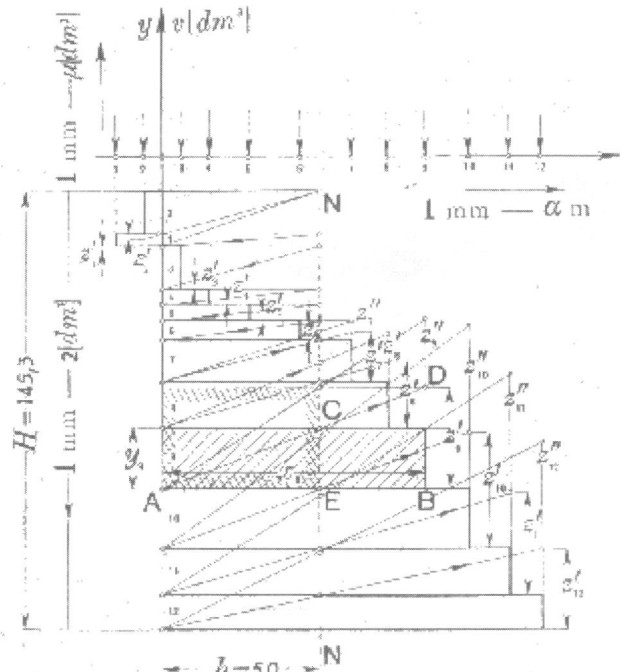

Fig.6

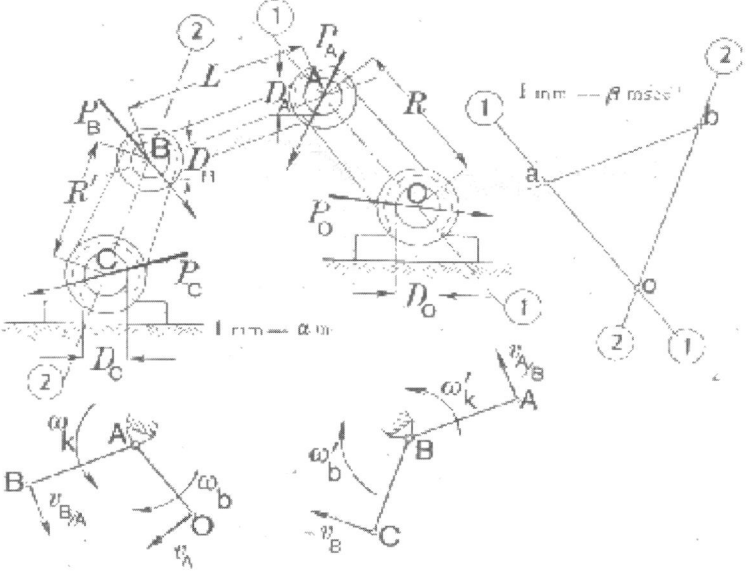

Fig.7

REFERENCES:

1. Scientific schools of Moscow State Technical University named after N. E. Bauman. The History of Development / I. B. Federov and K. S. Kolesnikov. BMSTU Publishers, Moscow, 1995. – 424 pp, il. (in Russian)

2. Mertsalov N. I. Dynamics of mechanisms / Lithographic course of lectures, read in Imperial Technical College. Processed and published by M. I. Phelinskii. – Moscow, 1914. – 610 pp. (in Russian)

3. Mertsalov N. I. Selected Creations / In 3 volumes, vv. 1, 3. Moscow, 1950 (in Russian)

4. Smirnov L. P. Kinematics of Mechanisms and Machines, Moscow-Leningrad, 1926. – 186 pp. (in Russian)

5. Smirnov L. P. Kinetics of Mechanisms and Machines, Moscow-Leningrad, 1926. – 206 pp. (in Russian)

6. Reshetov L. N. Involute Gearings. Moscow – Leningrad, 1935. (in Russian)

7. TMM Reference Lists / Edited by L. P. Smirnov. (in Russian)

8. Methodical Textbooks for Carrying-out Graphical Works in Applied Mechanics / Edited by L. P. Smirnov, - Moscow, 1940. (in Russian)

9. Gavrilenko V. A. The Basis of Involute Gearings Theory. – Moscow: Mashinostroenie, 1969. – 432 pp, il. (in Russian)

10. A. A. Golovin, C. B. Danilenko: Evolution of Theory mechanisms and machines from G. Monge to present day and modern problems of TMM training in Technical University / Proceeding of International Symposium on History of Machines and Mechanisms, Edited by Marco Ceccarelli, Kluwer Academic Publishers, 2000. pp. 263-270.

11. Mechanism's dynamics, Edited by A. Golovin, BMSU Publishers, Moscow, 2001. – 192 p. (in Russian)

12. Baranov G.G. Kinematics and Dynamics of Mechanisms, part 1. Moscow – Leningrad, 1932. (in Russian)

13. Gavrilenko V.A. and others, Theory of mechanisms and machines, text -book, Moscow, 1973 (in Russian)

HISTORY OF MACHINES AND MECHANISMS BY THE STEVENS FAMILY AND AT STEVENS INSTITUTE OF TECHNOLOGY

Richard S. Berkof
Department of Mechanical Engineering
Stevens Institute of Technology
Castle Point on Hudson, Hoboken, NJ 07030, USA
E-mail: rberkof@stevens.edu

ABSTRACT

Stevens, a name shared by three major inventors and a School of Technology, are all historically connected with Machines and Mechanisms.

The three notable inventors were John Stevens, Robert Livingston Stevens, and Edwin Augustus Stevens, all of whom made pioneering innovations in steam engines for ships and locomotives. Their other innovations in ship and rail transportation helped the worldwide commercialization of these technologies.

The Stevens' interest in technology led to the start of the Stevens Institute of Technology, the first college dedicated exclusively to Mechanical Engineering.

Innovative faculty and laboratories have earned a place for Stevens in education as well as technical advancements in the world.

This paper describes the major contributions of the original inventors, as well as the earlier days of the school.

KEYWORDS: History of TMM, Steam Engines, Steamships, Steam Railroads, Stevens

INTRODUCTION

The Stevens family has been called the America's family of invention. The three most notable inventors in this family were: **John Stevens,** "The Father of the Modern Steamship" and "The Father of American Railroads," and sons **Robert Livingston Stevens** and **Edwin Augustus Stevens.**

Their interest in technology also led to the start of the **Stevens Institute of Technology**, one of the first Schools of Mechanical Engineering in the world.

Herein are described the contributions of these inventors as well as the earlier days of the school. References [1] – [4] were used primarily; references [5] – [7] were supplementary.

1. JOHN STEVENS

Background

John Stevens III (1749-1838) of Hoboken, NJ was a developer of steamboats and a pioneer in mechanical transportation.

John's grandfather came to the US from England in 1699, became a lawyer, and purchased land in New Jersey. John's father, was a merchant and ship owner, and became prominent in politics. John thus grew up in a family of well-to-do landowners.

John was subsequently trained as a lawyer. He was treasurer of New Jersey during the American Revolutionary War, and achieved the rank of Colonel.

After the war, in the 1780's, he bought an elevated piece of land next to the Hudson River, surrounded by swamps, called the Island of Hoboken. He also became an 18th century gentleman scientist.

Crossing the Hudson River from Hoboken to Manhattan by barge was slow and often perilous. John became interested in better means of transportation, and steam power.

Steamboats

John watched a crude demonstration of steam power by John Fitch on the Delaware River around 1788, and this observation altered his life. After corresponding with James Rumsey, who had a crude steam engine, John designed a **new steam engine** with a vertical boiler.

He was concerned about protecting this novel invention. In order to do this, he helped bring about the first US patent laws, and earned one of the first patents in 1791. One of his early designs was for an alcohol-burning **internal combustion engine**.

John continued his developmental efforts, joining with others in 1797 to build the steamboat *Polacca*. This paddle wheel ship had the first non-condensing double-acting steam engine modelled on the Watt type in America. It was unworkable, due to boiler and other components leaking, as well as excessive vibrations of the wooden hull.

Developments continued, and John patented a boiler with multiple fire tubes that produced higher pressures without failure. He continued efforts to understand the problems, and developed practical solutions.

In 1803, John and son Robert designed an innovative stream engine using high pressure steam in a multitube boiler and single cylinder that delivered power to the screws via connecting rods. They built the boat with twin screw propellers and, in 1804, they sailed the **first successful propeller-driven steamboat in the world**. This revolutionary, steam-powered boat was small, only 20-odd ft. long, named *Little Juliana* [Figs. 1 & 2].

*Fig. 1: The **Little Juliana**, with original boiler & engine in a reproduction of the boat.*

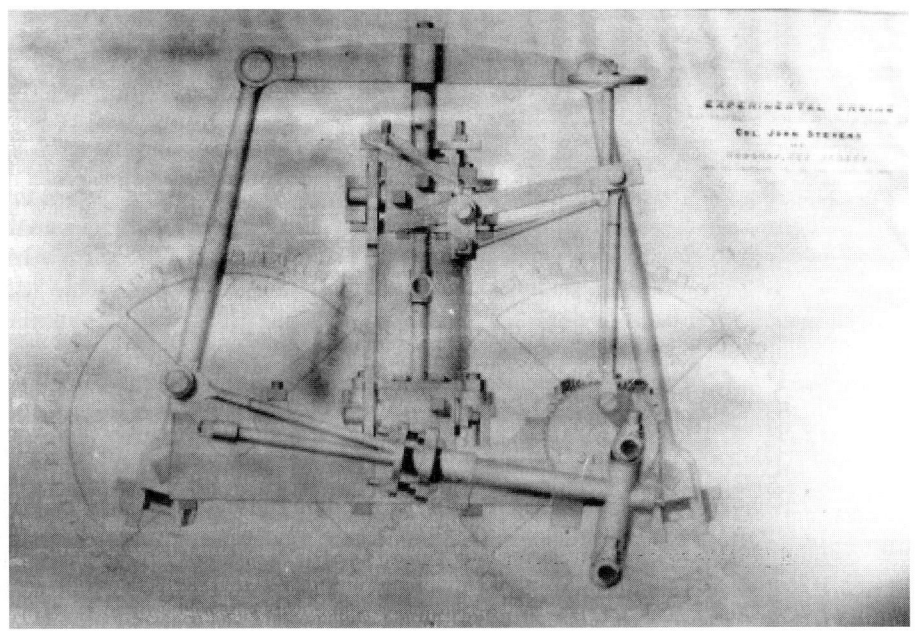

Fig. 2: Experimental engine for 1804 steamboat

The boat was operated by sons John Cox Stevens and Robert Livingston Stevens, who were to continue development efforts with their father in steam power.

In 1806, John started to build the 100 ft. *Phoenix* [Fig. 3]. Robert was the primary designer, with his father collaborating. This was a large ship, which had concave waterlines, the first such feature in a boat of this type, plus special bracing for the engine, to reduce strain on the hull. However, the boiler was not strong enough to make the high pressure needed for efficient propellers, so the ship used paddle wheels.

*Fig. 3 The **Phoenix**, which made the first ocean voyage by a steamboat (painting)*

The *Phoenix* was completed in 1808, and was propelled at over 5 mph, a fast rate at that time. There were plans to use this ship commercially on the Hudson River, but there were other limitations on rights to sail on this river. Another possible location for its use was the Philadelphia area. However, to get there from the New York City area, a ship had to go into the ocean, which was potentially very dangerous.

Danger not withstanding, in 1809 Robert sailed the *Phoenix* to Philadelphia. The voyage took 11 days and 150 miles, from the Hudson River into the Atlantic Ocean, around Cape May, to the Delaware River. This was the **first open ocean voyage by a steamboat**.

The *Phoenix* was then used on regular commercial routes between Philadelphia and Trenton.

John, "The Colonel," remained active for many years, working closely with Robert and Edwin on many new inventions and developments, especially in steamships and railroads.

Railroads

Railroads became a passion starting in 1812, and John was ultimately devoted to planning and building them. At that time, there were no true steam railroads anywhere.

In 1812, John published a widely distributed and controversial pamphlet entitled *Documents Tending to Prove the Superior Advantage of Railways and Steam Carriages over Canal Navigation.* In this, he considered all phases of railway transportation, with engineering and costs included.

After much difficulty and resistance, he initially obtained a New Jersey charter in 1815 to build a railway from Trenton to New Brunswick, but backers were hard to attract.

In 1820, he obtained a charter in Pennsylvania to build a railway from Philadelphia westward. His organization was called the Pennsylvania Railroad Company. Subsequently, his extensive development of commercial railroads in Pennsylvania ultimately enabled him to be called "The Father of the Pennsylvania Railroad," later **"The Father of American Railroads."**

John and Robert designed and built a demonstration railroad in 1826 in Hoboken, which had a circular track and "steam carriage" locomotive [Fig. 4]. The novel steam locomotive delivered its power though a cogwheel and cogged third rail, and moved at 12 miles per hour pulling a carriage with up to 12 passengers. This engine was the **first locomotive ever built in the U.S.**

Fig. 4 Demonstration railroad of 1826 (painting)

A model of the locomotive is seen in Fig. 5. The original engine is in the Chicago Museum.

Then, the Camden and Amboy Railroad was built, starting the era of commercial railroading. Robert was key in developing the railroads and many of the technical innovations needed for their commercialization.

Fig. 5 Model of original 1825 "Steam Waggon" demonstrated in 1928.

2.   ROBERT LIVINGSTON STEVENS

Steamship Development

Robert Livingston Stevens (1787-1856) was one son of John Stevens who worked closely with his father on technical developments. While John had many ideas, it was Robert who was the practical designer, innovator, and transportation entrepreneur in steamboats and railroads.

Robert helped design and build *Little Juliana,* then took the lead in large ship development with the 100 ft. *Phoenix* in 1808 (see above), which sailed the Philadelphia to Trenton route. Three years later, the *Philadelphia* was built and added to this important route.

Robert's marine engineering skills were prodigious, but he also became interested in other technical problems.

New Technical Challenges

Steamboats were just one area of technical interest to Robert (and John). In the 1790's, he helped design New York City's first water-supply system, using hollowed-out pine logs.

Robert suggested ideas for designing the **internal combustion engine**, which he never lived to see. He proposed concepts for bridges and tunnels. He had many innovative ideas.

During the war with England in 1812, Robert and John designed the **first ironclad gunship** for harbor defense. This was an early embodiment of principle later used for the ironclad *Monitor*. He also proposed another concept, which was also a precursor to the *Monitor*.

Another development in 1813 was the **elongated shell** fired from a cannon, and a forerunner of **armor-piercing projectiles**.

Steam Railroads

Robert was a brilliant engineer and visionary, especially in areas of transportation. With John, he became absorbed by railroads, which he foresaw as a future major force.

John and Robert built the demonstration railroad in 1826 (see above), which solved the problem of developing a steam engine on a locomotive. But, that was only one problem to solve.

Another was the rail needed for the train to move on, which until then consisted of iron straps attached to wooden rails. Robert solved this problem in 1830 by developing the **current iron rail cross section** as a modified "I," called a T-rail [Fig. 6].

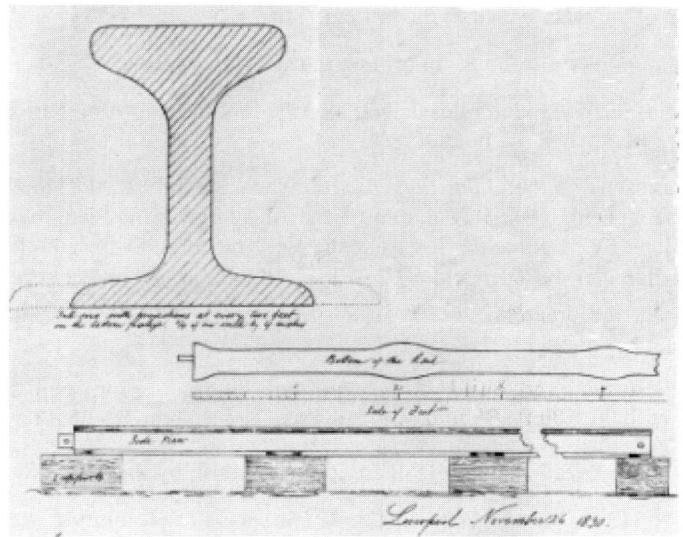

Fig. 6 Robert Stevens "T-rail" for railroads

He was able to have it manufactured by Welsh ironworkers, which in itself was a major innovation. He developed the **full rail construction system**, including the spike to hold rails in place, the "fish plate" to fasten rail ends to each other, and wooden ties embedded in crushed rock to stabilize the track bed.

Robert and family started the Camden and Amboy Railroad in New Jersey, one of the first in the U.S. His rail was used for this railroad, the locomotive *John Bull* that he brought back from England, and eventually all railroads in the world.

Another innovation incorporated into this railroad was the "cowcatcher," the sloping iron bars in front of the engine used to push animals away to avoid a derailment.

More Steamboats

Following a Supreme Court decision in 1824 that struck down a Livingston-Fulton monopoly on interstate ship traffic on the Hudson River, the Stevens family resumed heavy ferry operations on the Hudson.

Robert added improvements with each new boat. These included the following innovations:

- Substitution of lighter wrought iron for solid cast iron used for walking beams;
- Placement of boilers outside paddlewheels to increase deck space and improve safety;
- Guard beams supported by iron rods to protect the ship hull;
- **Balance valves** to help control the complex engine;
- Use of anthracite coal in the boilers;
- Building of a pilothouse to protect the steering operator;
- Use of deep wooden piles driven into the ferry slip bottom, which act as shock absorbers when the boat docks.

One notable vessel was the *John Stevens* side-wheel steamboat, designed and built by Robert at Hoboken in 1844. It was one of the first large iron steamboats built in the U.S., 245 ft. long by 31 ft. beam. It used a steeple-type engine with vibrating crosshead, with 75 inch diameter cylinder and 6 ft. stroke, and able to achieve a speed of 19 miles per hour.

Business Growth

The Stevens' technology innovations led to major growth in its combined railroad and ferryboat business from New York to Philadelphia:

- 1837 C&A Railroad had 15 locomotives (9 built by the company in Hoboken)
- 1840 $13 million investments, 8 steamboats, 17 locomotives, 71 passenger & baggage cars, and 55 freight cars

- 1869 $30 million investments, 100 trains a day carried 6 million passengers a year

Yacht Innovation

Yacht design was also of interest to Robert and his oldest brother, John Cox Stevens (1785-1857). John Cox formed the New York Yacht Club in 1844, becoming its first Commodore. In 1850, he formed a syndicate to race in a new event in English waters, establishing the **America's Cup** races. The now-legendary schooner *America* was built, and won the first America's Cup.

Many of the ten brothers and sisters in Col. Stevens' family retained a keen interest in technology. Working with his inventive father and innovative brothers was the youngest of the six sons, Edwin.

3. EDWIN AUGUSTUS STEVENS

Edwin Augustus Stevens (1795-1868) also pursued the family interest in technical innovations, displaying mechanical ingeniousness and an astute business sense. He was an equal partner with John and Robert in many engineering developments as well as business affairs. In 1821, at the age of 26, John turned over to him full responsibility for the family estate and other properties.

Also in 1821, Edwin designed and produced a successful plow. It was made of cast iron, and had an ingenious moldboard curved to keep dirt from sticking as well as a replaceable heel piece. This **"Stevens Plow"** was very popular on New Jersey farms.

Later, for New York City, Edwin designed a two-horse dump wagon with removable sides. This was used by the city for many years to haul refuse.

For the family's steamboat fleet, Edwin developed the **"closed fireroom"** system of forced draft air. This system used preheated air in the engine, greatly increasing efficiency. In 1842, Edwin was awarded a patent for the airtight ship fire room, subsequently used by navies throughout the world.

For the Camden and Amboy Railroad, Edwin designed the modern passenger car. Previously, all passenger railway cars had doors on the side. He developed the **vestibule car**, with a center aisle, and door and vestibule at each end for getting on and off or going to the next car.

When John passed away in 1838, at the age of 88, Edwin became responsible for one of the families most significant contributions to the nation. In 1841, because of a dispute with Canada, Edwin and Robert resurrected plans for ironclad naval vessels. They ran tests to determine necessary armor plate thickness, and then built the *Stevens Battery*. This was a huge ship, 410 ft. long and 45 ft. wide, with a flat deck designed to ride only 2 ft. above the water. It was covered in 4½ inches of iron armor. Guns mounted on the deck kept changing, because of new naval gunnery advances, and the ship was never completed. However, it still proved the concept of the modern armored warship, and was **the earliest American ironclad**

Robert passed away in 1856, leaving Edwin to carry on. During the Civil War, the *Stevens Battery* was considered too large, so Edwin built a new, scaled-down version at his own expense. This was the 100 ft. long *Naugatuck*, and it saw naval action alongside the *Monitor.*

Other Family Members

The family maintained a strong interest in technology.

Francis Bowes Stevens, son of James Alexander Stevens (another son of Colonel John), in 1839, designed the **"Stevens cut-off."** This device allowed the steam and exhaust valves of an engine to be separately but interdependently controlled by their own eccentrics. This enabled a single operator of a paddle wheel steamer to easily put the steam engine in reverse, and it became a standard feature of such boats.

Edwin's son, Edwin Augustus Stevens, Jr., developed the now standard **double-ended ferryboat**. While Colonel John had the idea in 1813 of a reversible propeller-driven ferryboat, it had never been built. Edwin Jr. worked out the mechanism of a single drive shaft running the length of the vessel, powered by a single engine, with a propeller at each end. This boat would go into reverse quickly, and be able to load vehicles at both ends. In 1881, the ferry *Bergen* was built, and essentially ended the use of paddlewheel ferries in New York harbor.

Vision for Education

Edwin also made a major contribution to education. He left a bequest to found the **Stevens Institute of Technology**, on the family estate in Hoboken.

4.   STEVENS INSTITUTE OF TECHNOLOGY

In 1867, one year before he died, Edwin decided to leave a portion of his estate for the establishment of a college. Edwin's wife, Martha Bayard Stevens, was instrumental in translating Edwin's wishes into the founding of an engineering college.

Martha's brother, Samuel Bayard Dod (1837-1906), was given the "details" of starting up the school. He became president of the trustees of the school. After extensive research, the trustees decided to establish "a school of technology with special reference to mechanical engineering."

The presidency of Stevens Institute was offered to **Henry Morton** (1836-1902) of the Franklin Institute of Philadelphia. The Franklin Institute had been founded in 1824 to promote advancement and research in the "mechanic arts." Until the American Society of Mechanical Engineers was founded in 1880, the Franklin Institute and its Journal was the only organization dedicated to promoting and sharing information about mechanical engineering.

Thus, **Stevens Institute of Technology** was founded in 1870 as a School of Mechanical Engineering based on scientific principles.

As part of the plan for development of the new school, the following was one key section: "The subject of mechanical engineering, in reference to the theory and practice of construction of machines, will form, like the others, a distinct department under the charge of a special Professor…"

The school opened to students in Sept. 1871. The first graduating class was in 1873, with one student graduating with degree of Mechanical Engineer.

Facilities

The main academic building at Stevens was constructed in 1870, and the **Mechanical Laboratory** was established in 1875. The laboratory included tools and equipment, as well as special machines constructed by students, including:

> Thurston autographic testing machine (1876), lubricant testing machine (1877), large oil tester, Prony dynamometer, small horizontal engine, and small oscillating engine (1878), autographic transmitting dynamometer (1879), 3½ hp compound condensing engine (1880).

Over the years, a great many of tools and equipment was contributed by companies, alumni, and other individuals.

One major gift was from Andrew Carnegie to build the Carnegie Laboratory of Engineering, dedicated in 1902. One large installation from the Stevens family was an Allis-Corliss Cross Compound Engine.

Faculty

The early days at Stevens had notable faculty members, some of whom included:

Prof. Robert Henry Thurston organized the Dept. of Mechanical Engineering in 1870. His many contributions to the field included investigations into materials of construction, especially iron & steel. He also did extensive work on steam engines and boilers.

Much of Thurston's research was accomplished in the Mechanical Laboratory, which he established as the first of its kind to conduct funded research in an American institution of higher learning. He achieved goals of performing government-sponsored research, commercial research for companies, and the training of undergraduate students on research.

Major laboratory efforts included the evaluation of boiler explosions, the testing of the strength of materials, particularly those used in boilerplate, tools, and engine parts, as well as friction and engine lubricant testing.

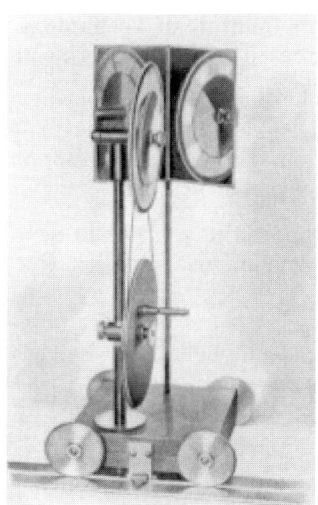

Figs. 7 and 8: Thurston's Autographic Tester and Mayer's Chromatic Photometer

He invented the magnesium-ribbon lamp, autographic recording torsion testing machine for materials [Fig. 7], a form of steam engine governor, and apparatus for determining the value of lubricants and friction of lubricated surfaces.

Thurston left an indelible mark on Stevens Institute and the engineering profession through his 574 articles and 21 books, and as a founding editor of the journal *Science*. He was the first president of the **American Society of Mechanical Engineers**, which was founded at Stevens in 1880.

Prof. Alfred M. Mayer, Physics, was well known for investigations in acoustics and vibrations, and was deemed "prince of experimenters." [See Fig. 8 for one experimental device, the Chromatic Photometer.] He had about 100 publications, including 6 standard books.

Mayer, along with Koenig, Helmholtz, and Rayleigh, were the first to devise methods to measure the intensity of sound by non-electrical means. He used the term "pendulum vibration" for simple harmonic motion, and he performed experiments with ticking clocks in soundproofed rooms. This led to the discovery of "masking," in which a lower-pitched sound could obliterate a higher-pitched sound of substantial intensity.

Prof. Charles William MacCord, Mechanical Drawing and Designing, developed many advancements of descriptive geometry and kinematics. He wrote many papers and books, including one on *Kinematics* [6], which documented many new developments in gears, wear measurement, and many other kinematic topics.

MacCord had been the draftsman for Ericsson for the ironclad *Monitor*, and was the chief draftsman under Gen. George McClellan for the *Stevens Battery*.

Prof. John Burkitt Webb, Mathematics and Mechanics, invented many power measurement devices. These included the "Floating Dynamometer" in 1888 (pat.) to measure power given out or absorbed by motors, dynamos, and other machines; the "Viscous Dynamometer" in 1892 (pat.) which was an absorption dynamometer using fluid friction; and the "Dynamophone" in 1900, which was a transmission dynamometer measuring the twist of a shaft carrying power using telephonic technology.

Prof. David Schenck Jacobus, Experimental Engineering, developed original apparatus for the illustration of physical laws and the testing of various mechanical devices.

Prof. William Bristol, Mathematics, was an inventive genius in instruments and received over 100 patents. He was a pioneering applicator of the bimetallic principle to the measurement of electrical quantities. He invented the helical pressure element used in pressure-indicating and pressure-recording instruments, and invented the first practical pyrometer for measuring high temperatures in combustion chambers. Later, in the 1900's, his Bristol Manufacturing Co. was a major world manufacturer and supplier of industrial instruments, ranging from steam engine indicators to electronics products, amplifiers, and speakers.

George Meade Bond of the Stevens Mechanical Laboratory designed the Rogers-Bond Comparator in 1880. This revolutionary device was conceived by William Rogers, a physicist, mathematician, and astronomer, who specialized in optical methods of measurement. This "optical calipering" comparator significantly increased the quality of measurements in manufacturing.

Bond was hired by Pratt and Whitney to supervise production of the comparator. This device led to a major increase in the ability to fabricate interchangeable parts, and was the standard industrial comparator used well into the 20th century.

5.   CONCLUSION

Stevens Institute of Technology has its roots in the innovations and developments of the Stevens family, who were pioneers in many key technological advances in the early days of the United States.

Col. John Stevens, together with sons Robert and Edwin, brought steam technology to ships and railroads, and built a substantial transportation business. They were very innovative engineers, inventing many machines and mechanisms to bring ideas to practical fruition.

Edwin also had a vision for a new school which taught Mechanical Engineering. Since 1871, Stevens Institute of Technology has prospered and grown into a strong university.

Today, Stevens provides a well-rounded engineering program at the undergraduate level, together with science, management, and humanities. There is an extensive graduate program leading to master's and doctoral degrees, with wide-ranging research activities in technical areas.

ACKNOWLEDGEMENT

Thanks to Doris Roberts and the S.C. Williams Library at Stevens Institute of Technology for their invaluable help. Thanks also to Dr. Constantin Chassapis, Director of the Department of Mechanical Engineering at Stevens Institute of Technology, for his strong support and encouragement.

REFERENCES

1. Allen, Oliver E., The First Family of Inventors, *Invention & Technology*, Fall 1987, pp. 50-58.

2. Clark, Geoffrey W., *History of Stevens Institute of Technology*, 2000, Jensen/Daniels.

3. Cunningham, John T., *Railroading in New Jersey*, Associated Railroads of New Jersey, based on 17 articles written for the magazine of The Newark Sunday News, January 7 to April 29, 1951.

4. Furman, Franklin De Ronde, Editor, *Morton Memorial, A History of the Stevens Institute of Technology*, The Alumni Association of the Stevens Institute of Technology, 1905.

5. Guthrie, John, "American Walking Beam Engines," *Marine Engineers Review*, Feb. 1979.

6. MacCord, Charles William, *Kinematics*, 4[h] Ed., John Wiley & Sons, New York, 1883. [Subtitle: "A Treatise on the Modification of Motion, as Affected by the Forms and Modes of Connection of the Moving Pats of Machines."]

7. Mitman, Carl W., *Steven's "Porcupine" Boiler, 1804: A Recent Study*, Presented at the Science Museum, London, and in New York, April 19th, 1939. [Regarding the Stevens 1804 boiler and engine located at the United States National Museum.]

RUSSIAN SCIENTIST – MECHANICS UFIMTSEV A.G.

Zaharov Ivan, Jatsun Sergey, Jatsun Svetlana
Kursk State Technical University
305000 Kursk 50let Oktyabrya, 94
jatsun@kursknet .ru

ABSTRACT The paper presents description of the some inventions of one talented Russian scientist – inventor who already one hundred years ago suggested new ideas of the development of the motors for aero plane, electrical power station and more than twenty different inventions.

KEYWORDS: birotary engine, sphere plane, aerospace centre, cycle shape wing

INTRODUCTION

Anatoly Georgievich Ufimtsev   is one of the talented Russian inventor and designer of the beginning of the last century. Four   years ago Jatsun Sergey has written one paper about some inventions of Ufimtsev concerning wind generator and inert ion accumulator [1]. But generally he had 22 inventions. And when we began to investigate different inventions, that had done this man, we understood we should describe   all of his inventions, because a lot of his ideas today remaining actual.

This very handsome man was well known during first Russian revolution time. When he was young, he close to Russian revolution movement. He wanted to help Russian revolution and those times he had done special printer and copy machine. Unfortunately detail description of these machines we didn't find.

He fought against Russian church at 1898. At this time he had done the bomb with a clock and destroyed one very famous Russian icon which was located in the one of the biggest Cathedral in Kursk. It is Znamenskiy Cathedral (see fig.1)

After this experiments he was in prison and only at 1905 this young man came bag to Kursk where continued his invent ional activity.

At 1909 Ufimtsev designed two cylinders two cycles rotary engine. The motor today is kept in the exposition of the inventor Ufimtsev in Moscow aerospace Centre (see fig.2).

164

Fig.1. Znamenskiy Cathedral in Kursk

Fig.2. General view on the exposition of the inventor Ufimtsev in Moscow aerospace
Centre.

Fig.3. General view on the birotative engine.

Size of the motor: d - 1 ? , lenth- 85 ?? , weith- 50 ??,power: 18kW. The motor was built by Ufimtsev and was investigated in a Schetinin factory at 1908-0909 in Forest university   where   was   register   power   around   30   kW   [2,3]. This   motor   is   kept   in   aerospace   Centre   in   very   good   condition. At 1910-1911 he has done two four and six cylinders birotative engine. The general view of this engine we can see on the fig. 3. In the 1912 Ufimtsev has got Silver medal on the3-d Russian conference.

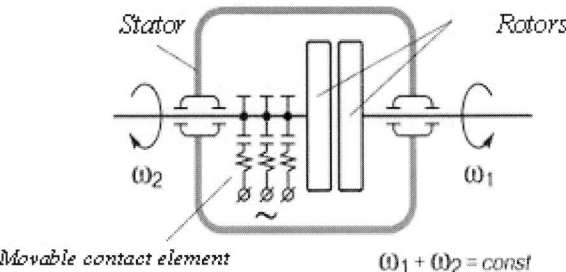

Fig.4. Scheme of the birotative engine.

166

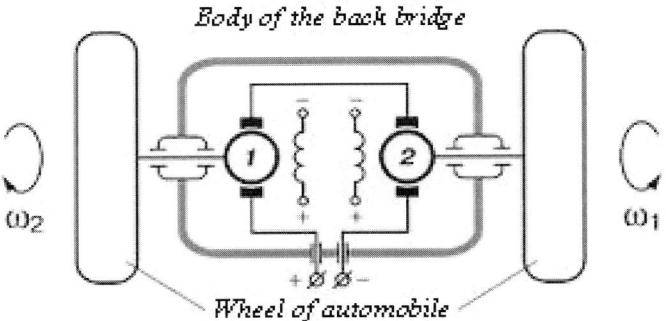

Body of the back bridge

Wheel of automobile

Fig.5. Birotative engine for wheel drive

The schemes of the birotative engine on the fig.4 and fig.5 are shown. For example on the fig 4 you can see electrical birotative engine. In one stator engine has two rotors that have only connection with electromagnetic field.

Axes of each rotor come in to opposite side of frame (stator) of engine. Electromagnetically torques is equal and has opposite direction. The sum of angle velocities is constant. Another scheme of the engine we can see on the fig. 5. There are two DC motors which connected each another.

Fig.6 General view on the Sphere plane of the Ufimtsev.

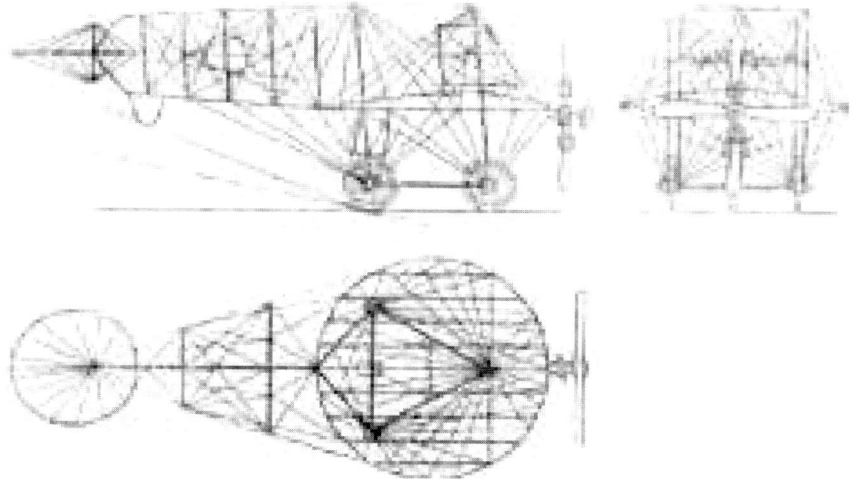

Fig.7. Scheme of a Sphere plane ? 1.

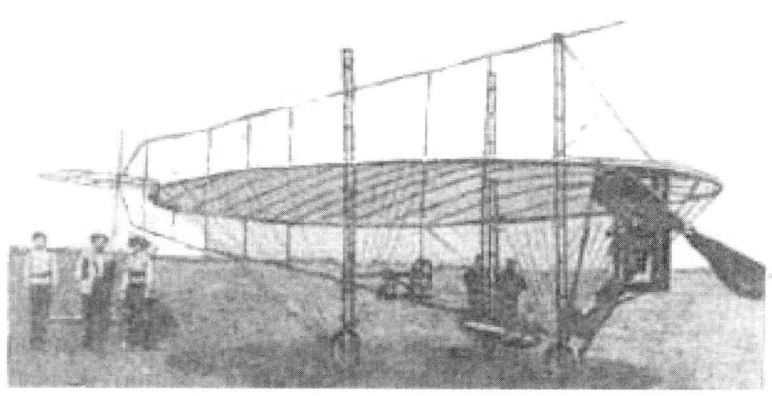

Fig.8. Scheme of a Sphere plane ? 2.

Sphere plan 1 (see fig.6). This picture shows inventor Ufimtsev in the moment of investigation of the sphere plane. This aero plan had original structure. The wing of the plan had cycle shape and three wheals one of them was maintained in a front of the plane [4]. This apparatus was arranged by engine with power around 15kW. This engine was designed by Ufimtsev the engine had four elements. Two of them were maintained to cylinders and two carter of engine. Three wheals chassis was built first time in the world simultaneously with American researcher Kertis. The wing consists from cycle frame and had diameter 3,4 m with area 9 ?^{2} and had 6 elements which provide durability of the wing. Steering system was maintained behind his screw. This sphere plan was investigated and tried flying. Sphere plan 2. The scheme and structure of this plan you can see on the fig. 4. General scheme of this plan were similar to first sphere plan but dimension of this machine were two time bigger than it was before. Area of the wing was 36 ?^{2} steering system –in the wing used 11 elements for durability.

CONCLUSION

The inventions that are described in this paper allowed creating modern motors and aero planes in last century. A lot of researchers in different countries continued this inventions in the field of aero plane design.

Ufimtzev A.G. (1880-1936)- He was the author of more than 22 inventions. The most interesting:

1896- electrical method of printing was invented;

1900- a project of engine without valves was developed;

1909 - the engine without valves was manufactured;

1918- inertia-kinetic accumulator was invented;

1924- wind electrical generator was suggested;

1927- wind electrical station was invented.

REFERENCES

1. S.Jatsun The windgenerator of electric power with inertia accumulator which was constructed by A.G. Ufimtsev, International Symposium on History of machines and Mechanisms - Proceedings HMM 2000,157-162. Cassino, Italy.
2 CGVIA, F. 802, op. 3, d. 1637, l. 202.
3. Biblioteka vozduxoplavania, 1910, ? 7, p. 1—11.
4. Vestnic vozduxoplavania, 1911, ? 10, p. 33; ? 11, p. 27
5. Aero and avtomobil life, 1910, ? 16, p. 20 -21.

THE KINEMATIC WORK OF W. WUNDERLICH

Manfred Husty

Inst. for Engineering Mathematics, Geometry and Computer Science,
University of Innsbruck, Austria
Technikerstr. 13e, 6020 Innsbruck, Austria
E-mail: Manfred.Husty@uibk.ac.at

ABSTRACT: Walter Wunderlich (1910-1998) was an Austrian geometrician who did a lot of work in kinematics of planar and spatial motions. As most of his important papers are in German, this contribution is aimed at introducing some of his results to a broader scientific publicity.

KEYWORDS: Wunderlich, scientific work, kinematics, geometry of surfaces, geometry of curves.

1. SHORT OVERVIEW OF W. WUNDERLICH'S LIFE

Figure 1: Walter Wunderlich (picture from [4])

In this section we give a short overview of the life of Walter Wunderlich. The data are taken from Stachel [4], where a detailed biography can be found.

W. Wunderlich was born on March 6th 1910 in Vienna. After graduation from the gymnasium(high school) he studied civil engineering, but after finishing the undergraduate courses he changed to a mathematics and descriptive geometry program. He wanted to become a high school teacher in these subjects. At University of Vienna he had excellent and famous professors as teachers: in mathematics P. Furtwängler (founder of the Vienna school of number theory), H. Hahn (one of the founders of functional analysis, "theorem of Hahn-Banach"), Mayrhofer and Wirtinger (functional analysis, Wirtinger was the third winner of the Cayley prize after Cantor and Poincaré) and in descriptive geomety L. Eckhart, J. Krames, E. Kruppa, Th. Schmid and L. Schrutka. His master thesis (Diplomarbeit) under supervision of E. Kruppa on "Nichteuklidische Schraubungen" ("Non-Euclidean Skrews") was finished in 1933

169

and already one year later he handed in his PhD. thesis ("Doktorarbeit") having the title "Über eine affine Verallgemeinerungen der Lyon'schen Grenzschraubungen" ("An affine generalization of Lyon's limit skrew"). With many problems mainly due to the political situation in Austria he started a scientific career and became assistant of L. Eckhart. 1939 he filed a habilitation thesis with the title "Darstellende Geometrie nichteuklidischer Schraubflächen" ("Descriptive geometry of non-Euclidean screw surfaces"). Immediately after the habilitation exam he was conscripted to the army. I mention this fact because during the war he wrote his first papers and three of his papers even bear the affiliation "in the battlefield". Later on he came to a military research institution and wrote a book titled "Introduction to under water blasting" which shows his versatility [1]. In 1946, after the war he became associate professor at the Technical University in Vienna and in 1951 he was appointed full professor. He retired in 1980, but wrote more than 40 high standard scientific papers as an professor emeritus. Ten years before his death he suffered a retina ablation and died almost blind in 1998.

In the frame of the workshop HMM 04 it must be mentioned that W. Wunderlich was more than 25 years honorary editor of the IFToMM journal Mechanism and Machine Theory.

2. SCIENTIFIC WORK

Walter Wunderlich's scientific work comprises of 205 papers including three books. Almost all of the papers are journal papers and single authored. The main part of his scientific work is devoted to kinematics. More than 60 papers including one book are from this field. His other interests were in descriptive geometry, the theory of special curves and surfaces, the classical differential geometry and the application of geometry in surveying, civil and mechanical engineering.

2.1 KINEMATIC PAPERS

It will not be possible to give a deep insight into the extensive kinematic work of W. Wunderlich, but we will try to explain the main ideas and most important achievements. In 1947 he published a paper with a generalization of cycloidal motions [5]. In this paper he introduced so called higher cycloidal motions. Cycloidal motions are obtained if both centrodes of a planar motion are circles. The paths of points are so called cycloids. The same motion can be obtained by the endeffector of a planar 2R-linkage when both revolute joints rotate with a constant angular velocity. Wunderlich's generalization is now that he allows n systems. Therefore he obtains the one parameter motion of a nR chain where all links rotate with constant angular velocity (see Fig. 2). The paths of points of the last system are called *cycloids of n-th stage*. The analytic representation of the paths is obtained easily when one uses complex numbers to describe planar motions:

$$\mathbf{z} = x + Iy = \sum_{i=1}^{s} a_\nu e^{I\omega_\nu t} \tag{1}$$

---

[1] He himself never mentions this book in his list of publications.

The description of planar motions with complex numbers requires a historical remark: Although R. Bereis [1] is often cited to be the founder of this method, W.
Wunderlich has used complex numbers (and also the so called isotropic coordinates) earlier. In private communications (Stachel [4]) he claimed that he was
the one who encouraged R. Bereis to develop a theory of planar kinematics using
complex numbers. According to Wunderlich it was the Italian geometrician G.
Bellavitis who underlined the vectorial interpretation of complex numbers using
the symbol "ramun" (radice di meno uno) instead of I to describe the $90°$ rotation.
The application of isotropic coordinates was introduced by A. Cayley(1868) and
E. Laguerre(1870).

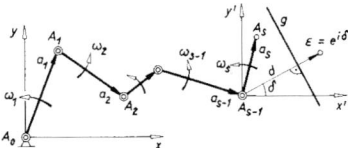

Figure 2: Higher cyloidal motions

The main result of Wunderlich concering generalized cycloidal motions is the generalization of Euler's theorem: *Every cycloid of n-th stage can be generated by a*
$2(n-1)$ *chain of circles in n! different ways.* He shows that the moving and the
fixed centrodes of a higher cycloidal motion are higher cycloids and discusses a lot
of geometric properties of the paths of points, and the envelops of lines. Later on
he used higher cycloids for curve approximation in the plane [6] .
In 1968 W. Wunderlich published his book on planar kinematics ("Ebene Kinematik", Fig. 3). This book contains all topics of planar kinematics treated from
point of view of geometry, but it has many links to the applications and some
unusual aspects. It starts with the classical basics of planar kinematics but note-
worthy, complex numbers are used continuously to describe the motions and their
properties analytically. The advantage of the method may be shown in Wunderlich's derivation of the equation of the coupler curve of a planar four-bar using
isotropic coordinates. Referring to the notation of Fig.4 he writes the bars LA, AB
and BM of the four-bar in complex numbers $\mathbf{u} = ae^{I\phi}, \mathbf{v} = ce^{I\psi}, \mathbf{w} = be^{I\chi}$. These
complex numbers have constant absolute value (modulus)

$$|\mathbf{u}| = a, \quad |\mathbf{v}| = c|, \quad \mathbf{w}| = b$$

but changing arguments (angles) and constant sum

$$\mathbf{u} + \mathbf{v} + \mathbf{w} = d.$$

An arbitrary complex number $\mathbf{m}$ is then describing a point of the coupler system:

$$\mathbf{z} = \mathbf{u} + \mathbf{m}\mathbf{v}.$$

W. WUNDERLICH
EBENE KINEMATIK
B·I·
HOCHSCHULTASCHENBÜCHER
447/447a

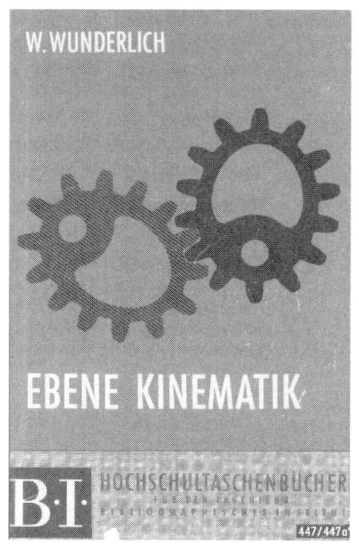

Figure 3: Book on Planar Kinematics

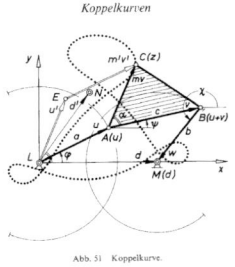

Koppelkurven
C(z)
B(u+v)
A(u)
M(d)
Abb. 51 Koppelkurve.

Figure 4: Four bar coupler curve

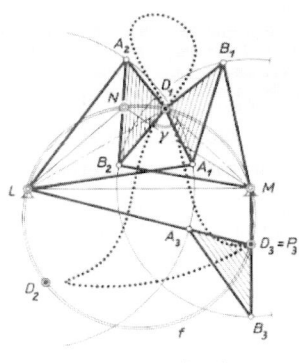

Abb. 55 Doppelpunkte der Koppelkurve

Figure 5: Double points of coupler curves and focal circle

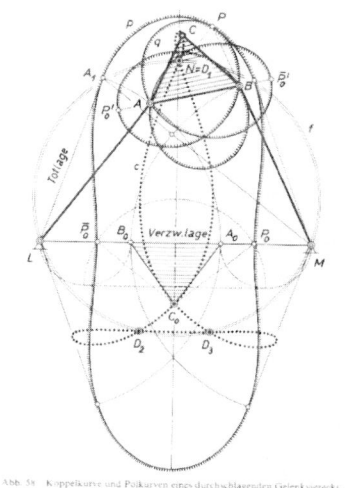

Totlage
Verzw. lage
Abb. 58 Koppelkurve und Polkurven eines durchschlagenden Gelenkvierecks.

Figure 6: Centrodes of four-bar motions

Using the relations:

$$\mathbf{u}\overline{\mathbf{u}} = a^2, \ \mathbf{v}\overline{\mathbf{v}} = c^2, \ \mathbf{w}\overline{\mathbf{w}} = w^2$$

to eliminate $\mathbf{u}, \mathbf{v}, \mathbf{w}$ and their conjugates he obtains a relation between $\mathbf{z}$ and the conjugate $\overline{\mathbf{z}}$ which is the equation of the coupler curve in minimal (isotropic) coordinates:

$$[\overline{\mathbf{n}}(\mathbf{z} - d)P - \overline{\mathbf{m}}\mathbf{z}Q][\mathbf{n}(\overline{\mathbf{z}} - d)P - \mathbf{m}\overline{\mathbf{z}}Q] + c^2R^2 = 0,$$

where P, Q, R are quadratic polynomials in $\mathbf{z}, \overline{\mathbf{z}}$ and $\mathbf{z} = \mathbf{m} - 1$. This concept of minimal coordinates is used in the book consequently up to the third differential order to derive e.g. the equation of center point curves and all the other well known curves in planar kinematics.

In Wunderlich's book on planar kinematics we can find a full chapter on four-bar mechanisms with applications and many examples. He treats the straight line mechanisms with help of curvature theory and differential geometry. Straight line mechanisms were also the topic of his first contribution [8] to the IFToMM journal Mechanism and Machine Theory (formerly called Journal of Mechanism). But through his whole scientific life he was interested in the topic of straight line mechanisms ([9, 10, 11]). The book "Ebene Kinematik" is also the only

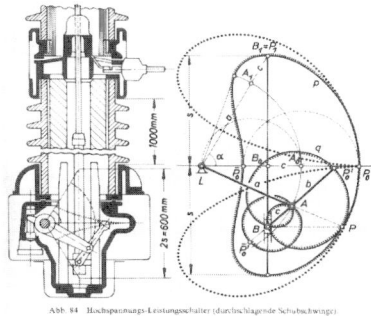

Figure 7: Four bar coupler curve - applications

book known to the author, which has a full chapter on the geometric theory of gears and cams. This is due to the fact that W. Wunderlich has made significant contributions to this theory. Starting with the geometric theory of constructing gear profiles for constant transmission ratio he develops the classical theory of evolute gears due to F. Reuleaux, the theory of cycloidal gears due to Ch. E. L. Camus based on the work of Ph. de la Hire. Remarkable is the paragraph on the geometry and kinematics of the Wankel motor. Here one can see clearly W. Wunderlich's approach to difficult kinematic problems. One of the main problems of this motor type is the geometric form of the moving system (piston, Σ_2 in Fig.10), where he shows that this problem can be solved as a gear profile problem. Moreover he uses the so called cyclographic mapping which maps circles into points of a three dimensional parameter space to investigate the geometric properties of the design curve of the piston. In the book he discusses also in detail motion transmission with changing transmission ratio. As examples we take from the book the Geneva wheel (Fig.9) and the involute gears with ellipses as base curves (Fig.8). Whereas the Geneva wheel is treated in many of the English textbooks on planar kinematics, the involute gear problem on non circular base curves can be found (to the author's best knowledge) in non of the English textbooks. Well known to the community of mechanical engineers is W. Wunderlich because of his contributions to the theory of cams. Therefore it is not surprising that a full chapter of Ebene Kinematik is devoted to this problem. In this field W.

174

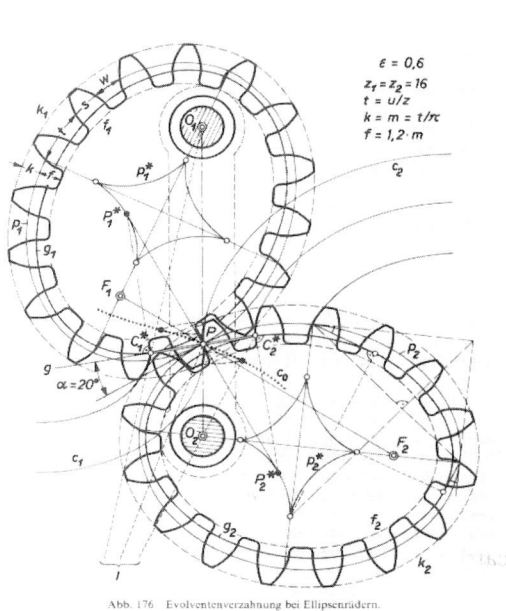

$$\epsilon = 0,6$$
$$z_1 = z_2 = 16$$
$$t = u/z$$
$$k = m = t/\pi$$
$$f = 1,2 \cdot m$$

$$\alpha = 20°$$

Abb. 176 Evolventenverzahnung bei Ellipsenrädern.

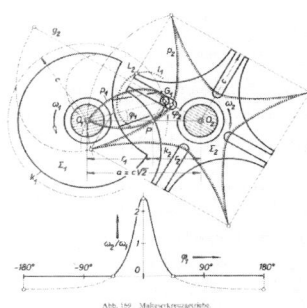

Abb. 180 Malteserkreuzgetriebe.

Figure 9: Geneva wheel

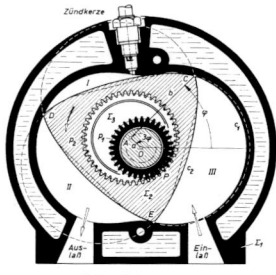

Abb. 167 Kreiskolbenmotor von Wankel.

Figure 8: Involute gears with elliptic base curves   Figure 10: Wankel motor

Wunderlich has made genuine contributions. Although two papers ([14, 18]) are published in English, the important papers showing the geometric basics to the cam problems are published in German. In these papers he shows that the problem of designing single cams, that steer a flat-faced follower pair, is closely related to the construction of curves having an isoptic circle [2]. In [15] he gives a complete solution to the problem which had been posed before (see [2]) but incompletely solved and additionally he shows algebraic examples of such curves.

The most important kinematic topic of his later work are his contributions to overconstrained and shaky (infinitesimal movable) structures. He shows many remarkable examples but also he develops methods that can be applied in today's research of singularity theory of nR-chains and parallel manipulators. Moreover he exhibits the close relation of shaky and pathologically movable structures to singularity problems in geodesy (see [22, 24, 26]). Already his first paper on "Rigid, snapping, shaky and movable octahedra" [19], where he gives geometric conditions for all phenomena mentioned in the title, can be viewed as a complete singularity analysis of 3-3 Stewart-Gough manipulators (see [3]). In [20] one can find a complete singularity analysis of closed serial $4R$-chains up to the conditions

---

[2] An isoptic curve is the locus of points from which two tangents to the curve subtend a constant angle ω.

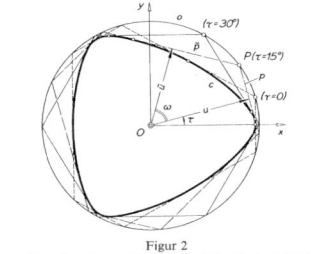

Figur 2
Konvexer Zug einer Kurve 12. Klasse mit isoptischem Kreis für 120°

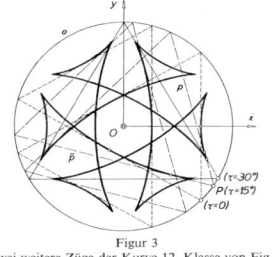

Figur 3
Zwei weitere Züge der Kurve 12. Klasse von Fig. 2

Figure 11: Curves with isoptic circle-convex part

Figure 12: Curves with isoptic circle

for movability. This analysis is of course mathematically identical to the inverse kinematics problem of serial 4R-manipulators. In [21] and [23] he corrects and improves an old result of Bricard concerning frameworks of rods connected by spherical joints: *Are $m \geq 4, 5, 6$ points P_i located on an ellipse and $n \geq 6, 5, 4$ points Q_i located on the respective focal hyperbola and are all points P_i connected to points Q_i by rigid rods, then this framework is movable even when all rods are connected by spherical joints.* Bricard claims that this framework has two degrees of freedom (dof) but this is not correct: Wunderlich proofs that the dof is three. Moreover he shows that one can take any pair of focal curves (not only ellipses and hyperbolas) and that the corresponding framework for arbitrary focal curves has 2 dofs.

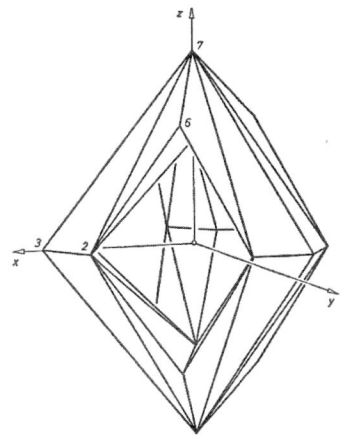

Figure 13: Shaky polyhedron of genus 2

In the papers [28]-[30],[32] and [34]-[38] he discusses the geometric conditions for infinitesimal movability or snapping (this means that two solutions of the assembly problem, are very close to each other) of polyhedra (icosahedra, dodecahedra, pyramids, prisms) and gives for each of the different cases examples. It must be mentioned that polyhedra having more then three edges in one face are considered as panel structures having revolute joints in the edges. Due to the basic theorem of Cauchy all of these polyhedra have to be non convex to allow infinitesimal moveability. Therefore for each of such polyhedra there exists more then one assembly mode.

Four of Wunderlich's papers on shakiness deserve special attention: In [31] (resp.

[33]) he proofs the projective invariance of shakiness of spatial frameworks. This means: if a framework is shaky, then any linear transformation of its design will not resolve the shakiness. In [36] and [40] (his last paper, written at the age of more than 85 years) he gives examples of shaky polyhedra with genus > 0 (Fig.13). The existence of such structures had bee doubted by all scientists working in the field.

This short overview was just the enumeration of the most important achievements of W. Wunderlich in kinematics. Many papers on small topics have been left out for sake of brevity.

2.2 SPECIAL CURVES AND SURFACES

To show Wunderlich's versatility a selection of other significant contributions will be presented. Through his whole scientific life he was working on the geometry of curves and surfaces. This interest was awakened at the beginning of his scientific career when he worked on the descriptive geometry of non-Euclidean screw surfaces and spiral surfaces. The theory of spiral motion and the curves and surfaces generated by this motion is treated extensively in his book "Descriptive Geometry I and II" ([41], [42]). The spiral motion is a generalization of a screw motion where instead of a rotation a planar spiral motion is concatenated with the translation.

The generation of special surfaces with special motions is a topic which has tradition in the Austrian geometry. Wunderlich contributed to this topic especially kinematic generations of J. Steiner's famous Roman surface ([43]-[45]). But there are also papers on kinematic generation of cubic ruled surfaces where a parabola is moved to describe the surface and a developable Möbius strip.

An other main area of Wunderlich's work is on surfaces with special fall lines (curves of steepest slope on a surface). Here he investigates surfaces with planar fall lines, Roman surfaces with planar fall lines or surfaces with conic sections as fall lines.

Concerning Wunderlich's contributions to the geometry of special curves one would have to write another overview article. But a short enumeration of some topics has to be made: Pseudo geodesic lines of cylinders and cones, D-curves on quadrics, principal tangent curves of special surfaces, loxodromic curves on different surfaces, irregular curves and functional equations, auto-involute curves, curves with constant global curvature, Zindler curves, Bertrand curves, spherical curves and auto-polar curves.

Concerning Wunderlich's versatile interest in geometry one has to mention his papers on "On the statics of the rope ladder", or "Geometric considerations on an apple skin" or "On the geometry of birds eggs" ([46]-[48]). This concludes the overview of W. Wunderlich's work. But we have to emphasize that this was only a small part of the results of a rich scientific life.

3. CONCLUSION

Walter Wunderlich was a great Austrian geometrician and kinematician. In this contribution we have tried to overview his papers in kinematics which are mostly written in German. The main purpose was to introduce some of his ideas to a broader scientific publicity.

REFERENCES

[1] Bereis, R., Aufbau einer Theorie der ebenen Bewegungen mit Verwendung komplexer Zahlen, Österr. Ing. Archiv, **5**, (1951), 246–266.

[2] Green, J. W., Sets subtending a constant angle on a circle, Duke. Math. J. **17**, (1950), 263–267.

[3] Merlet, J.-P., Parallel Robots, Klueer Akademic publishers, Dordrecht, The Netherlands, (2000).

[4] Stachel,H., Walter Wunderlich (1910-1998), Technical Report No. 65, TU Wien, July 1999.

[5] Wunderlich, W., Höhere Radlinien. Österr. Ing. Archiv, **1**, (1947), 277–296.

[6] Wunderlich, W., Höhere Radlinien als Näherungskurven. Österr. Ing. Archiv, **4**, (1950), 3–11.

[7] Wunderlich, W., Ebene Kinematik, Hochschultaschenbuch 447/447a, Bibliographisches Inst. Mannheim, (1970) 263 pages, 183 figures.

[8] Wunderlich, W., On Burmester's focal mechanisms and Hart's straight line motion. J. Mechanisms, **3**, (1968), 438-440.

[9] Wunderlich, W. Approximate optimization of Watt's straight line mechanism. Mech. Mach. Theory, **13**, (1978),156–160.

[10] Wunderlich, W. Über die Wattsche Geradführung. Math. stat. Sekt. Forschungszentrum Graz, **95**, (1978), 1–8.

[11] Wunderlich, W. Nomogramme für die Wattsche Geradführung. Mech. Mach. Theory. **15**, (1980),5–8.

[12] Wunderlich, W., Zur Triebstockverzahnung, Z. Angew. Math. Mech. (ZAMM), **23**, (1943), 209–212.

[13] Wunderlich, W., Über abwickelbare Zahnflanken und eine neue Kegelradverzahnung. Betrieb und Fertigung, **2**, (1948), 81–87.

[14] Wunderlich, W., Contributions to the geometry of cam mechanisms with ocillating followers. J. Mechanisms, **6**, (1971), 1-20.

[15] Wunderlich, W., Kurven mit isoptischem Kreis. Aquat. math., **6**, (1971), 71–81.

[16] Wunderlich, W., Kurven mit isoptischer Ellipse. Monathsh. Math., **75**, (1971), 346–362.

[17] Wunderlich, W. and Zenow, P. Contribution to the geometry of elliptic gears. Mech. Mach. Theory, **10**, (1975), 273–278.

[18] Wunderlich, W. Single-disk cam mechanisms with oscillating double roller follower. Mech. Mach. Theory, **109**, (1984), 409–415.

[19] Wunderlich, W., Starre, kippende, wackelige und bewegliche Achtflache. Elem. Math., **20**, (1965), 25–32.

[20] Wunderlich, W., Starre, kippende, wackelige und bewegliche Gelenkvierecke im Raum. Elem. Math., **26**, (1971), 73–83.

[21] Wunderlich, W., Fokalkurvenpaare in orthogonalen Ebenen und bewegliche Stabwerke. Sitzungsber., Abt. II, österr. Akad. Wiss. Math.-Naturw. Kl., **185** (1976), 275–290.

178

[22] Wunderlich, W., Über die gefährlichen Örter bei zwei Achtpunktproblemen und einem Fünfpunktproblem. Österr. Z. Vermessungswesen und Photogrammetrie, **64** (1977), 119–128.

[23] Wunderlich, W., Bewegliche Stabwerke vom Bricardschen Typus. Z. Angew. Math. Mech., **57**, (1977), 51–52.

[24] Wunderlich, W., Gefährliche Annahmen der Trilateration und bewegliche Fachwerke I, Z. Angew. Math. Mech., **57**, (1977), 297–304.

[25] Wunderlich, W., Gefährliche Annahmen der Trilateration und bewegliche Fachwerke II, Z. Angew. Math. Mech., **57**, (1977), 363–267.

[26] Wunderlich, W., Über gefährliche Annahmen beim Clausenschen und Lambertschen Achtpunktproblem. Sitzungsber. Bayer. Akad. Wiss., (1978), 23– 46.

[27] Wunderlich, W., Eine merkwürdige Familie von beweglichen Stabwerken. Elem. Math., **34**, (1979), 132–137.

[28] Wunderlich, W., Snapping and shaky antiprismas. Math. Magaz., **52**, (1979), 232–236.

[29] Wunderlich, W., Neue Wackelikosaeder. Anz. österr. Akad. Wiss., Math.-Naturwiss. Kl., **117**, (1980), 28–33.

[30] Wunderlich, W., Wackelige Doppelpyramiden. Anz. österr. Akad. Wiss., Math.-Naturwiss. Kl., **117**, (1980), 82–87.

[31] Wunderlich, W., Zur projektiven Invarianz von Wackelstrukturen. Z. Angew. Math. Mech. **60**, (1980), 703–708.

[32] Wunderlich, W., Wackeldodekaeder. Math. stat. Sekt. FZ. Graz **149**, (1980), 1-8.

[33] Wunderlich, W., Projective invariance of shaky structures. Acta Mechanica, **42**, (1982), 171–181.

[34] Wunderlich, W., Kipp-Ikosaeder I, Elem. Math. **36**, (1981), 153–158.

[35] Wunderlich, W., Kipp-Ikosaeder II, Elem. Math. **37**, (1982), 84–89.

[36] Wunderlich, W., Ringartige Wackelpolyeder. Anz. österr. Akad. Wiss., Math.-Naturwiss. Kl. **119**, (1982), 71–77.

[37] Wunderlich, W., Wackeldodekaeder. Elem. Math. **37**, (1982), 153–163.

[38] Wunderlich, W., and Schwabe, Ch., Eine Familie von geschlossenen gleichflächigen Polyedern die fast beweglich sind. Elem. Math. **41**, (1986), 88– 98.

[39] Wunderlich, W., Fast bewegliche Oktaeder mit zwei Symmetrieebenen. Rad Jugosl. Akad. Zagreb **428**, Mat. Znan. **6**, (1987), 129–135.

[40] Wunderlich, W., Shaky polyhedra of higher connection. Publ. Math. Debrecen **37** (1990), 355-361.

[41] Wunderlich, W., Darstellende Geometrie I. (Hochschultaschenbuch 96/96a), Bibiliograph. Inst. Mannheim (1966), 187 pages with 157 figures.

[42] Wunderlich, W., Darstellende Geometrie II. (Hochschultaschenbuch 133/133a), Bibiliograph. Inst. Mannheim (1967), 234 pages with 166 figures.

[43] Wunderlich, W., Über zwei durch Zylinderrollung erzeugbare Modelle der Steinerschen Römerfläche. Arch. Math., **18**, 325–336.

[44] Wunderlich, W., Durch Schiebung erzeugbare Römerflächen. Sitzungsber., Abt. II, österr. Akad. Wiss. Math.-Naturw. Kl., **176** (1976), 473–497.

[45] Wunderlich, W., Kinematisch erzeugbare Römerflächen. J. reine angew. Math., **236**, (1969), 67–78.

[46] Wunderlich, W., Zur Statik der Strickleiter, Math.Z., (1951), 13–22.

[47] Wunderlich, W., Geometrische Betrachtungen um eine Apfelschale. Elem. Math., **15**, (1960), 60-66.

[48] Wunderlich, W., Zur Geometrie der Vogeleier. Sitzungsber., Abt. II, österr. Akad. Wiss. Math.-Naturw. Kl., **187** (1978), 1–19.

5. Historical Development of Theories

THE ARCHIMEDEAN SCREW-PUMP: A NOTE ON ITS INVENTION AND THE DEVELOPMENT OF THE THEORY

Teun Koetsier & Hendrik Blauwendraat
Department of Mathematics, Faculty of Science
Vrije Universiteit, De Boelelaan 1081
NL-1081 HV Amsterdam, The Netherlands
e-mail: Teun Koetsier <teun@few.vu.nl>

ABSTRACT: Following Drachmann and others the authors argue that it is reasonable to assume that Archimedes invented both the infinite screw and the screw-pump. They argue that these inventions can be related to Archimedes' interest in the problem of the quadrature of the circle. Moreover, they discuss aspects of the development of the theory of the screw-pump.

KEYWORDS: screw pump, Archimedes, Galilei, Daniel Bernoulli, Hachette, Weisbach

INTRODUCTION

Some authors attribute the invention of the screw-pump to Archimedes; others believe that the screw-pump was invented earlier and only attributed to Archimedes because of his reputation.

Oleson [23] has given a survey of the data with respect to the origin of the screw-pump. There are several texts from Antiquity in which the screw-pump is attributed to Archimedes.[1] As for Archimedes' involvement with screws there is, next to these texts, another remark by Moschion, that is relevant. Moschion states that Archimedes launched a ship "by means of a screw, which was an invention of his own" ([7], p. 279). This seems to refer to an endless screw. Oleson describes the archeological evidence as well. The earliest representation of a water-screw is on a fresco from the Casa di P.

---

[1] The earliest evidence is a text by Moschion (after 241 B. C.) concerning the "Great Ship of Hieron of Syracuse". The text says: "And the bilge, although of a remarkable depth, was pumped out by a single man operating a water screw, an invention by Archimedes" ([23], p. 60). Then there is a statement from Agatharchides (floruit 180-116 B. C.) about the Nile delta: "the inhabitants easily irrigate the whole region by means of a certain device which Archimedes, the Syracusan, invented, called the 'screw' on account of its design" ([23], pp. 22-23) Then we have a text from Posidonius of Apamea (floruit ca 135-51 B. C.). The text describes the use of the water-screw in a series for mine drainage: "At a depth they [the miners] sometimes break in on rivers flowing beneath the earth, the strength of which they overcome by diverting their welling tributaries off to the side in channels [...] they draw off the streams of water with the so-called Egyptian screw, which Archimedes invented when he visited Egypt" ([23], pp. 92-93)

Cornelius Teges in Pompeii, obviously dating from before 79 A. D.[2] On the fresco an individual is moving a cylinder with his feet in a landscape that is allegedly Egyptian. Because water comes out of the cylinder it is generally assumed it must be a water-screw. From the imperial period we have two other Egyptian representations (in the British Museum and the Archeological Museum Cairo, respectively) and an Egyptian model of a water-screw (in the Hilton-Price collection). Moreover, remains of water-screws dating from the imperial period have been found in mines in Spain. None of these representations or remains of water-screws dates from before the time of Archimedes.

There is a very limited number of books on technical mechanics from Antiquity. For our purposes Vitruvius' book on Architecture [14] is important. It contains the oldest known description of a water-screw. The book dates from about 25 B. C.[3]

In this paper we will phrase a hypothesis with respect to the way in which Archimedes possibly invented the screw and the screw-pump. Moreover, we will discuss the way in which Cardano, Galilei, Daniel Bernoulli, Hachette and Weisbach studied the screw-pump.[4] The theoretical considerations concerning the screw-pump reflect the history of the theory of machines. Galilei was the first to give a complete and correct theory of the five simple machines (Cf. [21]). He was also the first to correctly explain the functioning of the screw-pump. In the course of the 17th and 18th century Newtonian mechanics and the calculus were the major new developments. The application of these new theories to machines was fragmentary and led to isolated results. Daniel Bernoulli's treatment of the screw-pump reflects this. Finally the treatment of Hachette and Weisbach of the screw-pump is characteristic of the 19th century approach to machines: geometric and graphical methods combined (in the case of Weisbach) with calculations. Judging on the basis of Rorres' remarks about the Archimedean Screw Pump Handbook ([22]) it seems that in 1968 the theory had not yet developed above the level reached by Weisbach and in practice Archimedean screw-pumps were built on the bases of rules of thumb ([26], p. 73). Rorres' treatment ([26]), that we will not discuss, represents the modern approach: he wrote a MathLab computer programme.

DRACHMANN'S RECONSTRUCTION

Dijksterhuis [10], Kellermann & Treue [18], Krause (in his contribution in [30]) prefer to assume that the screw-pump was invented before Archimedes and that Archimedes probably merely applied it or studied it theoretically. Others, like Drachmann ([7] and [8]), Oleson [23] and Rybczynski [27], argue that the available evidence, although rather limited, points clearly at Archimedes and that there is no evidence pointing elsewhere. The same argument applies to the endless screw. As for Archimedes and the

---

[2] [30] contains a complete reproduction of the (erotic) fresco.

[3] Also Heron's works on practical mechanics are important, in particular his textbook on Mechanics which has come to us in an Arabic translation only. Heron's books, written after 62 A. D. are by far the best source on ancient mechanical technology. Also Pappos wrote on mechanics; his work contains many fragments of earlier authors. Pappos lived at the time of Diocletian (285-305 A. D.). These books were all written more than about a century after Archimedes' death. Vitruvius describes the screw-pump without reference to Archimedes. Heron and Pappos describe other applications of the screw, but shed no direct light on Archimedes' role in this respect.

[4] The first sections of this paper are based on [20].

screw, we follow Drachmann and Oleson. Drachmann has argued ([8], p. 153) that Archimedes invented the screw-pump after having seen in Egypt the operation of a water-drum or tympanum (a water-lifting wheel with a body consisting of eight compartments, see Figure 1a). While the tympanum rotates, water enters a compartment through a hole close to the periphery of the drum, and after half a turn the water leaves the compartment again through a hole close to the axis. Oleson, who sympathises with Drachmann's reconstruction, described the moment of Archimedes' breakthrough as follows "if the tympanum were to be drawn slowly along the axis of its rotation as it turned, its compartment walls would describe the spirals of just such a screw" ([23], p. 298). Each of the eight compartments of the tympanum then generates one of eight spiral-shaped channels that together fill a cylinder (Figure 1b).

ANOTHER SOLUTION
If Drachmann is right, Archimedes must have seen a tympanum as described by Vitruvius in action and as a result of that imagined the screw-pump as described by Vitruvius. However, the insight that water can be lifted with the resulting object is far from immediate.

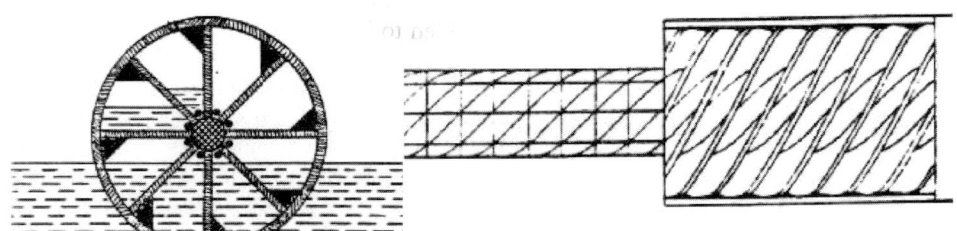

Fig. 1a Roman tympanum.   Fig. 1b Screw-pump.
In Drachmann's reconstruction Archimedes invented the screw-pump of Fig. 1b when he watched the tympanum of Fig. 1a in action (Illustrations taken from [8], p. 150 and 153)

If a tympanum in action is tilted and imagined to be moved during its rotation along its axis it seems natural that this movement will be upward, away from the water. If the thought-experiment is executed in this way the resulting screw-pump will only lift water if the direction of the rotation that generates the pump in the thought-experiment is *reversed*![5]

We would like to suggest another scenario. The Chinese are famous for their inventions. Yet they did not invent the screw. This suggests that something was missing in China which was present in the West. Greek mathematics was missing in China. My

---

[5] In order to execute the thought-experiment in such a way that a working screw pump is obtained without having to reverse the direction of the rotation it is necessary that in the though-experiment the tympanum is "screwed" downwards, into the water. This seems hardly a natural thing to imagine.

reconstruction of the invention starts with the problem in pure Greek mathematics of the quadrature of the circle, which led to the study of different kinds of spirals, among them the cylindrical helix. We will show that the cylindrical helix can be used to execute the quadrature of the circle, a construction that is immediately related to the fact that the helix can be obtained by wrapping a triangle around a cylinder. We are arguing that at this point Archimedes will have related pure mathematics to mechanics and Heron of Alexandria tells us how. In Heron's *Mechanics*, Book 2, this particular relation of cylindrical helix and rectangular triangle is not only used to design a screw, but Heron also explains the functioning of the screw as follows ([8], p. 76):

> "Now we must take the screw to be just a twisted wedge, for the triangle from which we draw the screw line is really a wedge, and ist head is the side which is equal to the height of the screw turn [...]. And so the screw becomes a turned, twisted wedge, which is worked not by a blow, but by turning, and its turning here replaces the blows on the wedge; and so it lifts the weight [...]".[6]

The wedge is a very old piece of equipment, dating from long before Archimedes. The first somewhat unsatisfactory theoretical remarks that we know of are in *Mechanical Problems* from say 280 B. C., attributed to the peripathetic school ([8], p. 12). It is inevitable that Archimedes studied the wedge. My hypothesis is that the revolutionary idea that we can wrap a wedge around a cylinder and get a screw, comes from Archimedes. His work on the quadrature of the circle combined with his vivid interest in applied mechanics made him see all of a sudden the relation between the wedge and the cylindrical helix and at that moment he realised that also the screw could be used to exert power. The first application will have been very simple: a spiral groove in a wooden cylinder will have been used to move a piece of wood that was forced to remain in the groove (See footnote 6). All other applications came later. Pondering on how water could be lifted by a screw, wondering how water could "stay securely and solidly in its place by a power which is in itself", to use Heron's words, he discovered the screw-pump. Gravity would hold the water in its place!

FROM THE QUADRATURE OF THE CIRCLE TO THE CYLINDRICAL HELIX

One of the famous problems that mathematicians in Antiquity struggled with was the problem of the quadrature of the circle: given an arbitrary circle, construct by means of compass and ruler a square with an area equal to the area of the circle. The circle quadrature is equivalent to the production of a straight line segment that is equal to the circumference of the circle. Circle quadrature and circle rectification are equivalent problems. The Greek mathematicians did not succeed in solving the problem of the quadrature or the rectification of the circle by means of compass and ruler.[7] For a

---

[6] The use of the screw (not the screw-pump) is actually described by Heron as well. Heron writes (in Drachmann's translation): "When the screw is turned, it moves the piece of wood that is called tulus, as I have said already, and it lifts the weight in a straight line; and this tulus must, when the screw is not moved, stay securely and solidly in its place by a power which is in itself, and it must not be so that, when the screw is at rest from turning, the weight overcomes it, I mean that when this block is fitted into the screw furrow and is like a support for it, then it does not slip out of the screw furrow, because if it does slip out, the whole weight will go down to the place from which it was lifted."([8], pp. 97-80).

[7] In the nineteenth century it became clear that the problem in this form is in fact insoluble.

rectification we need more than merely the use of compass and ruler. For example, in a famous treatise, *Spiral Lines*, Archimedes shows how any circle can be rectified provided one particular curve, a planar spiral is given and a tangent can be drawn to the spiral. Although Archimedes' argument is quite complicated, the result boils down to the situation of Figure 2. If PQ is the tangent to the spiral and OQ ⊥ OP, we have OQ=Circular arc PS.

Fig. 2 Rectification of a circle by means of a planar spiral

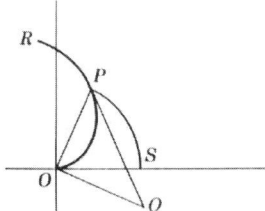

Archimedes undoubtedly knew the cylindrical helix and its properties.[8] It can be generated easily in a way analogous to the generation of the planar spiral by means of a superposition of a uniform rotation and a uniform translation: a segment of fixed length rotates uniformly in a plane about one of its endpoints, while at the same time this plane is moved uniformly in a direction perpendicular to the plane. The cylindrical helix can also be used to rectify a circle. Pointing out the analogy with the rectification by means of the planar spiral, Heath described the rectification as follows:

"if a plane be drawn at right angles to the axis of the cylinder through the initial position of the moving radius which describes the helix, and if we project on this plane the portion of the tangent at any point of the helix intercepted between the point and the plane, the projection is equal to an arc of the circular section of the cylinder subtended by an angle at the centre equal to the angle through which the plane through the axis and the moving radius has turned from its original position." ([16], Vol.I, p. 232)[9]

This rectification of the circle by means of the cylindrical helix boils down to unwrapping a rectangular triangle from a cylinder, precisely the reverse of Heron's above-mentioned construction of the cylindrical helix.

---

[8] A contemporary mathematician, Apollonius (c.262-c.190), is known to have written a treatise on the Cochlias, i. e. the cylindrical helix.

[9] In an interesting reconstruction Knorr ([19], 1986, pp. 166-167) argues that the rectification as described by Heath actually led Archimedes on a heuristic level to the rectification by means of the planar spiral.

THE SCREW-PUMP IN THE RENAISSANCE

The water screws from Antiquity that we know of are all based on one or more helical blades fitted inside a cylinder.[10] After Antiquity for many centuries there is no theoretical interest whatsoever in screw-pumps. Only at the end of the Middle Ages the interest returns. The first illustration after Antiquity is in Konrad Kyeser's *Bellifortis*, an early 15th century manuscript. It is obviously a pump with one or more helicoids inside (Figure 3a). [11]

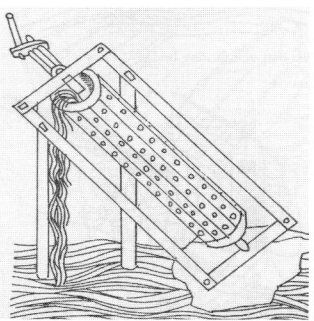

Fig. 3a Screw-pump in Kyeser's *Bellifortis*[12]

In Da Vinci's *Codice Atlantico* we find a screw-pump consisting of a helical tube wound around a central drum (Figure 3b), as well as a pump made by winding a tube around a central core with a triangular intersection. According to Da Vinci the triangular pump can lift much more water, but is less easily turned around.

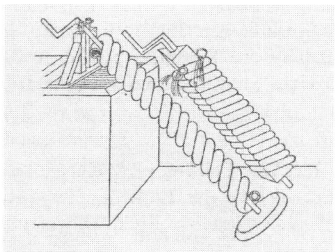

Fig. 3b Screw-pumps in Da Vinci's *Codice Atlantico*[13]

---

[10] In the Spanish copper-mines of the imperial period pumps with one helicoid were found ([18], p. 23).

[11] For technical reasons the illustrations 3a, 3b and 4a are taken from secondary sources.

[12] Illustration taken from [13], p. 64

[13] Illustration taken from [3], p. 468. Described by Cardano in[5].

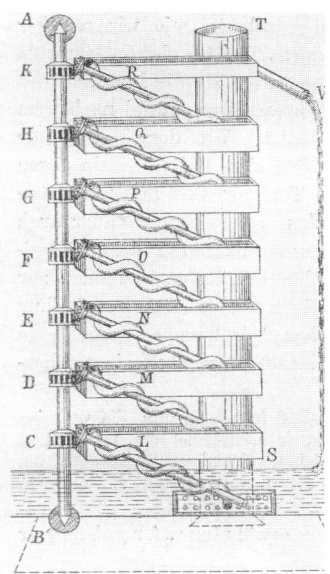

Fig. 4a The Augsburg Machine[14]

Fig. 4b Plate 64 from Ramelli [25]

---

[14] Illustration taken from [3], p. 180

Other 16th century authors in whose work we find the screw-pump are Cardano and Ramelli. In Cardano's work we have the Augsburg Machine (Fig. 4a), that actually existed. The pumps are based on helical tubes. In Ramelli's 1588 drawing (Fig. 4b) the pumps are based on two helical blades.

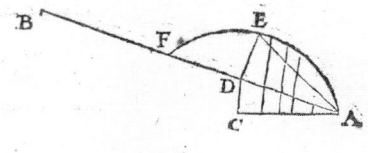

Fig. 5 From Cardano, [5], p. 19

it and so did Galilei. The first explanation of the functioning of the screw was given by Cardano. With respect to Figure 5 (in which AB is the axis of the pump, DC represents its elevation, and the curved segment AE represents the tube wound around the core of the pump). Cardano's argument is essentially this: He replaces the curved segment by the cord AE. He assumes that DE is longer than DC. Then, when E rotates around A, in its opposite position E will be under C, and a weight in A will fall towards E. Clearly this will also be the case when we consider the motion along the curved segment when it is in its lowest position. The fact that a weight continues to fall from E to F when the pump is rotated is explained by Cardano by the impetus the weight has in E and the fact that the situation repeats itself. Cardano in this context seems not to relate the cylindrical helix to a triangle wound around a core. It means that he understood how the screw pump works, but could not really make his insight precise. That is where Galilei came in in the 90s of the 16th century. Galilei wrote about the screw-pump "it is not only marvellous, but it is miraculous" (non solo è meravigliosa, ma è miracolosa – [12], p. 183), because in the srew-pump the water ascends by continually descending.

Galilei considered the cylinder MJKH with the winding line JLOPQRSH round it (Figure 6) The winding line is considered as a channel in which the water rises by descending. The winding line is generated by means of triangle ABC (drawn on the right side very small in Figure 6 just above the horizontal line), which means that the elevation of the channel is determined by angle CAB. Galilei now argues as follows.

> "Now it is clear that the rise of this channel will be taken away if the point C is dropped to B, for then the channel will have no elevation at all, and dropping the point C a little below B, the water would naturally run out downward through the channel AC from the point A to the point C."

Galilei then assumes that angle A is one third of a right angle and he continues (Figure 3).

> "These things understood, let us turn the triangle round the column, and let us make screw BAEFGHJD. If placed upward at right angles with the extremity B in water, this would not upon being turned draw up the water […]"

However, if we tilt the column through one third of a right angle "the water will move downward from the point J to the point L. And turning the screw round, its various parts successively displace one another and present themselves to the water in the same position as the part JL."[15]

Figure 6 From Galilei's *Mecchanice*,[12]

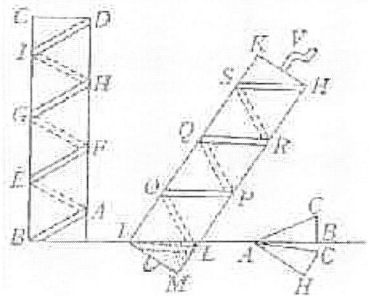

Galilei concludes: the water raising screw must be tilted a little more than the angle of the triangle that generates the screw. This is fundamental and Galilei seesm to have been the first to prove it and write it down.

THE CALCULATIONS OF DANIEL BERNOULLI

The Archimedean screw raises discrete scoops of water. Daniel Bernoulli (1700-1782), who studied the screw pump in his *Hydrodynamics* (1738),[16] seems to have been the first to study the size of the pockets that contain the water. He considered a tube of infinitely small diameter shaped as a cylindrical helix, generated by wrapping a triangle around a cylinder. The triangle is in Fig. 7b[17] under the horizontal line; in the upper part we have in fact an orthogonal projection of the tilted cylinder on a vertical plane. The projection of the cylindrical helix is a tilted sinusoidal curve. Bernoulli applied the differential- and integral- calculus (at the time less than half a century old). He first determined the height above the horizontal plane H(X) of a point P of the cylindrical helix on the tilted cylinder as a function of X, the angle of rotation during the generation of the helix. The radius of the cylinder is taken equal to 1 and then X equals the arc.
It is not difficult to check that

$$H(X) = X.\tan\psi \cos\varphi + \sin\varphi(1 + \cos X)$$

Here ψ is the angle that the pump makes with the horizontal and φ is the inclination of the cylindrical helix. It is clear from figure 7a that the pocket that can contain water has a deepest point corresponding to a minumum of H(X) and that one of its endpoints corresponds to a maximum of H(X) Putting the derivative of H(X) equal to zero yields

$$\sin X = \tan\psi/\tan\varphi$$

---

[15] The translation is Stillman Drake's ([9])
[16] For an English translation see [4] and for an annotated German translation see [11].
[17] Fig. 7b is our drawing, suggested by the drawings in the annotation by Flierl ([11])

On the interval $0 < X < \pi$ corresponding to the first section of the cylindrical helix we have no extrema when $\psi > \varphi$; the helix is then an ascending curve and water does not enter the tube. This is in fact Galilei's result, derived differently. However when $\psi < \varphi$ the equation has two solutions. There is a maximum corresponding to point o and a minimum corresponding to point p. Bernoulli remarked that in this case the quantity of water in one pocket is determined by the section opq of the helix; q being on the same level as o. He wrote that the length of this scoop cannot be determined algebraically, but can be approximated in every specific case.

It is interesting that Bernoulli also calculates the force needed to operate the pump. He imagined a weightless helix in which a masspoint is rolling without friction.

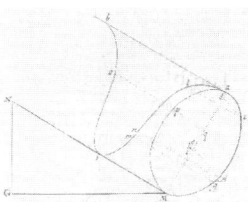
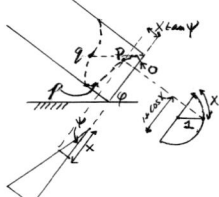

Fig.7a Bernoulli's drawing   Fig. 7b

Gravity pulls the mass point down and makes the helix turn. He imagined the masspoint at point p, corresponding to the above determined minimum, the lowest point of the pocket. The component of the weight directed along the axis of the cylinder does not have any effect, but the other component makes the cylinder turn. One can easily verify that if the weight is p, the moment about the axis equals

p.tanψ.cosφ.

In the case in which we are not dealing with one pointmass positioned at the lowest point, but in which the whole pocket opq is filled with water Bernoulli determines the moment that the weight of the water exerts about the axis of the cylindrical helix by means of an integration. The conclusion that he reaches is that the same formula applies: if the total weight of the water is p, the moment equals

p.tanψ.cosφ.

and we have the same formula as above.

In or some time before 1755 Leonhard Euler formulated fourteen mathematical problems (*Quaestiones Mathematicae*), that were read in 1757 to the members of the Petersburg Academy. The first problem concerned the need for a theory of the screw-pump. Although the pump was widely and successfully applied still Euler felt the need for a theory, he wrote. As far as we know Euler himself never developed such a theory.

HACHETTE'S APPROACH

One of the reasons why Gaspard Monge is famous, is the fact that he defined and developed the subject called "descriptive geometry". When the *Ecole Polytechnique* was founded in Paris in 1794, a course in descriptive geometry became obligatory for all students. Part of the course was devoted to (elements of) machines. This part was taught by Jean-Nicolas-Pierre Hachette (1769-1834), who later published the notes in the form of his *Traité des machines*. The goal of descriptive geometry is to make drawings of three-dimensional objects that are such that from the drawing, shape and relative positions of parts of the object can be deduced geometrically. The method boils down to the following. In space two perpendicular planes are chosen, Π_1 and Π_2. The 3-dimensional object is projected on both planes and in order to get a 2-dimensional picture Π_2 is rotated 90 degrees about the line of intersection of the two planes until it coincides with Π_1. Usually the projection on Π_1 is in the upper part of the drawing and the projection on Π_2 is in the lower part. Although the drawings in Hachette's *Traité élémentaire des machines* are all very well made, the use of descriptive geometry is limited, because most of the mechanisms are planar. In important exception is, however, Hachette's treatment of Archimedes' screw pump. Hachette's treatment of the screw is entirely based on descriptive geometry. The pump consists of a cylindrical core and a barrel with a helical blade between them. The blade can be generated by means of a straight line segment with its points describing cylindrical helixes. Hachette considers a case in which the angle between the tangents to the inner helix (described on the core) and the axis of the pump is 45 degrees. Fig. 8 shows the pump and above it the two projections. The projection in Π_1 is *Fig. 3* in Hachette's numbering and the projection in Π_2 is *Fig. 4* The sinusoidal projections of the inner and outer helixes are constructed in *Fig. 3* on the basis of Hachette's *Fig. 4* by constructing some discrete points and connecting them by means of a smooth curved line.

Suppose that G (*Fig.4*) and C (*Fig.3*) are the projections of a point on the inside of the barrel for which the tangent to the helix is parallel to the plane Π_1. GK (*Fig. 3*) and CM (*Fig.4*) are the projections of this tangent. When the pump is working this tangent is rotated about the axis of the pump and describes a hyperboloid of revolution. The question is whether there are positions of the tangent that are such that it points downwards, so that water present at point C would "fall" upwards in the pump. In order to treat this question Hachette considers an arbitrary point on the axis of the pump and considers the cone with this point as center consisting of all lines parallel to positions of the tangent during its rotation about the axis. If the horizontal plane through the center of this cone has two lines of intersection with this cone, the pump will raise water; if there is one or there are no lines of intersection, the pump will not raise water. The border case in which we have precisely one line of intersection leads Hachette to a problem in descriptive geometry: given a cylindrical helix and a plane, construct a tangent to the helix parallel to the plane. Hachette had already discussed this problem in his *Traité de géométrie descriptive* ([15], pp. 142-153).

Hachette points out that the cylindrical helixes that are generated by points on the generating segment closer to the axis are steeper. If the outer helix is such that water can be raised with it, the inner one may be too steep. In fact Hachettte proposes to define the optimal inclination as the one in which the cone corresponding to the inner helix is tangent to the horizontal plane ([16], p. 183).

192

Fig. 8 From Hachette's *Traité élémentaire des machines*

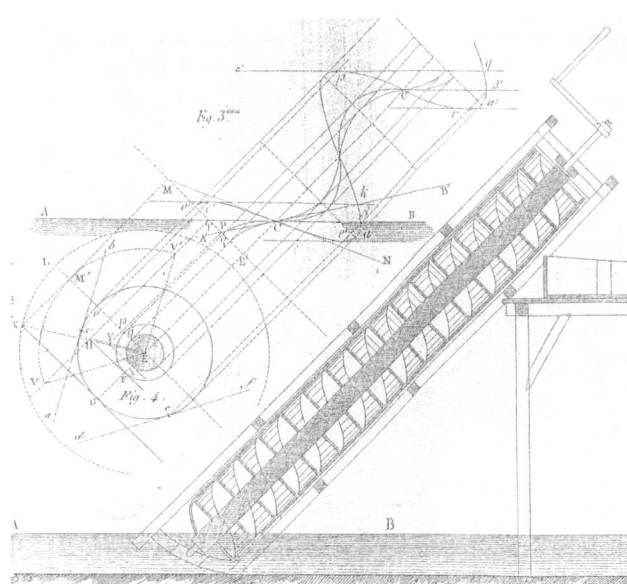

WEISBACH'S CONTRIBUTION

As far as we know the most complete 19th century treatement of the problem was given by Julius Weisbach ([29], 1851-1860, pp. 811-828). Figure 9 illustrates his approach. The central problem is the volume of water that can be lifted in one pocket. The integral involved cannot be calculated analytically (Cf. [26], p. 80). Weisbach determines the intersection of the water in one pocket with a number of cylinders, in fact slicing up the volume in curved slices, starting from the core of the pump and finishing with the inner cylinder of the outer barrel. He takes the average value of the areas of the slices, and he multiplies it with the distance between the core cylinder and the outer cylinder of the pump. This gives him a good approxiamtion of the volume.

He determines the areas of the curved slices as follows. On the left side of Fig. 9 we have a drawing of the pump in a vertical position in accordance with the rules of descriptive geometry. BFA is the projection on the horizontal plane of half of the core; $B_1F_1A_1$ is the projection of the outer cylinder. The area QSUT in Fig 9 is the intersection of the inner cylinder with the pocket of water. The surface of the inner cylinder is flattened and the intersection of the horizontal surface of the water with the cylinder becomes a sinusoidal curve QTRS. When the cylinder is flattened the cylindrical helix becomes a straight line: QS. Because Weisbach studies a pump with two helicoids, the water is trapped between them. TU in Figure 9 corresponds to the cylindrical helix of the second helicoid. The area of QSUT is determined with Simpson's rule

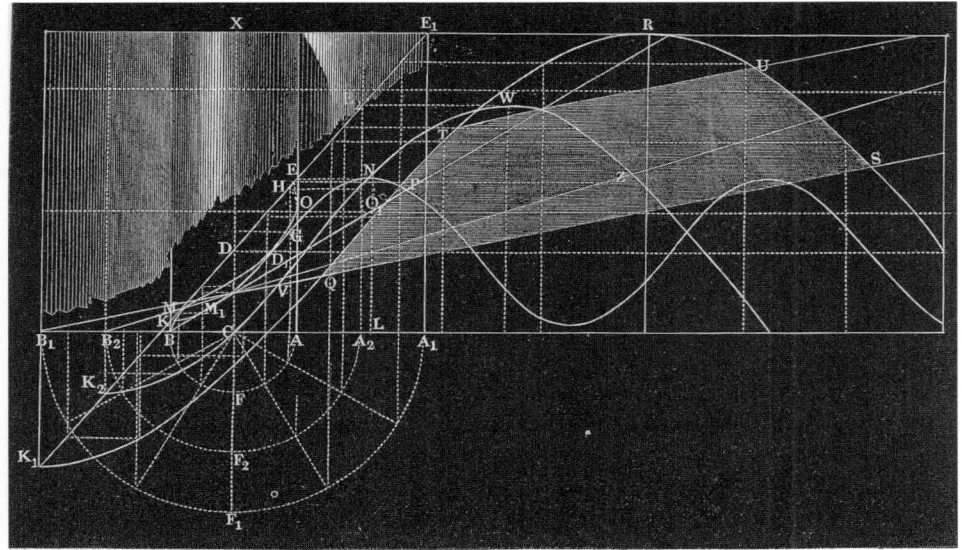

Figure 9 From Weisbach ([29])

CONCLUSION

In this paper we have concentrated on the theoretical considerations concerning the screw-pump. We have seen that the treatment of the screw in each of the cases we discussed reflected the state of the art in the theory of machines. For a modern treatment we refer to ([26], 2000)..

Inherently the determination of the volume of water that a screw pump can lift is a difficult problem that cannot be solved analytically. Understandably, because of the theoretical difficulties, the pump was also studied by means of experiments. Hachette reports about experiments by Touroude from 1766 ([16], p. 186-187). In the twentieth century experiments were executed by Addison ([1]). Addison concluded that the discharge at a given speed falls off rapidly as the inclination of the axis increases. Because the screw-pump was so popular, undoubtedly more experiments have been done. It seems that still in 1968 in practice rules of thumb determined the design of screw pumps.

ACKNOWLEDGEMENT

This research was supported by NWO, the Netherlands Organisation for Scientific Research.

REFERENCES

1. H. Addison, Experiments on an Archimedean Screw, *The Institute of Civil Engineers, Selected Engineering Papers*, No. 75, 1929

.2. Archimedes, *Werke*, (editor: A. Czwalina), Wissenschaftliche Buchgesellschaft Darmstadt, 1972

194

3. Theodor Beck, *Beiträge zur Geschichte des Maschinenbaues*, Julius Springer, Berlin, 1900
4. Daniel Bernoulli, *Hydrodynamics by Daniel Bernoulli & Hydraulics by Johann Bernoulli*, Dover Publications, New York, 1968
5. Girolamo Cardano, *De subtilitate Libri XXI*, Basel, 1584
6. G. Drachmann, The screw of Archimedes, *Actes du VIIIe Congrès international d'Histoire des Sciences Florence-Milan 1956*, Vol. 3 Vinci (Firenze) et Paris, 1958, pp. 940-953
7. G. Drachmann, How Archimedes Expected to Move the Earth, *Centaurus* 5, 1958, pp. 278-282
8. G. Drachmann, *The Mechanical Technology of Greek and Roman Antiquity*, Munksgaard (Publishers), Copenhagen, 1963
9. Stillman Drake & I. E. Drabkin,. *Galileo Galilei, On motion and mechanics*, The University of Wisconsin Press, Madison, 1960
10. E. J. Dijksterhuis, *Archimedes*, Ejnar Munksgaard, Copenhagen, 1956
11. Karl Flierl, *Annotated German translation of Daniel Bernoulli's Hydrodynamics of 1738*, Veröffentlichungen des Forschungsinstituts des Deutschen Museums für die Geschichte der Naturwissenschaften und der Technik, Reihe C, Quellentexte und Übersetzungen, Nr. 1a and 1b, 1963
12. Galileo Galilei, *Opere di Galileo Galilei*, A cura di Franz Brunetti, Volume I, Utet, Torino, 1996
13. Bertrand Gille, *The Renaissance Engineers*, Percu Lund, Humphries and company Ltd, London, 1966
14. Frank Granger, *Vitruvius on Architecture*, London, William Heinemann Ltd, Vol. 1, 1945, Vol 2, 1962
15. Jean-Nicolas-Pierre Hachette, *Traité de géométrie descriptive*, Paris, 1822
16. Jean-Nicolas-Pierre Hachette, *Traité élémentaire des machines*, Paris, 1828 (4th edition)
17. Thomas L. Heath, *Greek Mathematics*, Vol I and II, Oxford: Clarendon Press, 1921
18. Rudolf Kellermann und Wilhelm Treue, *Die Kulturgeschichte der Schraube*, Verlag F. Bruckmann, München, 1962
19. Wilbur Richard Knorr, *The Ancient Tradition of Geometric Problems*, Birkhäuser Boston 1986
20. Teun Koetsier, *Archimedes and the Invention of the Screw-Pump*, In Paipetis, 2003, pp. 24-37
21. Teun Koetsier, La théorie des machines au XVIe siècle: Tartaglia, Guidobaldo, Galileo, *Corpus, revue de philosophie*, 39, 2001, pp. 155-189
22. G. Nagel, *Archimedean Screw Pump Handbook*, Prepared for Ritz-Altro Pumpwerksbau GMBH Roding, Nürnberg, Germany, 1968
23. John Peter Oleson, *Greek and Roman Mechanical Water-Lifting Devices: The History of a Technology*, University of Toronto Press, Toronto, Buffalo London, 1984
24. S. A. Paipetis (ed.), *Extraordinary Machines and Structures in Antiquity*, Peri Technon Publishers, Patras, 2003
25. Agostino Ramelli, *Le diverse et arcificiose machine*, Paris, 1588 (References are to the English translation, John Hopkins University Press, 1976)
26. Chris Rorres, *The turn of the screw: optimal design of an Archimedes screw*, ASCE Journal of Hydraulic Engineering, Volume 126, Number 1, January 2000, pp. 72-80
27. Witold Rybczynski, *One good turn, a natural history of the screwdriver and the screw*, Simon & Schuster, New York etc., 2000
28. D. L. Simms, Archimedes the Engineer, *History of Technology* 17, 1995, pp. 45-111
29. Julius Weisbach, *Lehrbuch der Ingenieur- und Maschinen-Mechanik in drei Theilen. Dritter Theil: die Zwischen- und Arbeitsmaschinen enthaltend. Zweite Abtheilung: Die Arbeitsmachinen*, Braunschweig, Friedrich Vieweg und Sohn, 1851-1860
30. Reinhold Würth und Dieter Planck (Herausgeber), *Die Schraube zwischen Macht und Pracht, das Gewinde in der Antike*, Jan Thorbecke Verlag Sigmaringen, 1995

A NOTE ON THE HISTORY OF TRIGONOMETRIC FUNCTIONS

Jean-Pierre Merlet
INRIA, BP 93, 06902 Sophia-Antipolis Cedex, France

Abstract: Trigonometric functions appear very frequently in mechanism kinematic equations (for example as soon a revolute joint is involved in the mechanism). Dealing with these functions is difficult and trigonometric substitutions are used to transform them into algebraic terms that can be handled more easily. We present briefly the origin of the trigonometric functions and of these substitutions.

Keywords: trigonometric substitutions, robotics

1 INTRODUCTION

When dealing with the kinematic equations of a mechanism involving revolute joints appears frequently terms involving the sine and cosine of the joint angles. Hence when solving either the kinematic equations one has to deal with such terms. But the problem is that the geometric structure of the manifold build on such terms is not known and that only numerical method can be used to solve such equations. A classical approach to deal with this problem is to transform these terms into algebraic terms using the following substitution:

$$t = \tan(\frac{\theta}{2}) \Rightarrow \quad \sin(\theta) = \frac{2t}{1+t^2} \quad \cos(\theta) = \frac{1-t^2}{1+t^2}$$

This substitution allows one to transform an equation involving the sine and cosine of an angle into an algebraic equation. Such type of equation has a structure that has been well studied in the past and for which a large number properties has been established, such as the topology of the resulting manifold and its geometrical properties. Furthermore efficient formal/numerical methods are available for solving systems of algebraic equations.

But when asking in the community about the possible origin of this substitution I get only vague answer, most of them mentioning ancient Greek mathematicians. Hence the purpose of this paper is to investigate the origin of these substitutions.

2 THE ORIGIN OF THE TRIGONOMETRIC FUNCTIONS SINUS AND COSINUS

2.1 Origin for sinus and cosinus

The word "trigonometry" appears for the first time in the book *Trigonometria: sive de solutione triangulorum tractatus brevis et perspicuus* published by B.

Pitiscus (1561-1613) in 1595 and means "the study of trigons" in Latin (trigon being the word used for triangle) with a first appearance in English in the 1614 translation of the book of Pitiscus by Ra. Handson [Grattan-Guiness 94, Karpinski 46, Katz 87]. But *trigonometric functions* (a term that according to Cajori was introduced in 1770 by Georg Simon Klügel (1739-1812) while the term *trigonometric equations* can be found in 1855 as a chapter title of the book "A treatise on plane and spherical trigonometry" published by William Chauvenet) were in use well before: Hipparchus of Rhodes (190-120 BC), also called Hipparchus of Bithynia, called the founder of trigonometry, publishes a lost work on the chords of a circle in 12 books (although this number is contested) in 140 BC [Toomer 74] (the chord function Crd is related to the sine by $\sin(a) = Crd(2a)/120$), probably using Pythagora's theorem and the half-angle theorem. Despite the fact that Hipparchus was a major mathematician and astronomer it remains only one known of it writings, the *Commentary on Aratus and Eudoxus*, a minor work. One of the use of these tables was to tell the time of day or period of the year according charts of the angular measure between various stars compiled by astronomers.

Around AD 100 Menelaus (circa 70-130 AD) has published six lost book of tables of chords. In the two first books of its 13-books *Almagest* Ptolemy (85-125 AD) also gives a table of chords (note that Almagest is not the real name of the work of Ptolemy: originally the Greek title was *The Mathematical Compilation* that was soon replaced by the *Greatest Compilation* which was translated in Arabic as *Al-majisti* from which *Almagest* is derived).

The first appearance of the sine of an angle appears in the work of the Hindu Aryabhata the Elder (476-550), in about 500, that gives tables of half chords (that are 120 times the sine) based on the Greek half-angle formula and uses the word of *jya* to describe these quantities [Singh 39, Singh Chand 94, VdWaerden 86]. The same sort of table was presented by Brahmagupta (in 628) and a detailed method for constructing table of sine was presented by Bhaskara in 1150.

The Hindu word *jya* was phonetically reproduced by the Arabs as *jiba*, a word that has initially no meaning. But *jiba* became *jaib* in later Arab writings, a word that has the meaning of "fold". When Europeans translated the Arabic mathematical works into Latin, they translated *jaib* into *sinus* meaning fold, bay or inlet in Latin: especially Fibonnacci's use of the term *sinus rectus arcus* was one of the main step for the universal use of the word *sinus*. Note that the first appearance of *sinus* is still a subject of controversy:

- according to Cajori (1906) *sinus* appears in the 1116 translation of the astronomy of Al Battani (that formally introduces the cosine) by Plato of Tivoli that was published in 1537

- Eves claimed that *sinus* appears in the translation of the Algebra (*al-jabr w'al-muqabala*, the science of transposition and cancellation) of Al-khowarizmi (circa 780-850), a mathematician, astronomer and geographer, by Cremona (1114-1187)

- Boyer claims that *sinus* appears in 1145 in the translation of the tables of Al-khowarizmi provided in *Sindhind zij* by Robert of Chester (or Robert from Ketton) (his translation of the treatise on algebra starts with *Dixit Algorithmi: laudes deo rectori nostro atque defensori dicmus dignas*, "Algorithmi says: praise be to God, our Lord and Defender", and this was the first occurence of a sentence that will lead to the modern word of algorithm). Robert of Chester was a translator that was hired by the Castillian (Spanish) king Alphonso the 6th who captured Toledo from the Arabs and found a large library with many Arab manuscripts, including translations of Greek books unknown in the rest of Europe.

Georg Joachim von Lauchen Rheticus (1514-1574) published in 1542 some chapters of Copernicus's book giving all the trigonometry relevant to astronomy and produced accurate tables of the 6 trigonometric functions that were published after his death in *Opus Palatinum de triangulis* or *Canon of the Doctrine of Triangles* (probably written in 1551 but published in 1596).

Johann Müller of Königsberg also called Regiomontanus (1436-1476) writes the book *De triangulis omnimodis* that includes accurate data on the sine and its inverse that were done around 1464 but published only in 1533.

The word *cosinus* has a similar development: Viète (1540-1603) uses the term *sinus residae* while Edmund Gunter (1581-1626), a Rector and professor of astronomy, suggested the word *co-sinus* in 1620.

2.2   The sine and cosine abbreviation

In 1624 Edmund Gunter uses the abbreviation *sin* in a drawing (but the term was not used in the text) while it is claimed that it appears for the first time in the book *Cursus mathematicus* of the French mathematician Pierre Hérigone (1580-1643) published in 1634: this claim is controversial as many works relates that none of the notation proposed by Hérigone was used afterwards except for the ∠ one. Some other authors claimed that William Oughtred (1575-1660), the English rector of Albury, uses also *sin* in its book *Addition vnto the Vse of the Instrvment called the Circles of Proportion* published in 1632.

Other abbreviation were used: *Si* by Cavalieri (1598-1647), an Italian Jesuit mathematician, *S* by Oughtred (1574-1660), an English Episcopal minister who became passionately interested in mathematics, in its book *Trigonometrie* published in 1657. The term *sin.* (with a period) was proposed by Thomas Fincke (1561-1656), a Danish professor in rethorics, medicine and mathematics, in 1583 in his book *Geometriae rotundi*.

As for the cosine Cavalieri was using the notation *Si.2*, Oughtred using *s co arc* or *sco* and Sir Jonas Moore (1627-1679) proposes *Cos.* in *Mathematical Compendium* (1674). John Wallis (1616-1703) was using *S* while Samuel Jeake (1623-1690) used *cos.* in Arithmetick published in 1696.

The earliest use of *cos* is attributed either to Euler in 1729 in *Commentarii Academiae Scient. Petropollitanae, ad annum 1729* or to William Oughtred either in 1631 or 1657.

The modern presentation of trigonometry can be attributed to Euler (1707-1783) who presented in *Introductio in analysin infinitorum* (1748) the sine and cosine as functions rather than as chords.

2.3 Trigonometric substitution in sine and cosine

Ptolemy was aware of the formula

$$\sin(x + y) = \sin x \cos y + \cos x \sin y$$

and in 980 the Arab Abu'l-Wafa was using the formula

$$\sin 2x = 2 \sin x \cos x$$

in its book *Kitab al-Khamil*. This substitution was essential to calculate the tables that were used for astronomy and engineering.

Now consider the substitution:

$$\cos s \cos t = \frac{\cos(s + t) + \cos(s - t)}{2}$$

Tycho Brahe (1546-1601) uses this substitution to perform the multiplication of 2 number using an algorithm known as *prosthaphaeresis*. Assume that you want to multiply x times y. You first look a cosine table to look up the angle s whose cosine is x and the angle t whose cosine is y and then determines what are the cosines of $s + t$ and of $s - t$. If you average these two cosines you get the product xy.

3 THE ORIGIN OF THE TRIGONOMETRIC FUNCTION TANGENT

Tangent was initially not associated to angles or circles but to the length of the shadow that is projected by an object and that was used, for example, by Thales to measure the height of the pyramids.

The first known shadows tables were produced by the Arabs around 860 using both the tangent and the cotangent that were translated into Latin as *umbra recta* and *umbra versa*.

The first appearance of the term tangens (from the Latin *tangere*, to touch) is due to Thomas Fincke (1561-1656) used in 1583 in its Book 14 of *Geometrica rotundi* and was also used in 1632 by William Oughtred in *The circles of Proportion*. Viète was using the terms *amsinus* and *prosinus* and *sinus foecundarum* because he did not approve of the term tangent because it could be confused with the term in geometry..

The term cotangens was used by Edmund Gunter in 1620, *cot.* by Samuel Jeake in 1696. Finally *cot* was proposed by A.G. Kästner in *Anfangsgründe der Arithmetik*.

As for the notation Cavalieri was using *Ta* and *Ta.2*, Oughtred *t arc* and *t co arc* and Wallis *T* and *t*. The modern notation *tan* appears in a book of Albert Girard (1595-1632), a French musician settled in the Netherlands with an interest in algebra and military engineering, in 1626 and in the drawings of Edmund Gunter but was written as

$$\tan$$
$$A$$

The notation *cot* was proposed by Sir Jonas Moore in 1674.

4 THE HALF-ANGLE TANGENT SUBSTITUTION

As seen previously the half-angle sine formula was used very early. This is not the case of the transformation:

$$t = \tan(\frac{\theta}{2}) \Rightarrow \quad \sin(\theta) = \frac{2t}{1+t^2} \quad \cos(\theta) = \frac{1-t^2}{1+t^2} \tag{1}$$

All the authors seem to agree that this substitution was first used by Weierstrass (1815-1897) and is often called *Weierstrass substitution* of *Weierstrass t-substitution* [Stewart 94]. Weierstrass was interested in the integration of rational functions of $\sin(\theta)$ and $\cos(\theta)$. In addition to equation (1) it is indeed easy to prove that

$$d\theta = \frac{2dt}{1+t^2} \tag{2}$$

Combining equations (1) and (2) any integrand containing a rational function of $\sin(\theta)$ and $\cos(\theta)$ can be converted to an integrand containing a rational function of t. When considering integrals this substitution may lead to spurious discontinuities [Jeffrey 94]. For example the function $3/(5 - 4\cos x)$ is continuous and positive for all real x, and so its integral should be continuous and monotically increasing. Using the Weierstrass substitution we get

$$\int \frac{3dx}{5 - 4\cos x} = \int \frac{6du}{1 + 9u^2} = 2arctan(3\tan(x/2))$$

which is discontinuous at odd multiples of π.

5 CONCLUSION

Although trigonometric functions and substitutions are widely used in the mechanism community their history is not so much known. The purpose of this note was to emphasize some key points in their discovery. An open question remains: who is the first researcher that uses the Weierstrass substitution for the purpose of the kinematic analysis of a mechanism ?

Acknowledgment: Numerous Web resources were used to propose this paper, namely:

- MacTutor History of Mathematics archive:
  http://turnbull.mcs.st-and.ac.uk/~history/

- geometry: http://www.geometry.net/index.html

- History of Mathematics:
  http://aleph0.clarku.edu/~djoyce/mathhist/

- Math Archives:
  http://archives.math.utk.edu/topics/trigonometry.html

- Earliest Known Uses of Some of the Words of Mathematics :
  http://members.aol.com/jeff570/mathword.html

- The Galileo Project:
  http://es.rice.edu/ES/humsoc/Galileo/

References

[Grattan-Guiness 94] Grattan-Guiness I. *Companion Encyclopedia of the History and Philosophy of the Mathematical Science*. I. Grattan-Guiness Ed., Londres, 1994.

[Jeffrey 94] Jeffrey D.J. The importance of being continuous. *Mathematics Magazine*, 67:294–300, 1994.

[Karpinski 46] Karpinski L.C. The place of trigonometry in the development of mathematical sciences. *Scripta Math.*, 12:268–272, 1946.

[Katz 87] Katz V.J. The calculus of the trigonometric functions. *Historia Mathematica*, 14(4):311–324, 1987.

[Singh 39] Singh A.N. Hindu trigonometry. *Proc. Benares Math. Soc.*, 1:77–92, 1939.

[Singh Chand 94] Singh S.L. and Chand R. Hindu trigonometry. *J. Natur. Phys. Sci.*, 5/8:159–166, 1991/94.

[Stewart 94] Stewart J. *Single variable calculus*. Brooks/Cole, 1994.

[Toomer 74] Toomer G. J. The chord table of Hipparchus and the early history of Greek trigonometry. *Centaurus*, 18:6–28, 1973/74.

[VdWaerden 86] Van der Waerden B.L. On Greek and Hindu trigonometry. *Bull. Soc. Math. Belg. Sér.*, A38:397–407, 1986.

ON THE DEVELOPMENT OF THE CONSTRAINT MOTION THEORY OF FRANZ REULEAUX – AN OVERVIEW

Hanfried Kerle
Institute of Machine Tools and Production Technology
Technische Universität Braunschweig
Langer Kamp 19b, D-38106 Braunschweig, Germany
e-mail: h.kerle@t-online.de

ABSTRACT – The paper focuses on the constraint motion theory for mechanisms and machines defined by Franz REULEAUX, the founder and pioneer of the "geometry of motion" or "phoronomy". At his time machines with more than one degree of freedom were not controllable and therefore of minor or no importance in mechanical engineering science. Nevertheless the study of the question under which circumstances a mechanism is constrainedly running led to several degree-of-freedom equations for planar, spherical and spatial mechanisms, robots for example. Till today kinematicians and mechanism designers took a lot of advantages of the pioneering ideas of Franz REULEAUX.

KEYWORDS: History of Mechanisms and Machines, Constraint Motion Theory, Systematics of Mechanisms, Degree-of-Freedom Equations, Mechanism Design

INTRODUCTION
Franz REULEAUX, the founder of the constraint motion theory, lived from 1829 until 1905, within a period of a breath-taking industrial revolution. Industry discovered mass production, an important prerequisite for that was the mechanization of the production processes. Working machines were driven by steam engines or even already by combustion engines [1]. In spite of these immense engineering achievements a teachable mechanical engineering science was just at the beginning.

Franz Reuleaux, 1829 - 1905

201

It was the great merit of REULEAUX to have put in order the world of notions in mechanism theory, to have investigated and classified the elements of a machine systematically in his famous two-volume work "Lehrbuch der Kinematik", edited 1875 and 1900 [2, 3]. By means of his analytic-synthetic concept of the kinematic machine theory he created the fundamentals of a modern mechanism theory or technical kinematics theory [4].

What does constraint motion really mean? REULEAUX gave the following definition [3, 5]:

The constraint motion theory is that part in mechanics teaching us how a machine is arranged or should be arranged so that by means of exerting external forces all motions acting in it become determinate concerning all their paths.

This definition is equivalent to the statement that the motions of the links in a machine may only depend upon one single (drive) coordinate which means one d.o.f. (degree of freedom). More than one, for example two drives, were unimaginable at that time and led to undeterminate paths of points on moved links [6], say, such machines were not regarded to be very useful.

We owe a series of definitions and notions to REULEAUX starting with the kinematic chain including the frame as a link of the chain being the base for any mechanism [7]. A mechanism that transmits motion becomes a machine that transmits power by determining the drive and its time function. Here are some of the most important definitions given by REULEAUX [8]:

- A *kinematic chain* consists of rigid bodies (links) connected by *pairs of elements* forming reciprocal envelopes and thus allowing reciprocal relative motions. Such a pair, say joint, may be called a *lower pair* or a *closed pair*.
- Lower pairs of elements with reciprocal envelopes (surfaces) must be distinguished from *higher pairs* with mutual envelopes (curves or straight lines, points).
- Lower and higher pairs of elements may be called *complete pairs* if the closure is guaranteed solely by the form of the elements (*form-closure*).
- Lower and higher pairs of elements may be called *incomplete pairs* if the closure is guaranteed solely by opposite forces acting on the elements (*force-closure*).
- A *closed kinematic chain* of which one link is made stationary (frame) is called a *mechanism*.[1]
- The closed kinematic chain of a mechanism may be *simple* (one loop) or *compound* (more than one loop); the pairs of elements (joints) of it are characterized by the envelopes required for the motions which the bodies in contact must have so that all motions other than those desired in the mechanism are prevented (*constrained closed kinematic chain*).
- In general a *constrained closed kinematic chain* can be converted into a

---

[1] It has become customary to distinguish sometimes between a mechanism and a linkage, the latter being free from higher pairs of elements, cams for example.

mechanism in as many ways as it has links.

- A *machine* is a combination of resistant bodies so arranged that by their means the mechanical forces of nature can be compelled to do work accompanied by certain determinate motions (cf. also the definition of constraint motion).

Fig. 1 contains two models taken from the first volume of the main work of REULEAUX [2], a four-bar mechanism with marked paths of the moved rotary joints and a so-called "epicyclic reverted train" with coinciding input and output rotary axes. These models belong to the famous mechanism collection of REULEAUX which he started to establish at the former *Königliche Gewerbeakademie* (Royal Trade Academy) in Berlin and which later became the Royal Polytechnic.

Nowadays the models of REULEAUX and many replicas of them are spread almost all

Fig. 1: Two models of mechanisms belonging to the collection of Franz REULEAUX.

over the world, at universities in Germany (e.g. Dresden), Russia, USA (e.g. Cornell University, Ithaca, NY) and Canada [9, 10]. Two machines of outstanding historical meaning in the industrial revolution period are based on compound six-link kinematic chains leading to mechanisms with one d.o.f.; it is the steam engine of James WATT in 1769 and the locomotive steam control of Robert STEPHENSON in 1846. We find sketches of them in the well-known book of Ludwig BURMESTER, fig. 2 [11].

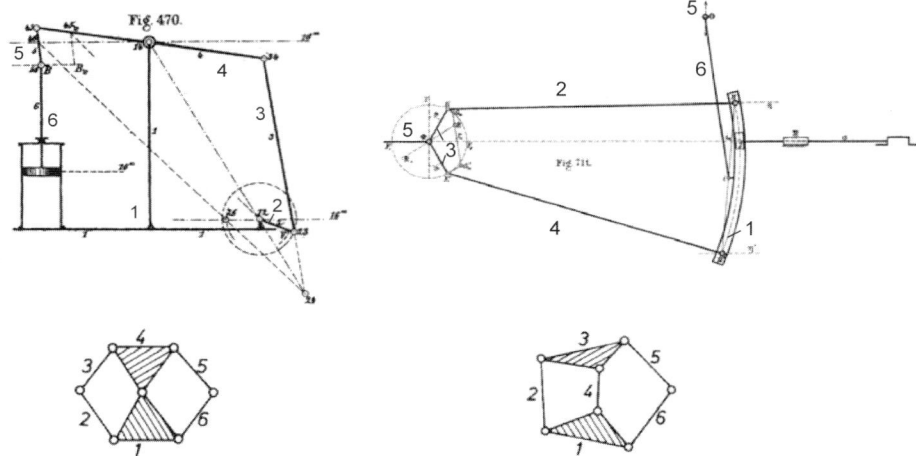

Fig. 2: WATT (left) and STEPHENSON (right) mechanism and related kinematic chains

Six-link kinematic chains may also lead to simple machine tools where no or only a small flexibility is necessary. Kurt HAIN gives two interesting examples [12, 13]: On the left side of fig. 3 we see a lathe for cutting straight twin edges at circular disks. The driving crank A_0A is attached to a first gearwheel. The crank moves a slider with the cutter by means of the coupler AB, the workpiece is turned by a second gearwheel meshing with the first one. On the right side of fig. 3 a machine tool for cutting rotary-symmetric shapes is shown. The linear axes X (drive) and Y are coupled by the trans-mission function of the four-bar linkage A_0ABB_0 with adjustable link lengths.

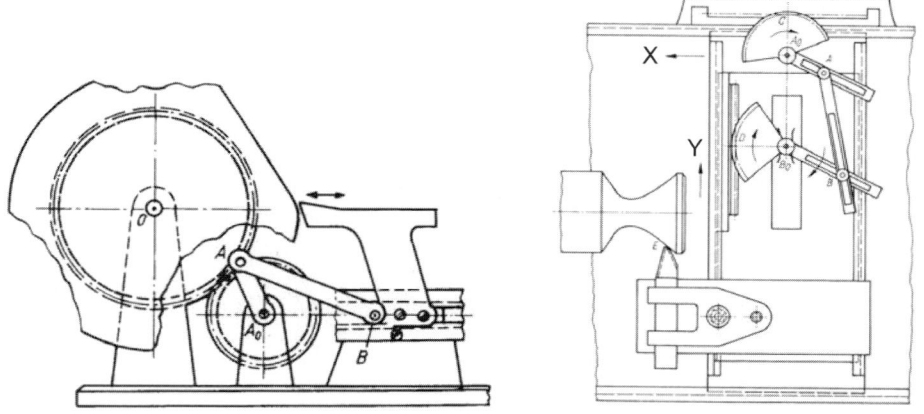

Fig. 3: Two simple machine tools based on six-link kinematic chains.

NUMERICAL RELATIONS IN MECHANISMS

In 1883 and 1885 Martin GRÜBLER developed his formula for planar mechanisms (with all joint axes being parallel or having one common point of intersection in infinity) calculating the degree of freedom, say, the number of drives necessary to move a mechanism in a determinate way [14, 15]. This well-known formula runs

$$F = 3 \cdot (n-1) - \sum_{i=1}^{g}(3 - f_i)$$ (1a)

or

$$F = 3 \cdot (n-1) - 2 \cdot g_1 - g_2$$ (1b)

where

F   resulting degree of freedom (d.o.f) of the mechanism
f_i d.o.f. of the ith joint
g   number of joints
g_1 number of joints with one d.o.f.
g_2 number of joints with two d.o.f.
n   number of links including the frame

Eq. (1a) is also valid for spherical mechanisms with all joint axes having one common finite point of intersection. The extension to spatial mechanisms is also possible, i.e.

$$F = 6 \cdot (n-1) - \sum_{i=1}^{g}(6 - f_i)$$ (2a)

[15] or respectively

$$F = 6 \cdot (n-1) - \sum_{i=1}^{5}[(6 - i) \cdot g_i]$$ (2b)

where g_i designates the number of joints with i d.o.f. [16]. A very interesting historical overview over different forms concerning the d.o.f. equation from authors in their early works who in most cases did not know much about one another is given by Ferdinand FREUDENSTEIN and his co-author [17].

A difficult situation arises in case there is a combination of a spatial and one or more planar mechanisms. In fig. 4 a spatial four-bar F_0FED_0 with arbitrarily oriented joint axes is coupled to a six-link mechanism (STEPHENSON chain) moving in the plane P1. The coupling is done in point E by a universal or cardanic joint ($f_E = 2$). The planar mechanism has only rotary joint axes (f = 1) being all parallel to each other (marked by a dot in the joint circle) in direction of the vector n_P perpendicular to P1. On the base of eqs. (1a) and (2a) H. BODEN developed a formula taking the whole mechanism as a spatial one. It runs [19]

$$F = \sum_{i=1}^{g}(f_i) - 6 \cdot (g - n + 1) + \sum_{k=1}^{z}(D_k)$$ (3a)

where

$$D_k = (g_P)_k - (F_P)_k, \quad k = 1, 2, ..., z$$ (3b)

is part of a sum giving the difference of d.o.f. between the spatial mechanism and the z planar mechanisms in arbitrarily located planes P1 to Pz coupled to it. Additionally to the nomenclature already introduced we have

$(F_P)_k$ d.o.f. of the planar mechanism part of number k,
$(g_P)_k$ number of joints of the planar mechanism part of number k.

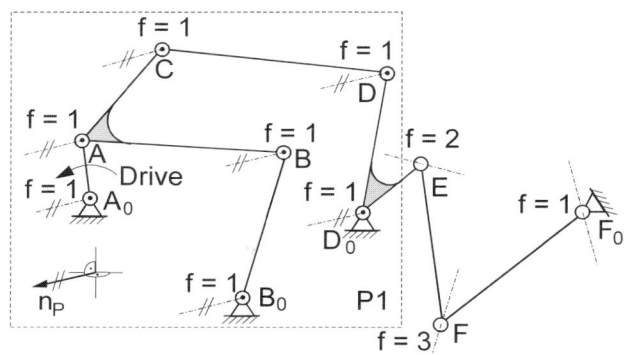

Fig. 4: Combination of a planar and spatial mechanism.

If we apply this formula to the combined mechanism of fig. 4, we get

$$F = \sum_{i=1}^{10}(f_i) - 6 \cdot (10 - 8 + 1) + D_1 = 1$$

where the sum of joint d.o.f. amounts to $8 \cdot 1 + 1 \cdot 2 + 1 \cdot 3 = 13$ and $D_1 = 7 - 1 = 6$. The mechanism as a whole proves to be constrainedly running transmitting the rotary drive motion of the crank A_0A of the planar part to the rocker F_0F of the spatial part around a fixed rotary axis arbitrarily located. Similar extensions exist for "branched" mechanisms, especially used with "parallel structures" [20, 21].

The GRÜBLER relations are only valid for mechanisms with arbitrary lengths of links or arbitrarily located joint axes in space, etc. If, however, there are geometric equalities, the mechanism is called "overclosed" or "compatible". In addition, some links may have *partial* or *passive* d.o.f. [18], say, they can be moved locally without moving the whole mechanism. We remember that in such cases kinematicians introduced a "corrected" formula to classify mechanisms with $F \le 0$ or simply to avoid $F \le 0$ or $-$ on the other hand $-$ to do not count partial d.o.f., for example

$$F = 6 \cdot (n-1) - \sum_{i=1}^{g}(6 - f_i) + c_{over} - f_{partial} \qquad (4)$$

where c_{over} means the number of overclosing constraints and $f_{partial}$ the number of partial or passive d.o.f. Omitting first c_{over} and $f_{partial}$ eq. (4) leads to table 1.

$F = m < 0$	Force-overclosed (overconstrained) mechanism of order $|m|$
$F = 0$	Form-overclosed mechanism (rigid framework)
$F = 1$	Constraint motion mechanism
$F = 2$	Two-drive mechanism, differential mechanism
$F = m > 2$	Multiple-drive mechanism, multiple differential mechanism of order m [23]

Table 1: F-classification of mechanisms.

Eq. (4) can now be exploited in three different ways:

1. ***Type synthesis:*** Calculation of F and classification of kinematic chains or mechanisms derived from them according to table 1
2. ***Number synthesis:*** Putting F = ..., -1, 0, 1, 2, ... into it to find out the correct number of links n and joints g taking account of the joint d.o.f.
3. ***Constraint synthesis:*** Calculation of c_{over} and creating mechanisms for special transmission of motions and forces

In the following chapters these three alternatives shall be discussed in detail with the help of some vivid examples.

TYPE SYNTHESIS OF SIMPLE BAND MECHANISMS

A simple band mechanism originates from the well-known four-bar mechanism with F = 1. This also includes the possibility to apply methods of kinematic analysis of the four-bar mechanism to simple band mechanisms. Because of the fact that the contact point of the band and a curved segment now replaces a rotary joint of the four-bar mechanism, a kinematic analysis program must take account of variable link lengths during motion cycles.

Kurt HAIN takes advantage of this substitute idea and presents six different forms of simple band mechanisms, fig. 5 [22].

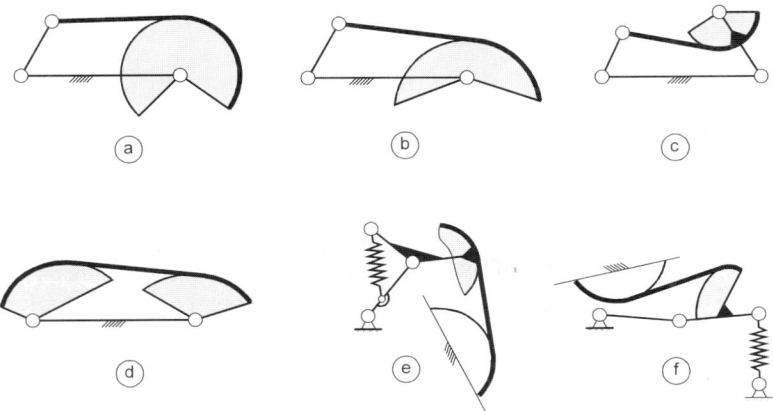

Fig. 5: Simple band mechanisms based on the four-bar kinematic chain.

208

The first form (a) is a centric version with coinciding pivot in the frame and centre point of a circular segment; (b) and (c) are eccentric versions of the same kind, but the pivot in the frame does not coincide with the centre point of the circular segment, and the two points lie either on the same side (b) or on opposite sides (c) with reference to the contact point. The form (d) is the most general one with two arbitrary segments for guiding the band, one segment may even be fixed. The last forms (e) and (f) include a spring for holding the band tight.

NUMBER SYNTHESIS OF SPATIAL SHAFT-COUPLINGS
If we want to find all possible solutions for spatial shaft-couplings consisting of four links and including the input and output axes with arbitrary location and orientation we put F = 1 and $c_{over} = 0$ into eq. (4) and get

$$\sum_{i=1}^{g}(f_i) = 6 \cdot (g - n) + 7 + f_{partial} \qquad (5)$$

from which we conclude that a four-link spatial coupling with n = g = 4 in general must have seven d.o.f. of the joints in series, partial d.o.f. not included. Thus, with two cranks hinged on a baseplate by simple rotary joints the two coupler-rod joints must have five d.o.f. in series [27, 28]. Six design examples of four-link shaft-couplings are presented in fig. 6. Only for ease of drawing the input/output axes are coincident, in general they are not necessarily. The last two couplings e) and f) have coupler-rods with one partial d.o.f. additionally.

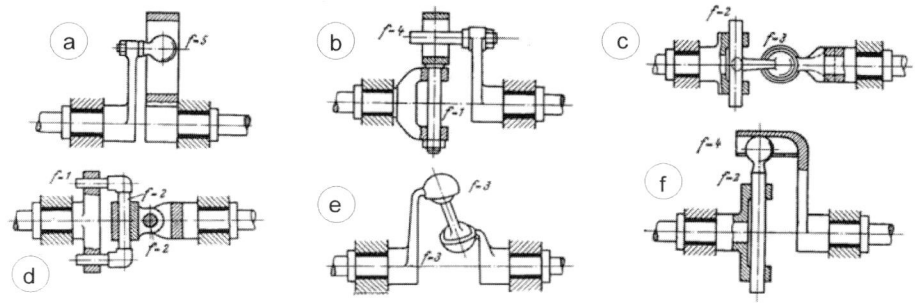

Fig. 6: Spatial four-link shaft-couplings with arbitrarily oriented input/output axes.

CONSTRAINT SYNTHESIS

ROBERTS´ theorem
In 1875 not only REULEAUX wrote his famous book about the phoronomy of mechanisms, but also S. ROBERTS published his famous theorem about the triple generation of coupler-curves [29]. The theorem states that the curve of a coupler-point K of a four-bar mechanism A_0ABB_0 (fig. 7, left) can be reproduced by two other four-bar mechanisms $A_0A^*B^*C_0$ and $C_0A^{**}B^{**}B_0$. Such linkages are called "cognates" [30]. From any given four-bar the two cognate four-bars may be uniquely determined. For example, it can be shown that the well-known straight-line mechanisms of ROBERTS, EVANS and

CHEBYCHEV are all ROBERTS´ cognates of the WATT four-bar [31].

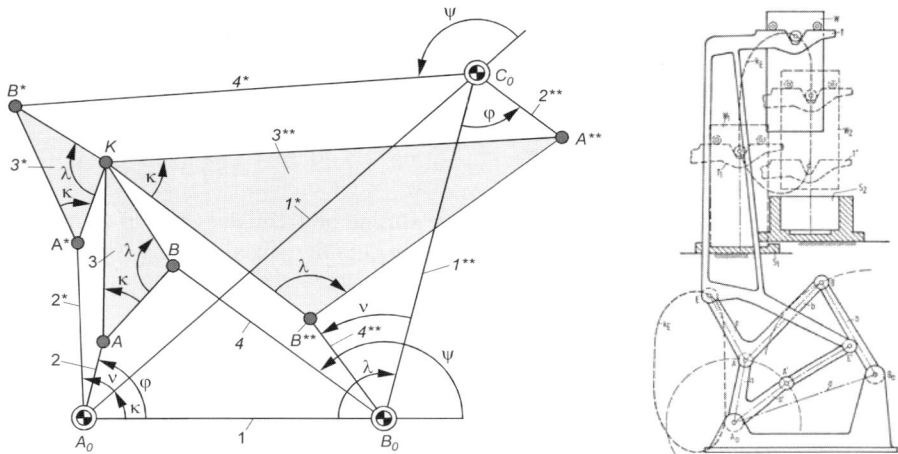

Fig. 7: Triple generation of a coupler-curve (left) and example of application (right).

If we calculate the d.o.f. of the mechanism on the left side of fig. 7 taking account of the double rotary joints in A_0, B_0, C_0 and K, we get from eq. (1b) $F = -1$. According to table 1 the mechanism is force-overclosed, in spite of that we know that the mechanism can be moved. This is due to the fact that the mechanism as a whole consists of three parallelograms of related lengths (table 2) which means $c_{over} = 2$ and leads to the triangles ABK, A^*B^*K, $A^{**}B^{**}K$ and $A_0B_0C_0$ (frame link 1) being similar to each other [32].

	Initial linkage	1. alternative linkage	2. alternative linkage
Length of crank (rocker) 2	$l_1 \equiv \overline{A_0 A}$	$l_1^* \equiv \overline{A_0 A^*} = l_2 \cdot \frac{k}{l_2} = k$	$l_1^{**} \equiv \overline{C_0 A^{**}} = l_1 \cdot \frac{l}{l_2}$
Length of coupler 3	$l_2 \equiv \overline{A B}$	$l_2^* \equiv \overline{A^* B^*} = l_1 \cdot \frac{k}{l_2}$	$l_2^{**} \equiv \overline{A^{**} B^{**}} = l_3 \cdot \frac{l}{l_2}$
Length of rocker (coupler) 4	$l_3 \equiv \overline{B B_0}$	$l_3^* \equiv \overline{B^* C_0} = l_3 \cdot \frac{k}{l_2}$	$l_3^{**} \equiv \overline{B^{**} B_0} = l_2 \cdot \frac{l}{l_2} = l$
Length of fixed link 1	$l_4 \equiv \overline{A_0 B_0}$	$l_4^* \equiv \overline{A_0 C_0} = l_4 \cdot \frac{k}{l_2}$	$l_4^{**} \equiv \overline{C_0 B_0} = l_4 \cdot \frac{l}{l_2}$
Defining parameters of the coupler point K	$k \equiv \overline{A K}$	$k^* \equiv \overline{A^* K} = l_1$	$k^{**} \equiv \overline{A^{**} K} = l_3 \cdot \frac{k}{l_2} = l_3^*$
	$l \equiv \overline{B K}$	$l^* \equiv \overline{B^* K} = l_1 \cdot \frac{l}{l_2} = l_1^{**}$	$l^{**} \equiv \overline{B^{**} K} = l_3$

Table 2: Length relations of ROBERTS´ cognates.

On the right side of fig. 7 an application example is given: A block W is lifted along an approximately straight line and guided downwards again by a special handling device, a six-link mechanism (WATT chain) driven by the ternary crank-link AA_0A'. Because of the congruent curves k_E of the coupler-points E and E' the block W is always kept parallel to the ground. Three different positions of the block W are shown. Point E belongs to the initial four-bar A_0ABB_0 with lengths $A_0A = A_0A'$ and $AB = B_0B$, the alternative cognate reduces to $C_0 = A_0$ and $A^{**} = A'$ with a new coupler-rod EE' [33].

Linear guides and bearings

Linear guides and bearings are used to realize sliding or prismatic pairs of elements, especially in cases where one of the elements belongs to the (fixed) frame. Nevertheless it is possible to treat such linear guides as spatial mechanisms consisting of at least two links and having one d.o.f.

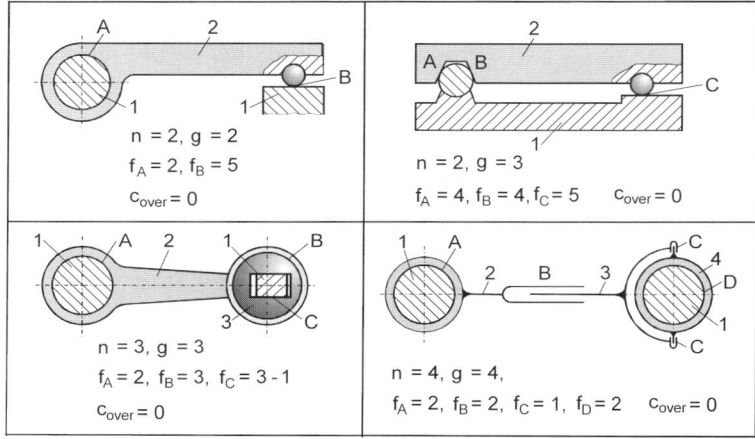

Table 3: Linear guides forming spatial mechanisms with F = 1.

Putting F = 1 into eq. (4) we calculate the number of overclosing constraints[2] from

$$c_{over} = 7 - 6 \cdot (n - g) - \sum_{i=1}^{g}(f_i) + f_{partial} \qquad (6)$$

If a linear guide is expected to work under general conditions regardless of special dimensions of links and/or specially located joint axes, the joint d.o.f. should be chosen or distributed respectively in such a way that $c_{over} = 0$. Examples of design are given in table 3. On the other hand we can have linear guides of high precision and stiffness with no backlash the higher the number of c_{over} or order $|m|$ is chosen or results. This is won by a high manufacture quality of the pair elements in contact as regards geometry, surfaces and adjustment of axes. Some examples are given in table 4.

---

[2] c_{over} corresponds to the order $|m|$ in table 1 according to $c_{over} = |m| + 1$, m < 0.

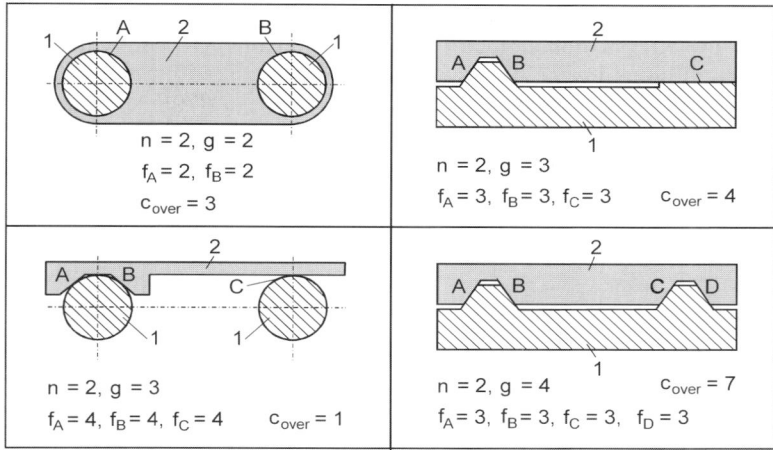

Table 4: Linear guides forming spatial mechanisms with F < 1.

Clamping devices and pliers

Force-overclosed mechanisms of order 1 having at least two binary links may be used for clamping devices [34], for instance in machine tools. One binary link transmits the driving force, by means of a screw-spindle for example, the remaining binary links are foreseen to exert the clamping forces. Putting $F = -1$ into eq. (1a) and setting $g_1 = g$, $g_2 = 0$ we get six-link kinematic chains ($n = 6$) with eight joints of one d.o.f. ($g = 8$) for example, from which proper mechanisms can be derived.

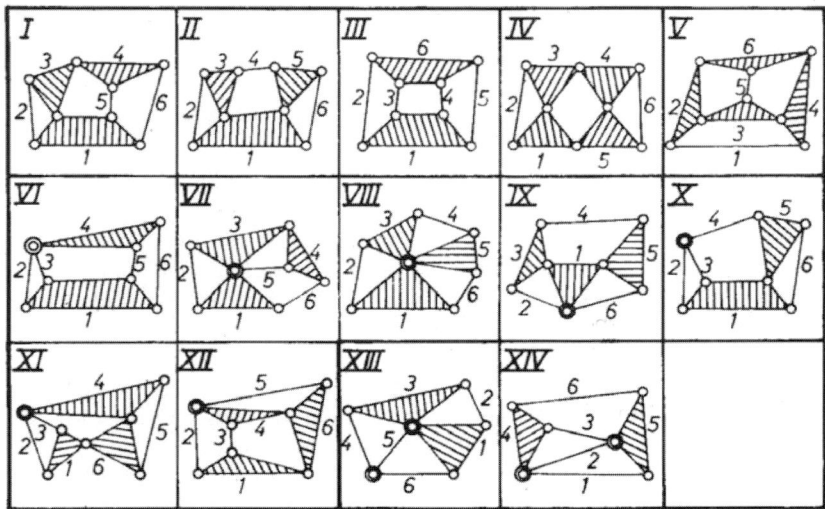

Table 5: Six-link kinematic chains for clamping devices with $F = -1$.

212

Table 5 presents 14 kinematic chains: The chains I to V have only single rotary joints, the chains VI to XII one double rotary joint and the chains XIII and XIV two double rotary joints.

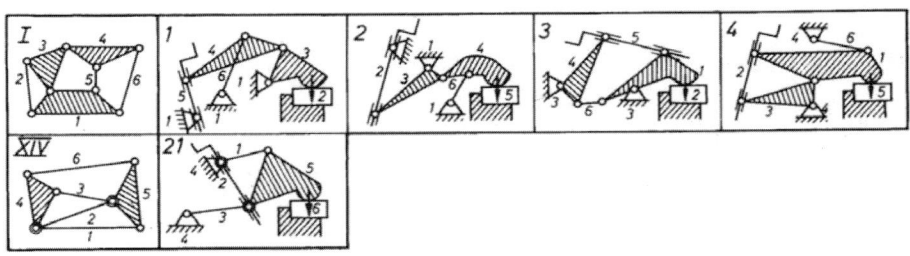

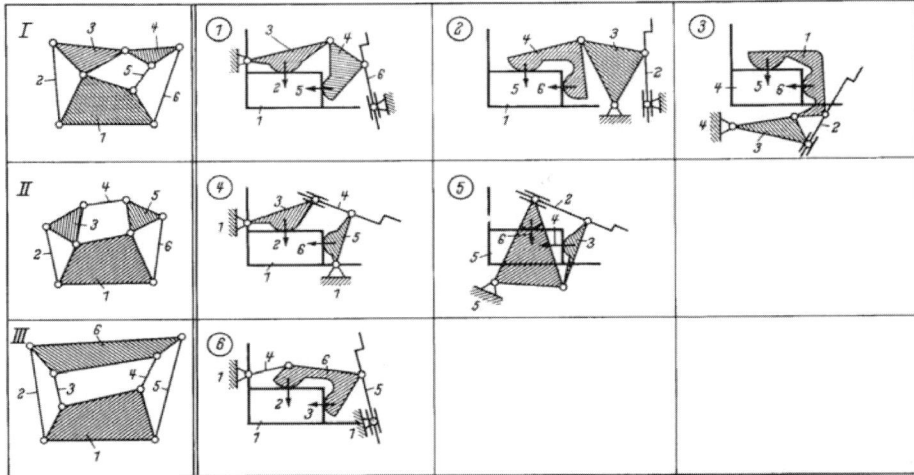

Table 6: Clamping devices taken from six-link kinematic chains of table 5.

All these chains form the base of some clamping devices after having selected the frame-link and the drive-link (screw-spindle): The upper part of table 6 contains clamping devices with one clamping location taken from the chains I and XIV, in the lower part of table 6 there are clamping devices with two clamping locations each taken from the chains I to III.

Increasing the number of links and joints leads to clamping devices with additional possibilities of application [35]. For example, if we want to clamp a flat, rectangular block (workpiece) from two sides and at the same time from above by means of a screw-spindle (with the possibility to drill a hole into it, for example), we can find proper solutions on the base of eight-link kinematic chains with eleven joints. The partial task "clamping from above" needs the frame as a force counterpart and a linear guide of the pressing slider in the frame. The partial task "clamping and holding the workpiece" can be done relatively inside the mechanism represented by a binary link in the kinematic chain. The gripping jaws belonging to this link must work in a way centering the work-piece with respect to the central axis of the pressing slider. There are only seven chains A to G (out of 372!) with eight links and eleven joints that meet these conditions, table 7. The chains H and I have one double rotary joint each and originate from chain E or F.

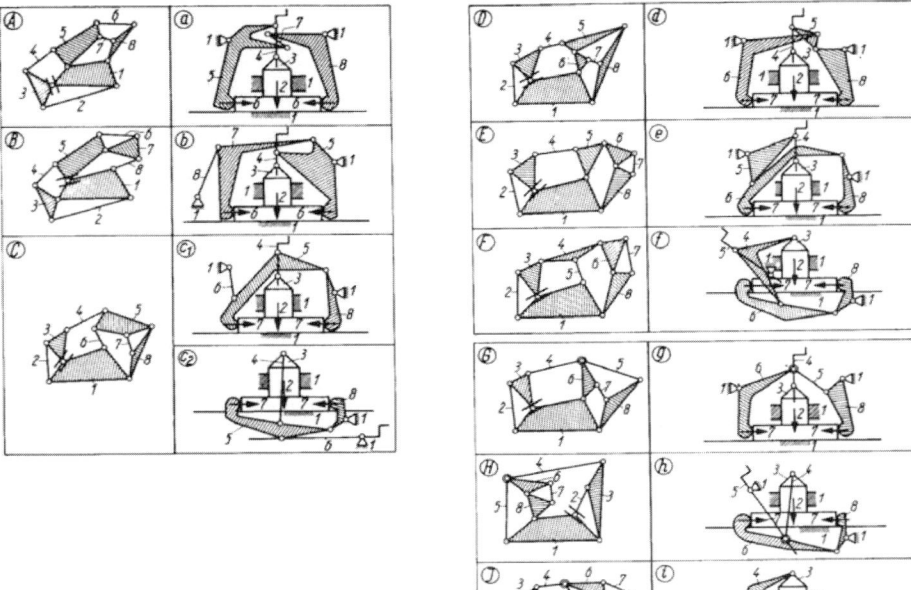

Table 7: Clamping devices taken from eight-link kinematic chains with F = -1.

Another interesting field of using force-overclosed mechanisms of order 1 are force-amplifying pliers [37]. Table 8 shows some examples of pliers consisting of four links connected by rotary joints. These pliers are based on the six-link kinematic chains already introduced with table 5. Two binary links of the chains on the left side in table 8 are replaced by forces, for example by P_{ij} as the output force between links i and j and by Q_{kl} as the corresponding input force between links k and l. Because of the fact that the two binary links do not contribute to the actual number of links of the mechanism itself, a four-link mechanism results from the six-link mechanism. There are several possibilities to select appropriate binary links in order to distribute the output and input forces P and Q respectively, symmetric kinematic chains lead to symmetric pliers where the output and input forces can be exchanged.

214

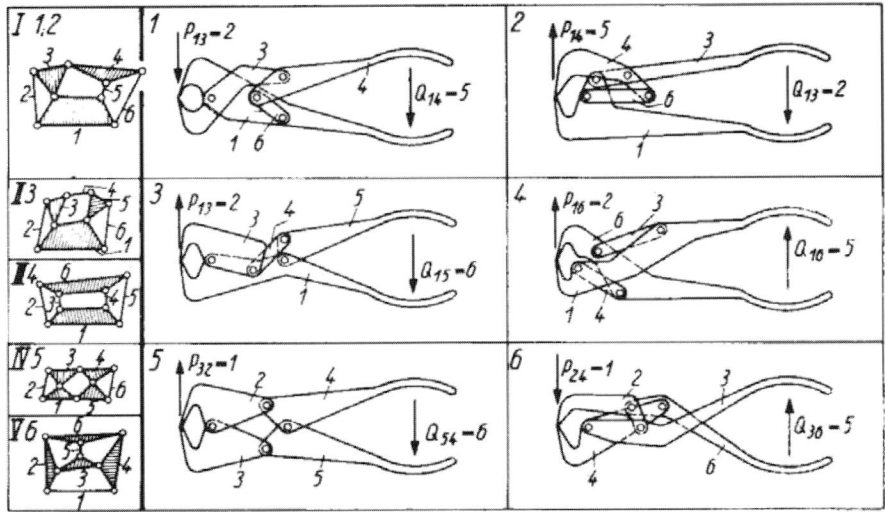

Table 8: Pliers mechanisms derived from six-link kinematic chains with $F = -1$.

CONCLUSION

The systematic approach of Franz REULEAUX laid the base for classifying mechanisms according to their types and numbers of links and joints. He defined a kinematic chain, a mechanism and a machine which at the beginning of mechanical engineering science was strictly constrained, say, had only one degree of freedom. But on this base other famous kinematicians like GRÜBLER, KUTZBACH and KRAUS could develop the degree-of-freedom equations of mechanisms and the type and number synthesis. Especially Kurt HAIN dedicated a big number of his publications to forward the systematic ideas of REULEAUX for new creative insights into mechanism theory till the end of the last century.

REFERENCES
1. Mauersberger K., Franz Reuleaux – Begründer der Kinematik, Feingerätetechnik, Vol. 37, 1988, No. 1, pp. 38-41.
2. Reuleaux F., Lehrbuch der Kinematik – Vol. 1: Grundzüge einer Theorie des Maschinenwesens, Verlag Friedrich Vieweg & Sohn, Braunschweig, 1875.
3. Reuleaux F., Lehrbuch der Kinematik – Vol. 2: Die praktischen Beziehungen der Kinematik zu Geometrie und Mechanik, Verlag Friedrich Vieweg & Sohn, Braunschweig, 1900.
4. Mauersberger K., Exkurs in die Geschichte der Maschinenlehre, Wissenschaftliche Zeitschrift der TU Dresden, Vol. 41, 1992, No. 4, pp. 6-15.
5. Grodzinski P., Was ist Zwanglauf?, Reuleaux-Mitteilungen – Archiv für Getriebetechnik, Vol. 2, 1934, No. 5, pp. 35-36.
6. Jahr W., Knechtel P., Grundzüge der Getriebelehre, Vol. 1: Allgemeine Grundlagen – Schraubentriebe – Kurbeltriebe, Verlag M. Jänecke, Leipzig, 1930.
7. Grodzinski P., M'Ewen E., Link Mechanisms in Modern Kinematics, Proc. of the Institution of Mechanical Engineers, London (UK), Vol. 168, 1954, No. 37, pp. 877-896.
8. Reuleaux F., The Kinematics of Machinery – Outlines of a Theory of Machines (translated into English by A. B. W. Kennedy), Macmillan & Co., London (UK), 1876.
9. Mauersberger K., Die Getriebemodellsammlung der Technischen Universität Dresden, Wissenschaftli-

che Zeitschrift der TU Dresden, Vol. 46, 1997, No. 3, pp. 103-106.

10. Bragastini R., Contributo per una interpretazione filosofica dell'opera di Franz Reuleaux, Tesi di laurea, Università degli Studi di Milano (Italy), 2003.

11. Burmester L., Lehrbuch der Kinematik, Vol. 1: Die ebene Bewegung, Verlag Arthur Felix, Leipzig, 1888.

12. Hain K., Die mechanischen Möglichkeiten des Unrunddrehens, Das Industrieblatt, Vol. 58, 1958, No. 12, pp. 530-537.

13. Hain K., Formdrehen mit zwangläufig geführten Werkzeugen, Technische Zeitschrift für praktische Metallbearbeitung, Vol. 66, 1972, No. 10, pp. 447-451.

14. Grübler M., Allgemeine Eigenschaften der zwangläufigen ebenen kinematischen Ketten, Part I: Civilingenieur, Vol. 29, 1883, pp. 167-200, Part II: Verhandlungen des Vereins zur Beförderung des Gewerbefleißes, Vol. 64, 1885, pp. 179-223.

15. Grübler M., Getriebelehre, Verlag Julius Springer, Berlin, 1917.

16. Malyshev A. P., Analysis and Synthesis of Mechanisms from the Viewpoint of Their Structure (in Russian), Izv. Tomsk Tekh. Inst., 1923.

17. Freudenstein F., Alizade, R., On the Degree of Freedom of Mechanisms with Variable General Constraints, Proc. 4th World Congress on the Theory of Machines and Mechanisms, Newcastle upon Tyne (UK), 1975, pp. 51-56.

18. Bagci C., Degrees of Freedom of Motion in Mechanisms, Trans. ASME, Journal of Engineering for Industry, Series B, Vol. 93, 1971, pp. 140-148.

19. Boden H., Zum Zwanglauf gemischt räumlich-ebener Getriebe, Maschinenbautechnik (Getriebetechnik), Vol. 11, 1962, No. 11, pp. 612-615.

20. Corves B., Simulation des kinematischen und dynamischen Verhaltens von Handhabungsgeräten mit geschlossenen kinematischen Teilketten, PhD thesis RWTH Aachen, VDI-Verlag, Düsseldorf, Fortschritt-Berichte VDI, Series 18, No. 71, 1989.

21. Frindt M., Modulbasierte Synthese von Parallelstrukturen für Maschinen in der Produktionstechnik, PhD thesis TU Braunschweig, Vulkan-Verlag, Essen, 2001.

22. Hain K., Die Synthese der viergliedrigen Bandgetriebe, Schriftenreihe des Verlags Technik, Vol. 193, 1954, pp. 40-51.

23. Kutzbach K., Mechanische Leitungsverzweigung – Ihre Gesetze und Anwendungen, Maschinenbau / Der Betrieb, Vol. 8, 1929, No. 21, pp. 710-716.

24. Beyer R., Technische Kinematik, Verlag J. A. Barth, Leipzig, 1931.

25. Beyer R., Ueber den Zwanglauf räumlicher und ebener Getriebe – Ein Beitrag zur Zahlsynthese, Reuleaux-Mitteilungen – Archiv für Getriebetechnik, Vol. 2, 1934, No. 5, pp. 37-40 and table 9.

26. Kraus R., Zur Zahlsynthese ebener kinematischer Ketten, Reuleaux-Mitteilungen – Archiv für Getriebetechnik, Vol. 3, 1935, No. 12, pp. 707-708 and Vol. 4, 1936, No. 4, pp. 335-336.

27. Kraus R., Zur Zahlsynthese der räumlichen Mechanismen, Reuleaux-Mitteilungen – Archiv für Getriebetechnik, Vol. 8, 1940, No. 1, pp. 33-39.

28. Altmann F. G., Raumgetriebe, Feinwerktechnik, Vol. 60, 1956, No. 3, pp. 83-92.

29. Roberts S., On Three-bar Motion in Plane Space, Proc. of the London Mathematical Society, Vol. 7, 1875, No. 14, pp. 14-23.

30. Roth B., On the Multiple Generation of Coupler-Curves, Proc. Mechanisms Conference, Lafayette, Ind. (USA), Paper No. 64—Mech-3, 1964, pp. 1-7.

31. De Jonge A. E. R., The Correlation of Hinged Four-bar Straight-line Motion Devices by Means of the Roberts' Theorem and a New Proof of the Latter, Annals of the New York Academy of Science, Vol. 84, 1960, pp. 75-145.

32. Association of German Engineers (VDI), Guideline No. 2740, Mechanical Devices for Automation Equipment, Part 2: Guidance Mechanisms, Beuth Verlag GmbH, Berlin, 2002.

33. Hain K., Getriebekonstruktionen mit Parallel-Geradführungen, technica, Vol. 11, 1982, pp. 982-987.

34. Hain K., Die Entwicklung von Spannvorrichtungen mit mehreren Spannstellen aus kinematischen Ketten, Das Industrieblatt, Vol. 59, 1959, No. 11, pp. 559-564.

35. Hain K., Achtgliedrige kinematische Ketten mit dem Freiheitsgrad F = -1 für Spannvorrichtungen, Das Industrieblatt, Vol. 62, 1962, No. 6, pp. 331-337.

36. Hain K., Spannvorrichtungen mit selbsttätiger Mitteneinstellung, Werkstatt und Betrieb, Vol. 97, 1964, No. 5, pp. 353-357.

37. Hain K., Entwurf viergliedriger, kraftverstärkender Zangen für gegebene Kräfteverhältnisse. Das Industrieblatt, Vol. 62, 1962, No. 2, pp. 70-73.

THE CONTRIBUTION OF COULOMB TO APPLIED MECHANICS

Agamenon R. E. Oliveira
Federal University of Rio de Janeiro
Polytechnic School
Applied Mechanics and Structures Department
P. O. Box 68536-21941, Ilha do Fundão, Rio de Janeiro, Brazil
agamenon.oliveira@gbbo.com

ABSTRACT- Coulomb's major work on friction forces appeared in his *"Théorie des Machines Simples"* in [3] which won him the Grand Prix from the "Académie des Sciences" in 1781. He investigated both static and dynamic friction of sliding surfaces, as well as friction in bending of ropes and in rolling. Because of this contribution to the science of friction one can say that he created it. This paper is an attempt to highlight the importance of Coulomb's work in the scientific context of eighteenth century, especially in relation to the birth of applied mechanics.

KEYWORDS: History of mechanics, History of applied mechanics, History of machines, History of work, History of industrial mechanics.

INTRODUCTION

Alongside Lazare Carnot (1753-1823), Coulomb can be considered the most important founder of applied mechanics. Carnot, as is well known, created the first general theory of machines in [12, 13]. Carnot's theory of mechanics emerged in the context of the development of Rational Mechanics under the influence of d'Alembert's principle, [4, 5]. Coulomb, through the discovery of laws of friction, created a new science to study the phenomenon of friction in machines. In addition, Coulomb provided solutions for many of the important problems of applied mechanics, such as his elegant solution to the problem of torsion in cylinders and his use of torsion balance in physical applications for physicists in successive years, [1].

Coulomb was the first to show how torsion suspension could provide physicists with a method of accurately measuring extremely small forces. By using torsion balance he achieved remarkable results: the law of attraction and repulsion, electric point charges, magnetic poles, distribution of electricity on the surface of charged bodies, etc. Based on experimental investigations Coulomb developed a theory of attraction and repulsion between bodies with the same and opposite electrical charges. He also demonstrated an inverse square law for such forces and went on to examine perfect conductors and dielectrics. He suggested that there was no perfect dielectric, proposing that every substance has a limit above which it will conduct electricity, [1, 2].

These works on electricity were also important for the mechanist view of the world because they confirmed for the field of electricity a similar law to Newton's theory of gravitation, both based on action at distance between masses and an inverse square law of the same distances.

This paper presents the fundamental aspects of Coulomb's investigations of friction forces, his experimental findings, and his discovery of the laws of friction, as well as the importance of these achievements for engineering sciences.

COULOMB'S BIOGRAPHY

Charles Augustin Coulomb was born in Angoulême, France on June 14, 1736. His father was Henry Coulomb and his mother was Cathérine Bajet. Both his parents came from families, which were well known and important in their regions. After being brought up in Angoulême, the capital of Angoumois in southwestern France, Coulomb's family moved to Paris. In Paris he entered the "Collège Mazarin", where he received a good classical education in language, literature and philosophy. He received the best available education in mathematics, astronomy, chemistry and botany.

After a period in Montpellier, Coulomb went to Paris in October 1758 to receive the tutoring necessary to take the examinations to enter the "École du Génie" in Mézières. He studied Camus's famous book, "Cours de Mathematique", for several months. In the same year, Coulomb took the examinations set by Camus himself. He passed and he entered the "École du Génie" in February 1760, the same school from which Lazare Carnot graduated in 1773.

Fig.1. Charles Augustin Coulomb

Coulomb graduated in November 1761, now a trained engineer with the rank of lieutenant in the "Corps du Génie". Over the next twenty years he lived and worked in several places where he was involved in engineering, structural design, fortifications, soil mechanics and many other areas. After his first posting in Brest in 1764 he was sent to Martinique in the West Indies. However, Martinique was attacked by a number of foreign fleets over the following years and finally captured by England in 1762.

On his return to France, Coulomb was sent to Bouchain. He now began to write important works on applied mechanics and in 1773 he presented his first work to the "Académie des Sciences" in Paris. This was entitled: "*Sur une Application des règles, de maximis et minimis à quelque problème de statique, relatifs à l'architecture*". The most significant aspect of this work is Coulomb's use of the calculus of variation to solve the problem under consideration.

In 1779 Coulomb was sent to Rochefort. During his time there he began his investigations into mechanics, using shipyards in Rochefort as laboratories for his experiments. His studies of friction forces led him to write the major work: "*Théorie des Machines Simples*".

This 1781 memoir changed completely Coulomb's life. He was elected to the mechanics section of the "Académie des Sciences", and moved to Paris where he now held a permanent post. He devoted the following years mainly to physical problems rather than engineering. He wrote treatises on electricity and magnetism, also submitted to the "Académie des Sciences" between 1785 and 1791.

He presented twenty-five *memmoires* to the "Académie des Sciences" between 1781 and 1806. He also carried out several investigations with Bossut (1730-1814), Borda (1733-1799), de Prony and Laplace (1749-1827) over this period. He participated in the work of 310 Academy committees. Besides these engineering projects, he also undertook services for the French government in several fields from education to hospital reform. His educational activities were largely carried out between 1802 and 1806, when he was inspector general of public instruction He was also mainly responsible for setting up the *Lycées* across France. Coulomb died in Paris on August 23, 1806, [1, 2].

SCIENCE AND FRENCH REVOLUTION

Although the names of Bacon (1561-1626) and Newton (1642-1727) ensure to England an important place in the history of scientific revolution, it was France which first recognized the real significance of Newton' s work and which first put into practice Bacon' s precepts. An explanation for this is the creation in France of a new type of organization for systematic scientific research, [17].

The political transformations in France, which led to French Revolution, were numerous and profound. The intellectual and cultural movement of this period known as Enlightenment influenced several other countries in Europe and led to many debates in France. The changes included the general reform of the French system education, resulting in the establishment and reorganization of several educational institutions.

When the French Revolution began in 1789, Coulomb was deeply involved with his scientific work. He retired from the " Corps du Génie" in 1781 and about the same time the "Academie des Sciences" was abolished in August 1793.

In 1794, the new government created the " École Polytechnique", allotting 12,000 pounds to it. It opened with 400 students and its staff included the mathematicians Lagrange (1735-1813), Laplace (1749-1827), the chemist Berthollet (1748-1821) and many others famous scientists.

By 1800 the scientific spirit was firmly established in France. In addition, as well as being the country with the most organized scientific research, France was the first to encourage the practice of experimental science systematically and on a large scale and the first country to realize that scientific work must also be summarized and disseminated. The French Academy and scientific schools closely associated with it, mobilized the intellect of the nation. Another important characteristic of this period was that France led the way in the application of scientific research to industry.

The authors that have analyzed the development of sciences in the second half of the eighteenth century agree that a remarkable characteristic of French science was the presence of " analysis"- a method that could be applied to a great number of physical problems, by using algebraic tools, some general principles and a deductive approach. This approach to problems enhanced the capacity of scientific research at the same time as it also increased the power of generalization and the formalization of mechanics and correlated sciences, [18].

In the period under consideration not only did Rational Mechanics reach a high degree of sophistication with the Lagrangian Mechanics in 1788, but astronomy, acoustic, electricity, optics, the theory of elasticity and probabilistic calculus were systematically modified, [11]. It is important to mention that in this period Condorcet (1743-1794) conceived a project for a mathematical social science.

THE PIONEERING WORK OF GUILLAUME AMONTONS

Amontons performed the first experimental investigation of friction forces, [19]. This study changed the way friction was interpreted as a physical phenomenon. Before Amontons (1663-1705), friction forces were treated either as a resistance to be overcome, resistance at the beginning of a relative motion between two bodies, or as a kind of adherence between surfaces in contact. In his famous study, Amontons showed that friction is independent of surface extension and stated: " If we think carefully about the nature of friction forces, we can concluded that it is an action in which a body under pressure from another can be moved through the common surface between them".

Amontons belonged to the founding group of scientists of the French Academy. In 1699 he published two memoirs, the first " Moyen de substituer commodement l' action du feu à la force des homes et des chevaux pour mouvoir les machines", deals with a rotating machine that uses hot air as a kind of "fire windmill". The second is called " De la resistance causeé dans les machines tant par les frottements des parties qui les composent, que par la roideur des cordes qu' on y emploie, et la manière d' en calculer l'un et l'autre". It had the same objective that Coulomb proposed to solve later. In other words, the last investigations tried to understand the nature of friction forces and separate the effect of bending stiffness and friction. To do this, Amontons used various materials, as copper, iron, lead and wood, in his experiments. Unlike Coulomb, however, Amontons used lubricating substances between the samples with relative motion and established rigorous control over two parameters, the pressure between bodies and the force to maintain motion. Obviously he made variations in the sample material. The experimental set up used clearly identified that the phenomenon is strongly dependent of the weights involved, as well as the nature of common surfaces in relative motion. Amontons did not published numerical results of this investigation but summarized his achievements as follows:

1) The resistance caused by friction forces varies in the same proportion as pressures between the parts in relative motion of more or less the same size.

2) The resistance caused by friction forces is nearly the same in iron, copper, lead or wood for any variation introduced once the same lubricant material is used.

3) The resistance of friction forces is approximately one third of pressure.
4) The resistance of friction forces depends not only on the weights and pressures involved but also on time and velocities.

The above postulations can be seen as Amontons´s laws of friction. Obviously this study influenced Coulomb's investigations. As is shown in next section, Coulomb introduced a great deal of rigour into his study of friction forces.

With respect to the force necessary to bend a rope round a cylinder, Amontons showed that it is inversely proportional to the radius of the cylinder and directly proportional to the tension and the diameter of the rope.

SOME CONSIDERATIONS ON *"THÉORIE DES MACHINES SIMPLES"*

The complete title of Coulomb's major work is: *'Théorie des Machines Simples, en ayant égard au Frottement de leurs Parties et a la Roideur des Cordages"*,[3]. It has 254 pages and is divided into two parts. The first part, which we are particularly concerned, has 99 pages and is subtitled: *"Du Frottement des Surfaces Planes qui Glissent l'une sur l'autre"*. It is divided into two chapters: Chapter I is entitled *"Du premier effort pour vaincre le frottement, ou pour faire glisser une surface après un temps de repos donné"*, and Chapter II *"Du frottement des surfaces en mouvement."* Obviously chapter one deals with static friction and chapter two with the dynamic case. With respect to the first case Coulomb discuss the friction dependence of four types of causes:
1) The nature of surfaces in contact and their surface finishing
2) Surfaces extensions
3) The pressure between the surfaces in contact
4) The elapsed contact time between the surfaces

The experimental apparatus used by Coulomb is shown in Fig. 2 and is suitable for sliding pairs of surfaces by means of a weight and a cable passing through a pulley. The table, which supports the moving parts, is stiff enough to maintain the whole system isolated from the ground. Coulomb made the following remarks after three tests:

"We have seen in previous tests that friction resistance was smaller after one second in contact than after one or two minutes; but after one or two minutes, friction reached the maximum possible value. After this result we tried to determine a rate between pressure and friction as soon as friction limit, or the maximum growth of friction, was reached; we obtained the following values for this ratio:

First test ------------------ 74/30 --------------------- 2.46

Second test ---------------- 877/406 ------------------ 2.16

Third test ------------------ 2474/1116 --------------- 2.21

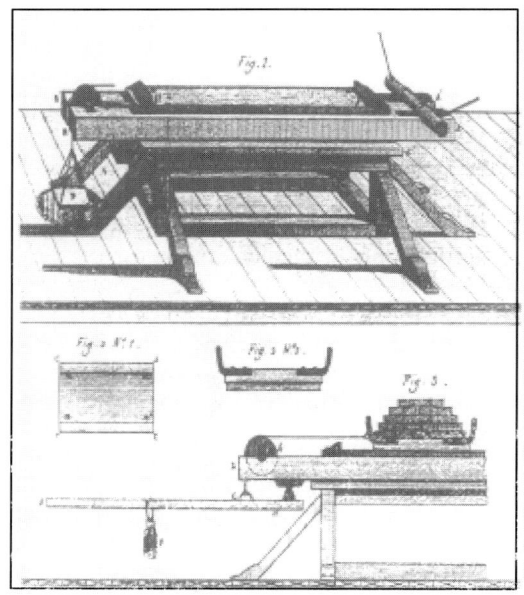

Fig.2: Coulomb's experimental apparatus

As the tests indicated, an approximate contact rate between pressure and friction force is obtained. This in spite of large differences in pressure, while I have tried to reduce the area between the surfaces to verify if this ratio is still the same."

The three tests were made with samples of wood without any kind of lubricant between the pieces in contact and changing the dimensions of the samples, the contact pressure and the elapsed time at rest. The numerical values showed in the experiments refer to the weights of samples in pounds.

Looking at the dynamic case, the same experimental apparatus was used, the contact areas were reduced and after several tests Coulomb concluded that friction is independent of relative velocities between the contact surfaces, as well as the magnitude of their contact areas.

In chapter three Coulomb summarizes the results of these investigations as follows:

1) With two pieces of dry wood in contact, after some time together without relative motion, the resistance increases significantly at the beginning before the relative motion arises; after some minutes the friction reaches its maximum, which is the limit for resistance.

2) Since relative motion initiates with two pieces of dry wood, friction is still proportional to pressure; however, the intensity of friction is smaller than in the static case.

3) If we look at metal surfaces sliding in relative motion without any lubrication material, friction is proportional to pressure; but the intensity of friction is the same, irrespective of how the motion is obtained after some time at rest.

4) Heterogeneous materials in relative motion such as wood and metal without any lubrication substance lead to different results; friction intensity in relation to time at rest increases slowly and reaches a limit after four or five days.

This summarizes the well-known laws of friction, laws as we know them nowadays. In the second part of his book, Coulomb studies the forces needed to bend ropes and friction in rotating shafts. The experimental structure consists of two suspended weights on each side of a pulley. In this type of device forces to compensate friction effects are added to forces to bend rope elements. Obviously it is necessary to separate both effects. Coulomb uses the same method and the experimental apparatus used by Amontons. In Fig. 3 Amonton's machine is shown. Other investigations were performed as that carried out by Desagulliers (1683-1743) and published in his "Cours de Physique Experimentale", 2 Vol., appeared in 1751. Basically it was the same experiment as Amontons and Coulomb conducted.

Coulomb generalized the effect of stiffness of ropes by means of the formula:

$$\frac{A+BT}{R}$$

Where $A = h r^q$, $B = h r^s$, R is the radius of the pulley, r the radius and T the tension of the rope. The exponents q and s are approximately equal.

Other generalization made by Coulomb refers to the force T to move a weight P along a horizontal plane:

$$T = A + P/u$$

Where A is a small constant depending of the "coherence" of the surfaces and u is a coefficient that is the reciprocal of the coefficient of friction, which is now commonly used. This coefficient depends on the nature of surfaces. For a more general situation, Coulomb calculated the force necessary to hold a body on an inclined plane, [9].

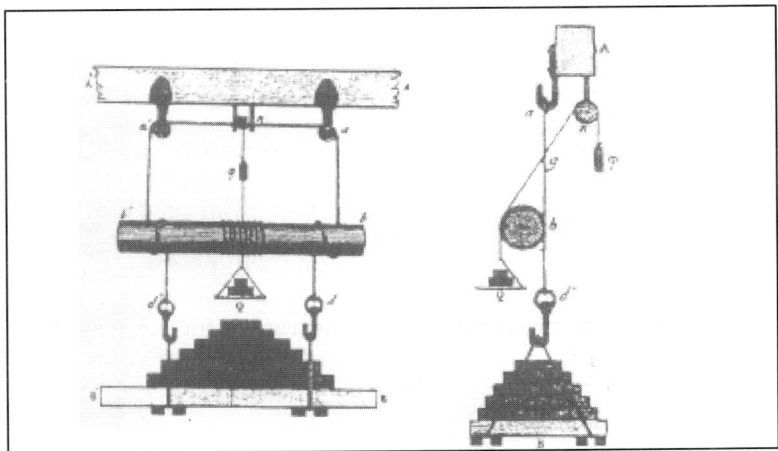

Fig.3: Amontons's machine

COULOMB AND APPLIED MECHANICS

Applied mechanics and further developments of industrial mechanics are a result and a heritage of the mainly French polytechnic engineers, associated with the knowledge of machine constructors since the end of Middle Ages in Europe. The successive creation of engineering schools in France, such as the School of Bridges and Highways (1747), the Royal School of Mézières (1748), the School of Mines (1781) and finally the Polytechnic School (1794), provided a homogeneous and a high level scientific education. In this context, the contributions of Coulomb and Lazare Carnot appeared as fundamental for the development of a new science of machines. As mentioned above, in addition to his proposed laws of friction, Coulomb solved several other important problems. Carnot created the first general theory of machines, thereby providing a tool to analyze any kind of machine.

Coulomb is also responsible by the introduction of economic studies in relation to machines and can be considered as a precursor of industrial mechanics. This task was done by mechanically measuring the human body as a kind of machine in order to measure economically a machine. The memoir *"Sur la Force des Hommes"* is an attempt to understand human work mechanically and is the first publication on physiology and ergonomics, [6, 8]. The question introduced by Coulomb appears from a pragmatic point of view. He proposed to measure the quantity of action (work) that a man can expend in a day of work by different ways of applying the forces, [6]. To do this, Coulomb studied an old problem that is the human mechanical capacity, also studied by Desagulliers and Daniel Bernoulli (1700-1782).

The heritage of Coulomb and Carnot was subsequently developed and transformed into applied and industrial mechanics by the French Polytechnics, Navier (1785-1836), Coriolis (1792-1843), Poncelet (1788-1867), Augustin Cournot (1801-1877), etc, [7]. The most important work in the sense of a transition to applied mechanics came up from Coriolis, in his book: *'Du calcul de l'effet des machines'*, published in 1829. In the twenty-five years which separate it from Carnot's *"Principes"*, no important intermediate work was published, except for the notes and additions made by Navier to the *"Architeture Hidraulique"* by Belidor (1693-1761). Navier is also important because of his introduction of economic ideas to study man-machine relationships. He uses the concept of work (the product of a force by a displacement) as a machine production measurement, attributing to work in this sense, the name of "mechanical currency", [6, 10].

Corioli's book is a complete manual on mechanics where the problems raised by the construction and use of machines are treated with rigour and theoretical refinement. The results of mechanics and differential and integral calculus are used with great skill.

Besides Coriolis, Poncelet's works represent the biggest effort towards effectively consolidating the discipline of applied mechanics. In 1825 he was admitted as a professor in Metz and there, using the work of Navier and Coriolis, he gave the first course on "Applied Mechanics to Machines", which would be transformed into a book in 1826. In 1829, Poncelet published the "Course of Industrial Mechanics", which after his death became an important reference and fundamental textbook.

Augustin Cournot is known as a founder of the probabilistic thought and also a precursor of the mathematization of economic theory. Before attempting to produce a broad synthesis of contemporary knowledge, he worked on mathematics and mechanics, being a disciple of Poisson. In 1834, he translated both Herschel's *"Treatise on Astronomy"* and Lardner's *"Elements of mechanics"* into French, [7]. After this period as mathematician and expert in mechanics, he moved into the application of probabilistic methods to natural sciences. Finally he dedicated himself to the philosophical and economics analysis of machine science.

CONCLUSION

Coulomb's work is closely related to the beginning of Polytechnic School education, especially the improvement in and new kind of teaching of the machine sciences. This change was conceived by Gaspar Monge (1746-1818) and his disciple Pierre Hachette (1769-1834).

In 1821, Coulomb's mechanical studies were published, including the *"Memoire sur la force des hommes"*. By introducing the question of human work as a machine operation, it was possible to study the work carried out by a machine economically. This double analogy created the conditions to construct the concepts of net work and global work applied to a given machine in order to measure its efficiency, costs, energy consumption, etc., providing the necessary theoretical tools to compare two different machines.

Coulomb's investigations of friction forces were of decisive importance for engineering and physics because all types of movement inside a machine are associated with friction and a loss of energy. In other words, the mechanical energy transformations by parts of a machine and its communication of movements degrade that quantity yet unknown to Coulomb - energy. It was only in 1847 that the principle of energy conservation was discovered by several independent investigations, [14, 16]. This fundamentally important achievement for the history of science did not appear in a strictly mechanical field. It was necessary to achieve a new synthesis of scientific knowledge by means of modern developments in electricity, magnetism, optics, heat, etc. This new situation also represents the birth of a new machine science: thermodynamics, [15].

Another remarkable consideration concerns Coulomb's method. It is, in fact, a combination of in-depth experimental research with the application of mathematics, mainly differential and integral calculus, to generalize the results obtained experimentally, by means of simple algebraic formulas. This was done for example for friction in bending of ropes and in rolling. From examination of many physical parameters, he developed a series of two-term equations. The first term a constant and the second term varying with time, normal force, velocity, or other parameters, [9].

The famous scientist Biot wrote: "It is to Borda and to Coulomb that one owes the renaissance of true physics in France, not a verbose and hypothetical physics, but that ingenious and exact physics which observes and compares all with rigour".

REFERENCES

1. O'Connor J. J., and Robertson E. F., Charles Augustin Coulomb, article, 2000.
2. Falomo L., Bevilacqua F. and Montalbetti C., Charles Augustin Coulomb, article, 2000

226

3. Coulomb C. A., Théorie des machines simples, Librairie Scientifique et Technique Albert Blanchard, Paris, 2002.
4. Oliveira A. R. E., Lazare Carnot and his general theory of machines, ABCM review, 2003.
5. Oliveira A. R. E., A short history of applied mechanics, COBEM, Brazilian Congress of Mechanical Engineering, S. Paulo, 2003.
6 Vatin P., Le travail: economie et philosophie- 1780-1830, PUF, Paris, 1993.
6. Vatin P., Économie politique et économie naturel chez Antoine-Augustin Cournot, Presses Universitaires de France, Paris, 1998.
8. Vatin P., Le travail, sciences et societé, Editions de l'Université de Bruxelles, 1999.
9. Dugas R., A history of mechanics, Dover Publication, N. Y., 1988.
10. Daumas M., Histoire general des techniques, Quadrige/Presses Universitaires de France, 1996.
11. Lagrange J. L., Mécanique analytique, Chez la Veuve Desaint, Librairie, Paris, 1788.
12. Carnot L., Príncipes fondamentaux de l'équilibre et du mouvement, de l'imprimerie de Crapelet,1803.
13. Charnay J. P., Lazare Carnot ou le savant citoyen, Presses de l'Université de Paris - Sorbone, Paris, 1990.
14. Kuhn T. S., La tension esencial, Fondo de Cultura Economica, Mexico, 1996.
15. Locqueneux R., Pre-histoire & Histoire de la thermodinamique classique (Une histoire de la shaleur), Societé Française d'Histoire des Sciences et des Techniques, Librairi A. Blanchard, Paris, 1996.
16. Paty M., A matéria roubada, Editora da Universidade de S. Paulo, 1995.
17. Ashby E., Technology and the academics, Macmillan, St. Martin' s Press, New York, 1996.
18. Rashed R., Science a l'époque de la Revolution Française, Librairie Scientifique et Technique, Paris, 1988.
19. Séris J. P., Machine et communication, Librairie Philosophique J. Vrin, Paris, 1987.

6. Mechanical Designs

Zrnic N., Hoffmann K.: *Development of Design of Ship-To-Shore Container Cranes: 1959-2004*

Strizhak V., Penkov I., Pappel T.: *Evolution of Design, Use and Strength Calculations of Screw Threads And Threaded Joints*

Bautista Paz E., Munoz Sanz J.L., Leal Wina P., Echavarri Otero J.: *Industrial Archaeology: Form the 17-th to the 21-st Century. Reconstruction of Herschel's Telescope*

Mauersberger K.: *Piston Engine versus Rotary Engine – Possible Structural Solutions in the Invention and Development of the Steam Engine before 1900*

DEVELOPMENT OF DESIGN OF SHIP-TO-SHORE CONTAINER CRANES: 1959-2004

Nenad Zrni ü
University of Belgrade, Faculty of Mechanical Engineering, Dep. of Mechanization
11000 Belgrade, 27 marta 80, Serbia and Montenegro
E-mail: nzrnic@mas.bg.ac.yu
Klaus Hoffmann
Vienna University of Technology, Faculty of Mechanical Engineering
Inst. for Engineering Design and for Transport, Handling, and Conveying Systems
A-1060 Wien, Getreidemarkt 9, Austria
E-mail: hoffmann@ft.tuwien.ac.at

ABSTRACT- The paper presents the historical development of mechanical and structural design of ship-to-shore (STS) container cranes, from 1959, when the first crane was built, up to now. The paper gives a short survey of the evolution of the container crane industry, the state of the art in modern container cranes, and focuses particular attention on mechanical design of trolleys and evaluation of the existing structures. The analysis of historical development and state of the art in modern container cranes enables us to analyze future trends in mechanical and structural design.

KEYWORDS: History of STS container cranes, design, trolley, construction

INTRODUCTION

The method of handling ship cargo in the early 1950s was not very different from that used during the time of the Phoenicians, Figure 1 [6]. The time and labor required to load and unload ships increased substantially with the size of the ship, requiring more time in port than at sea, Figure 2 [6]. The problem "how to lift a load" is as old as humankind. From the earliest times people have faced this problem. The first written information on the use of hoisting mechanisms appeared around 530 BC, mainly concerning the construction of the first temple of Artemis in Ephesus [13]. The forerunners of modern cranes in the ports were the wooden slewing cranes developed in the middle Ages, Figure 3 [20]. The slewing level luffing crane was the main means of cargo loading and unloading between ship and shore up to the end of 1950s, Figure 4 [20].

Through the 1950s, general cargo continued to be handled as break-bulk (i.e. palleted) cargo. Pallets were moved, generally one at a time, onto a truck or rail car that carried them from the factory or warehouse to the docks. Each pallet was unloaded and hoisted, by cargo net and crane, off the dock onto the ship. Once the pallet was in the ship's hole, it had to be positioned precisely and braced to protect it from damage during the voyage. This process was then reversed at the other end, making the marine transport of general cargo a slow, labor-intensive, and expensive process. All this began to change in 1955. Believing that individual items of cargo needed to be handled only twice - at their origin when stored in a standardized container box and at their destination when unloaded, Malcolm McLean purchased a small tanker company, renamed it SeaLand, and adapted its ships to transport truck trailers. The first voyage of a SeaLand container ship commenced on April 26, 1956 between Newark, New Jersey, and Puerto Rico. In the years that followed, standardized containers were constructed, generally twenty or forty feet long without wheels, having locking mechanisms at each corner that could be secured to truck chassis, a rail car, a crane, or to other containers inside a ship's hole or on its deck. The idea of shipping cargo in locked containers has been widely accepted, resulting in an uninterrupted worldwide growth of about 8% a year at the beginning of this century [16], [21].

Fig.1: Beginnings of handling ship cargo.

Fig.2: Past of handling ship cargo.

Fig.3: Wooden slewing crane.

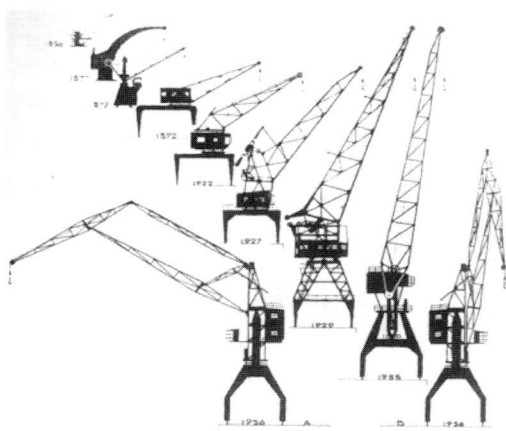

Fig.4: Development of slewing cranes 1856-1956.

One of the major problems facing the containerization concept was that during the mid-Fifties most ports were not equipped to handle the heavy containers except by mobile revolving cranes and even then many of the cranes did not have the capacity to lift containers. These cranes were extremely inefficient, in that at least two to three minutes of loading cycle was lost due to poor control at the points of pickup and discharge. In July 1957 the engineering staff of Matson company, under the leadership of Mr. Les Harlander, commissioned a study of existing crane types, to determine the state of the art and identify the type which best met the general requirement for loading containers between ship and shore and keep turnaround time of container ships to a minimum. The study concluded that no crane then on the market satisfactorily filled this requirement, and that an ore-unloading type crane with a horizontal boom and through-leg trolley came closest to meet this requirement. Early in 1958, performance specifications were finalized and put out for bid [6].

PACECO, one of eleven biddres, pioneered the first container crane for Matson in 1958 [4]. PACECO philosophy was that the best design has the fewest number of pieces, and developed the conceptual drawings paying particular attention to aesthetics. Trusses, used at that time by most manufacturers, were replaced with all-welded box girders wherever possible. This resulted in unique and extremely clean-looking A-frame configuration. The A-frame gantry crane takes its name from it's "A"-shaped noticeable when observed from the bridge. On 7 January 1959 the world's first container crane was put into service at the Encinal Terminals in Alameda, California, Figure 5. The original container crane set the standards for dozen of manufacturers around the world. Although there have been many significant improvements, all modern cranes are direct descendants of this first crane, and the design of later cranes has reminded relatively unchanged [6].

In the meantime, to keep up with the growth of container traffic, container ships and container cranes are getting bigger and bigger. During the first 45 years since the first container crane was designed, the size of container cranes and lifting capacity have more than doubled, Figure 6 [5].

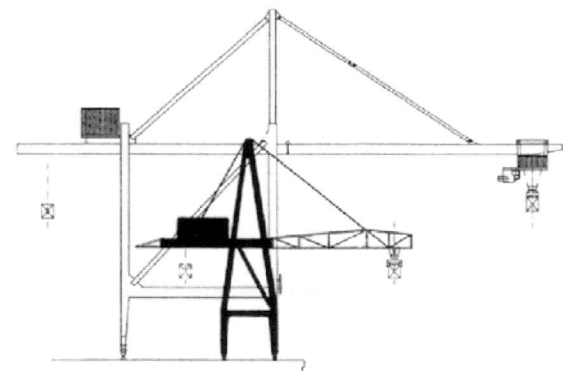

Fig.5: First container crane.   *Fig.6: Growth of container cranes 1959-1995.*

EVOLUTION OF CONTAINER CRANE INDUSTRY

At these early 1960s a crane container purchaser could call up a PACECO (at that time a US company) representative in any part of the world and order STS cranes. Around the same time, European manufacturers entered the market, offering improved and standard design, good quality and competitive prices. The Japanese entry into the container crane industry in the late 1960s presented some opportunities and, for the first time, some challenges for the purchasers. Japanese consumer and industrial products were introduced to the world market at significantly lower prices. The first reaction in the Western world was "No such thing as a free lunch – you will pay for it one way or another" [4]. But, some large shipping lines saw opportunity in the new competition and decided to give the Japanese a try. The Japanese provided good quality cranes, and within a short time they were in the same category as the Americans and Europeans. With the growth of the economy and domestic demand increasing, Japanese cranes ceased to be a bargain. The concept of purchasing cranes with the "Tailor-Made" specification was born. The "Tailor-Made" philosophy requires the detailed performance specifications and very competitive bids. A larger number of cranes brings economy of scale and favors this concept. Tailor-made cranes require high expertise, whether in-house or through outside consultants [4].

Korean STS container cranes manufacturers were next with the lower-priced cranes, starting from 1970s. Now, at the beginning of 21st Century the Chinese manufacturer ZPMC, Shangai, has made the most significant impact on the container crane industry, and may further increase its share of the world market during the coming years. The actual situation is that PACECO could no longer compete with the overseas suppliers, some European suppliers have merged or quit manufacturing STS cranes, and the Japanese suppliers retracted from the international market and remained focused on their protected domestic market. To compete in the current market, established cranes builders have shifted fabrication and assembly to remote plants with cheaper labor, purchased electrical components and integrated them in-house, used standardized components and reduced profit margins. The use of standardized components favors the "Off-the-Shelf" approach in design of STS cranes. The aim of this strategy is to emulate the early purchasing strategy of issuing a brief technical outline and inviting proposals from two or more crane suppliers [4].

DEVELOPMENT OF DESIGN OF STS CONTAINER CRANES

The basic structural form of STS container cranes (A-frame) remains practically unchangeable compared with the first structure built in 1959. Of course some modifications are done. The basic structural shape of STS cranes can be divided into two groups:

1. Conventional or modified conventional cranes, Figure 7 [15];
2. Low profile cranes, Figure 8 [1].

Conventional STS cranes are used for servicing the following ships:

1. **Panamax ships:** III-generation ships with a beam of less than 32.3 m (width of Panama Canal) – their structure was developed from end of 1960s up to early

1980s, and they operate with up to 13 containers abeam on deck and with maximum capacity of 4,700/4,900 TEU (Twenty Equivalent Unit);

2. **Post-Panamax ships:** IV-generation ships, developed from 1984, whose beam is greater than the width of the Panama Canal – they operate with up to 20 containers abeam on deck and with a maximum capacity of 7,000 TEU;

3. **Mega ships or Jumbo ships:** Ships of the most recent generation, pioneered in the last years of the 1990s, they operate with more than 20 containers abeam on deck and with a capacity of more than 7,000 TEU.

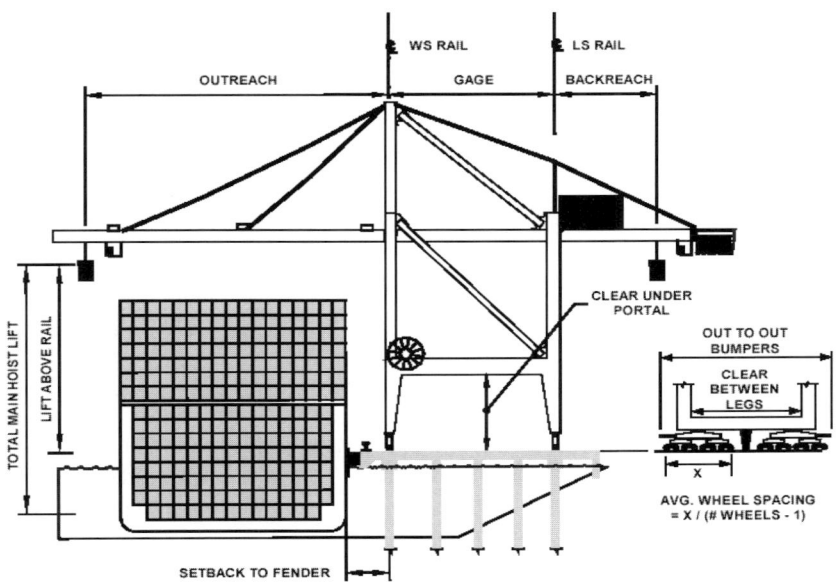

Fig.7: Conventional Post-Panamax container crane.

Fig.8: Low profile container crane.

During this time there was an incredible growth of overall dimensions of STS cranes, and an increase in drive performances such as speed, acceleration, etc [21]. The outreach of the first container crane was 23.8 m (70 ft), and the gauge was 15.24 m (50 ft). The later Panamax cranes reached the gauge of 30.48 m (100 ft), maximum outreach 36/42 m, trolley travel 120/150 m/min, and hoisting with rated load 20/50 m/min. The state of the art of modern Post-Panamax and Mega cranes is given in Table 1 [15, 17, 18].

Features of Post-Panamax and Mega cranes	16 wide	18 wide	20 wide	22 wide
Gantry rail gage (m)	30.48	30.48	30.48	30.48
Clear between legs (m)	18.3	18.3	18.3	18.3
Lift above rails (m)	34	34	36	≥36
Total main hoist lift (m)	50	52	54	≥60
Clear under portal (m)	12	12-15	15	12-18
Out-to-out bumpers (m)	27	24-27	24-27	24-27
Outreach from waterside rail	45-47	50-52	56	≥60
Hoisting with rated load m/min	50	60	75	70-100
Hoisting with empty spreader m/min	120	130	150	180
Trolley travel m/min	200	245	245	250

Tab.1: State of the art of Post-Panamax and Mega cranes.

The conventional and modified A-frame crane with single trolley and one operator is still the workhorse of container cranes industry [7]. In the last two decades there have been many attempts to improve the productivity of cranes by introducing the second hoist, or second trolley [23]. Dual hoist cranes have a second hoist located over the quay. They are conceived in the early 1980s, first by ECT (European Container Terminal) Rotterdam, and then by Virginia Port Authority [11]. These cranes can increase productivity by about 50%, but also increase the initial costs by 30 to 50%, require an additional operator, and increase operating costs. In practice they were not economic, but they may be making a comeback, as only one operator is needed on the ship trolley, and the second trolley may be fully automated. The newer solution of dual hoist crane combined the solution of dual hoist crane and elevating platform, Figure 9 [14]. Dual hoist elevating cranes are dual hoist single cranes, but the shuttle runway elevates to the ideal elevation. These cranes cost more than dual hoist cranes but have higher productivity [14].

The double trolley also boosts productivity of the crane, Figure 10 [8]. The suggestion is that two trolleys operate on the same runway. The SeaLand Ansaldo cranes in Taiwan are designed to carry two trolleys. For double trolley operations, the landside trolley transfer containers from the landside half of the ship to extreme landside lanes, and the waterside trolley operates from the waterside half of the ship to the waterside lanes. Boom deflections could cause problems when two trolleys are operating simultaneously over the ship. Since a trolley should not carry over the personnel, some of the transfer vehicles need to wait for the waterside trolley to pass. This reduces production, but not significantly. The second trolley requires a second operator. Nowadays, a dual hoist crane will be more productive than a double trolley crane. The initial and operating costs of two cranes solutions are the same, so there is no apparent advantage in the double trolley crane [8].

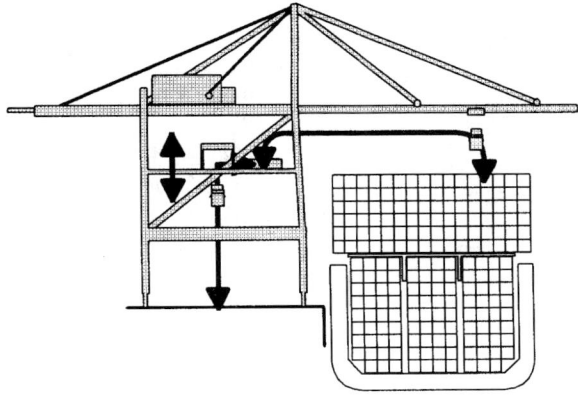

Fig.9: Dual hoist elevating cranes.

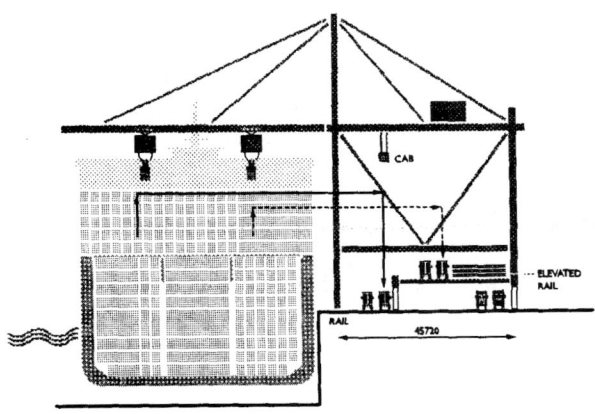

Fig.10: Double trolley cranes.

The first low profile cranes emerged in the late 1960s. The boom is supported by a hanger system including support truck and wheels; it rides on the wheels at all times. For the PACECO standard, all machinery and electrical equipment is on the boom. The machinery travels with the boom and causes significant increases in the wheel loadings [12]. Low profile cranes cost about 15% more than A-frame cranes and were used when it was necessary to restrict the height of the crane (under max. 46 m), i.e. because they have the advantage of keeping cranes profile below aircraft clearance lines, Figure 8. During the period 1970-1999 approximately 40 of these cranes were built, sized to service vessels between 13 and 16 containers wide. Except for the US, they are now found only in Italy.

DEVELOPMENT OF MECHANICAL DESIGN OF TROLLEY

The selection of a crane's trolley system type is significant for the structure of crane, for wheel loads, and for maintenance considerations. The trolley can be rope towed (RTT) or machinery type (MOT). A hybrid of the two systems, commonly known as a fleet-through machinery trolley (or semi-rope trolley [9, 22]), was adopted by some manufacturers. For the fleet-through machinery trolley the main hoist machinery is placed on the gantry frame, but the trolley is self-driven [5]. Since the machinery in the machinery house tow the trolley and hoist the load by a system of wire ropes, this system is called a RTT system. The main hoist ropes run from the machinery house to the landside of the crane, through the trolley and head block, and usually dead end at the waterside tip of the boom, Figure 11 [2].

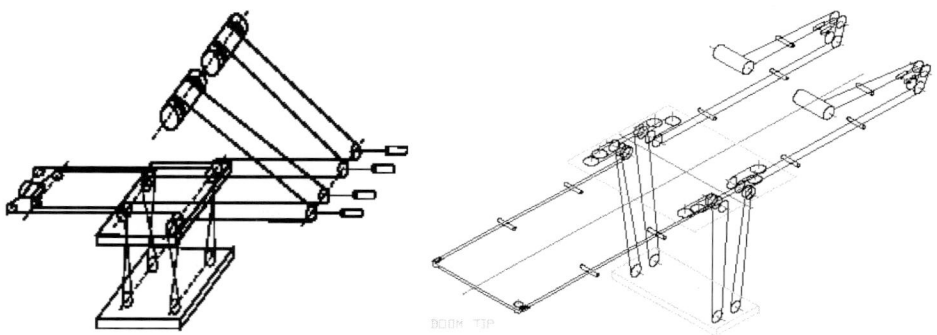

Fig.11: Main hoist reeving, RTT system.

The trolley tow ropes run from the machinery house to the sheaves at the landside of the crane, through the trolley to the tip of the boom, and back to the house, Figure 12 [2]. This arrangement allows the trolley to be shallow and lightweight, permitting greater lift height and smaller loads on the crane structure and wharf [15].

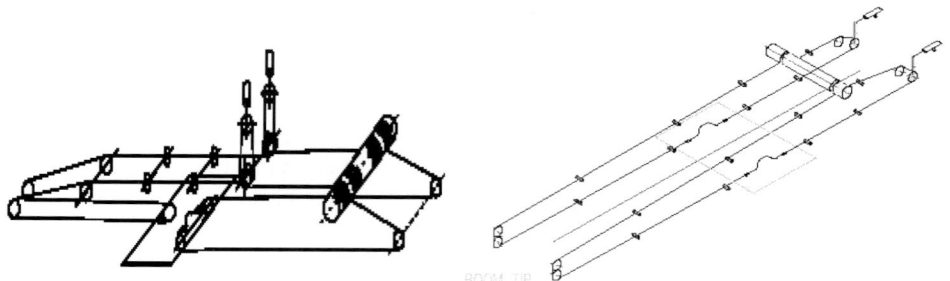

Fig.12: Trolley reeving, RTT system.

With the RTT design, there was concern that the rope would stretch and that catenary effect would reduce productivity. The auxiliary catenary trolley is the typical solution for reducing the catenary effect due to greater outreach of modern cranes and longer trolley travel, Figure 13 [2]. But even with the catenary trolley, the long runway will result in a significant catenary effect [3].

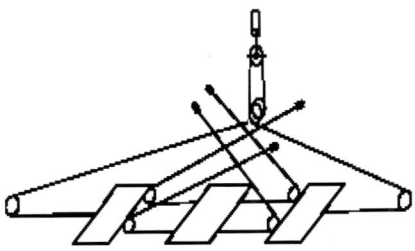

Fig.13: Catenary trolley.

A machinery trolley (Figure 14, [10]) has the trolley and main hoist machinery on board. No trolley drive ropes are required, and the main hoist ropes (Figure 15, [2]) are shorter than for a rope-towed trolley [15].

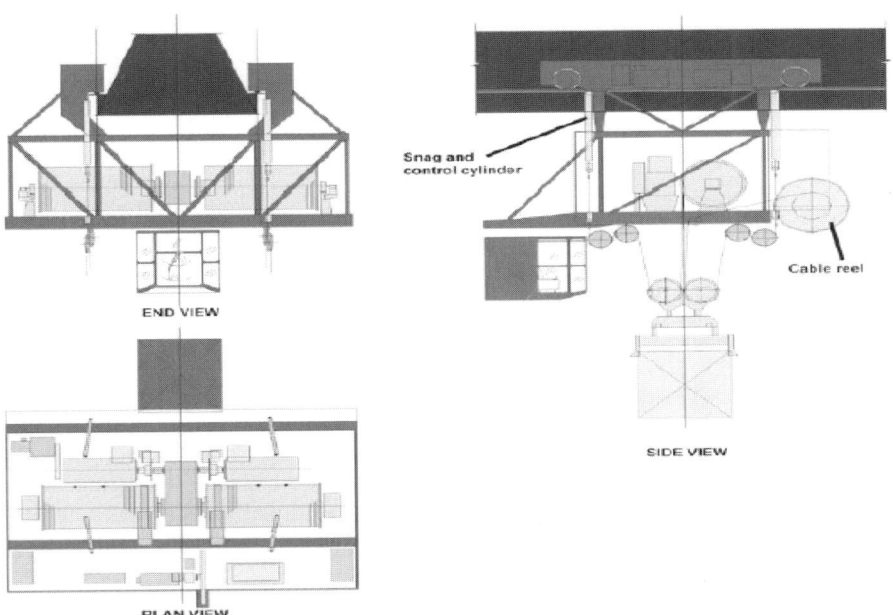

Fig.14: Construction of MOT.

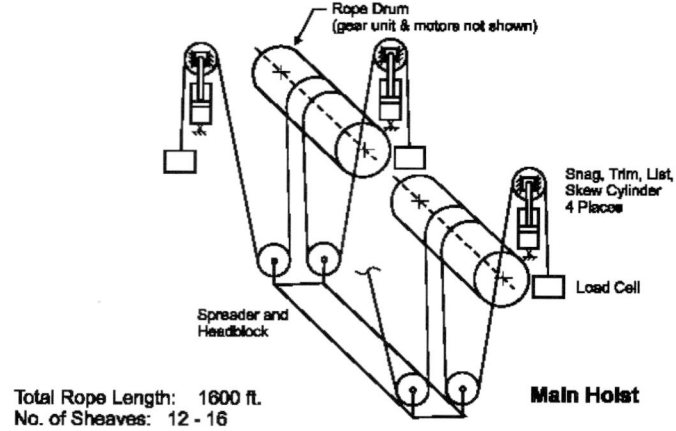

Fig.15: Reeving diagram for MOT.

For the structural design, weight is the main difference between the two types of trolleys. The weight of the rope-towed trolley is approximately one-third that of a machinery trolley. The main disadvantage of the machinery trolley is the increase in crane weight and wheel loads on the wharf.

The first dockside container handling crane, built by PACECO in 1959, utilized a RTT. This crane was conceptualized for operation on the existing wharf. This was common in 1959 [15]. The minimization of wharf loads was the primary factor favoring the RTT cranes for other existing facilities. This design became the model for the next generation of container handling cranes for most ports in the world, except for a few European ports. These RTT cranes have provided excellent service over the years and, with proper maintenance, have exhibited high reliability. However, the resulting maze of ropes, sheaves and trolleys has become complex. Some European crane manufacturers adopted the MOT design concept for most of their cranes. Kocks introduced the first European machinery trolley container crane in 1968 and have since used the same basic philosophy for most of their cranes worldwide. Another German manufacturer, Noell, later introduced their machinery trolley cranes [15].

Less than 10 years ago, American President Lines (APL) concluded negotiations with the Port of Los Angeles for construction of a new port facility. APL developed a new Post-Panamax crane specification written for the traditional RTT design. Bids were received from many international crane manufacturers. But one manufacturer, Noell, offered an MOT-type design. At first, APL did not consider changing from the traditional RTT – perhaps this was because the traditional design was familiar and worked well. But on second sight, the design began to intrigue the evaluation team. APL's in-house evaluation team looked at it this way [2,5]:

If no container cranes had ever been built and there were no dock wheel load constraints, would the team recommend a crane with a RTT system or as a MOT? Interestingly, the group was inclined toward the MOT. As a next step, APL invited a small group of crane experts to join the evaluation team. The team was asked to keep in mind the basic criteria: that no container crane had ever been built, no design constraints exist, and the goal is to optimize efficiency, reliability and maintenance of the crane. The interesting result was how easily the group concluded that, with these criteria, the MOT design was the logical choice. Prior to making the final decision, the APL team visited sites where Noell had installed cranes of similar design. Why was the MOT system chosen?

1. Depending on the design, approximately 1,650 m of wire rope is eliminated from the main hoist, trolley drive, and catenary trolley.
2. Approximately 36 sheaves of various sizes are eliminated.
3. Hydraulic rope tensioning devices are eliminated.
4. The spare parts inventory is reduced.
5. The intensity of maintenance is reduced.
6. Up-time reliability is increased because of the reduced number of crane components.
7. Wire rope lubrication is reduced.

Table 2 [15, 22], presents a comparison of both mechanical systems.

	Rope-towed Trolley RTT	Machinery On Trolley MOT	Advantage
Reeving Assemblies	Main Hoist, Trolley Drive Catenary Trolley Drive	Main Hoist	MOT
Trolley Positioning	Movement due to trolley travel rope stretch	Movement due to skidding	MOT
Festoon	Spreader power only	Power for main hoist (including trim, list, skew, and snag device), trolley drive, and spreader	RTT
Trolley Accelerations	0,6 m/s^2	0,6 m/s^2	---
Rope Lubricant	Exposed to environment. Oil spillage on ground.	Enclosed, spillage contained	MOT

Tab.2: Comparison of mechanical systems of trolley.

AUTOMATION OF STS CONTAINER CRANES

The development of efficient, automated, high-technology loading/unloading equipment has the potential of considerably improving the performance of terminal operations. The construction industry is relatively still slow in implementing advanced technology to improve safety. Current practice requires that control of the STS cranes dynamic behavior is the responsibility of a skilled operator. The operator applies corrective measures based on experience when any undesirable swaying is detected. The absence of automated sensing and control not only leave room for accidents arising because of human error and/or delayed response of the operator, but also can greatly reduced the productivity of the crane's operation, also as the productivity of a whole port terminal. There is also a potential danger of an exaggerated response, which will lead to an uncontrollable load swing. The biggest source of dynamic forces is the pendulum motion of the loaded spreader suspended by cables. Advances in STS container cranes

240

technologies have a significant effect on the efficiency of port terminal operations once properly implemented. Precise control of the spreader and load is only possible using mathematically correct algorithms, and properly implemented sensor systems. For the years many authors have been researching and developing ways to make STS cranes transfer containers faster and more safely through computerized anti-sway and automatic controls. Automation continues to evolve and will continue to improve productivity of STS cranes. The fact that the cranes should be quicker, larger and more efficient, force both the manufacturer and terminal operator to incorporate to the equipment some automation for the repetitive process of handling containers. Automation is also another important aspect of container crane becoming a conglomerate of sophisticated elements of high added value consisting of specialized software and hardware. The pressure on the port terminal by the shipping companies to release vessels as fast as possible is used by port operators to specify that the cranes shall be supplied with an antisway system. The behavior of the crane with antisway device is completely different from that the cranes without it. The outline of the classification of existing STS cranes due to their degree of automation is shown in Figure 16.

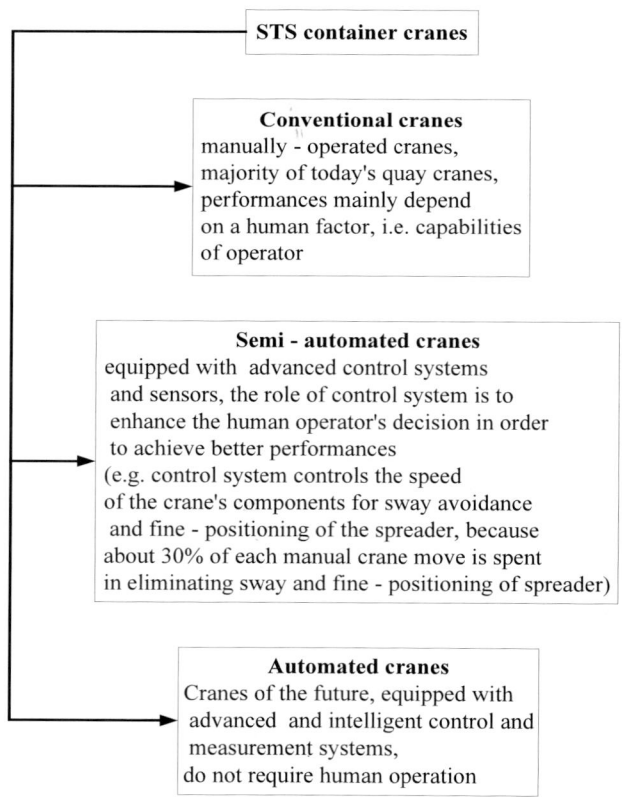

Fig.16: Classification of STS container cranes by their degree of automation.

CONCLUSION

The future of STS cranes starts today and will last until 2030, what is the reasonable life of a container crane. The analysis of the past and state of the art of STS container cranes may help us to predict the future trends in their development. New cranes should be designed that are big and fast enough to keep up with the demands of new larger ships. Some expected performances of future cranes are as follows [10,19]:

1. **Crane rail gage:** 30,48 m (100 feet); although there are some good arguments for increasing the gage to as much as 45,72 m (150 feet), the cost of shipping the erected crane will be much greater.

2. **Lift above rail:** 39,92 m (131 feet) now, 47,55 m (156 feet) future. The higher the trolley is above the wharf, the more difficult is to control the load. Therefore, the current height should be kept to a minimum.

3. **Outreach:** 64/69,2 m (210/227) feet; the increase of the outreach will follow the increase in the size of ships. Current New Standard Maersk crane orders provide for vessels with 22 containers on deck. Ships with 23 containers on deck are referred to as Suezmax, and with 24 containers across as Malacca-max.

4. **Trolley travel speed:** 250/300 m/min, or even more. In 2002 ZPMC made fundamental improvement on the trolley starting up, braking and traveling, and the trolley speed has successfully reached 350 m/min, instead of previous 240 m/min. This crane is now undergoing industrial tests.

5. **Trolley tipe:** Both types of trolley systems, RTT and MOT have beneficial site-specific applications, and are viable. For each crane purchase, the owner have to evaluate each design and then choose the design which best suits the site and the all-round operational needs. For large super productive cranes, the MOT will be the choice of the future.

An effective crane must be designed to suit the present and future needs of the end user. But if today's crane is built large enough to serve tomorrow's ships using future technology, the crane will not perform well on today's ship with today's technology. For years design team members - mechanical, structural, automation, and electrical engineers have worked together to produce economical design that meet operational demands and that can be efficiently fabricated and erected. As always, the "best" ("optimum") design requires balance. The cost and benefits of each alternative should be considered in respect of the specific case.

REFERENCES
1. Bhimani, A. K., Low Profile Cranes - New Applications, AAPA Facilities Engineering Seminar, American Association of Port Authorities, Baltimore, Maryland, USA, April 28-30, 2003.
2. Bhimani, A. K., Hoite, S., Machinery trolley cranes, Proceedings of the Conference "PORTS '98", American Society of Civil Engineers, Long Beach, CA, USA, 1998, pp. 603-613.
3. Bhimani, A. K., Hsieh, J. K., Cranes to serve ship in the slip Ceres Paragon terminal Amsterdam, Proc. of the Conference "PORTS '01", ASCE, section 30, chapter 3, Norfolk, VA, USA, 2001.
4. Bhimani, A. K., Jordan, M. A., Crane Purchase Specifications: Tailor-Made or Off-The-Shelf, Terminal Operations Conference "TOC'01", Lisbon, Portugal, June 20, 2001.

5.   Bhimani, A. K., Kerenyi, J., Rope towed or machinery trolley – which is better?, AAPA Facilities Engineering Seminar, American Association of Port Authorities, San Pedro, CA, USA, November 15-17, 1995.

6.   12th International Historic Mechanical Engineering Landmark: PACECO Container Crane - The World's First High Speed, Dockside, Container Handling crane, American Society of Mechanical Engineers, Encinal Terminals - Alameda, California, USA, May 5th, 1983.

7.   Jordan, M. A., Dockside container cranes, Proceedings of the Conference "PORTS '95", American Society of Civil Engineers, Tampa, Florida, USA, 1995, pp. 826-837.

8.   Jordan, M. A., Super productive cranes, Terminal Operations Conference "TOC '97", Barcelona, Spain, June 3-5, 1997.

9.   Jordan, M. A., Purchasing cranes in a changing world, Proceedings of the Conference "PORTS '98", American Society of Civil Engineers, Long Beach, CA, USA, 1998, pp. 591-602.

10. Jordan, M. A., Future-Proof Your Crane, Terminal Operations Conference "TOC Americas '01", Miami, FL, USA, October, 2001.

11. Jordan, M. A., Quay crane productivity, Terminal Operations Conference "TOC Americas '02", Miami, FL, USA, November, 2002.

12. Jordan, M. A., Davis Rudolf III, C., New container crane concepts, AAPA Facilities Engineering Seminar, American Association of Port Authorities, Savannah, Georgia, USA, April 14-16, 1993.

13. Lazos, C. D., Engineering and Technology in Ancient Greece, Laboratory for Automation and Robotics, University of Patras, Greece, 1998.

14. Morris, C. A., Hoite, S., The future of quayside container cranes, China Ports '97, Post Conference Workshop, China, March 26, 1997.

15. Morris, C. A., Mc Carthy. P. W., The impact of jumbo cranes on wharves, Proceedings of the Conference "PORTS '01", ASCE, section 30, chapter 4, Norfolk, VA, USA, 2001.

16. Payer, H. G., Technological and Economic Implications of Mega-Container Carriers, SNAME Annual Meeting, Society of Naval Architects and Marine Engineers, Orlando, USA, October 24-27, 2001.

17. Petkovi üZ., Zrni üN., Last trends in design of ship-to-shore container cranes, Proc. of the 4th Inter. Conf. on Heavy Machinery, Faculty of Mechanical Engineering Kraljevo, Kraljevo, 2002, pp. A.51-A.54.

18. Petkovi üZ., Zrni üN., State of the art of Post-Panamax ship to shore conventional and modified A-frame container cranes, Proceedings of the XVII International Conference on Material Flow, Machines and Devices in Industry, Faculty of Mechanical Engineering Belgrade, Belgrade, 2002, pp. 1.78-1.81.

19. Petkovi üZ., Zrni üN., The expectations in the trend of development of container cranes for the XXI century, Proceedings of the II Yugoslav Conference - Water transportation in 21st century, University of Belgrade, Beograd, 2002, pp. 121-128.

20. Verschoof, J., Cranes - Design, Practice, and Maintenance, 2nd edition, Professional Engineering Publishing Limited, London and Bury St Edmund, UK, 2002.

21. Zrni ü N., The influence of some container cranes design characteristics on terminal system performances, In: Modelling and Optimisation of Logistic Systems (edited by T. Banayai and J. Cselenyi), University of Miskolc, Hungary, 2001, pp. 159-171.

22. Zrni ü N., Petkovi ü Z., Evaluation of design solutions for trolley of quayside container cranes, Proc. of the Conf. "IRMES 2002", University of Srpsko Sarajevo, Bosnia and Herzegovina, 2002, pp. 99-104.

23. Zrni ü N., Dragovi ü B., Petkovi ü Z., Survey of some new concepts that increase STS container cranes productivity, Seminarband zu den Neuesten Ergebenisse auf dem Gebiet Foerdertechnik und Logistic "Miskolcer Gespraeche 2003", University of Miskolc, Miskolc, Hungary, 2003, pp. 133-138.

EVOLUTION OF DESIGN, USE AND STRENGTH CALCULATIONS OF SCREW THREADS AND THREADED JOINTS

Viktor Strizhak , Igor Penkov and Toivo Pappel
Tallinn Technical University, Department of Mechatronics
Ehitajate tee 5, 19086 Tallinn, Estonia
e-mail: strizhak@staff.ttu.ee, ipenkov@staff.ttu.ee ,
tpappel@staff.ttu.ee

ABSTRACT

Threaded joints are in wide use in different structures, machines and devices and these facts lead to new designs of threaded parts and joints having new useful properties for different conditions of loading. Therefore, in spite of the fact that there are a lot of different literature where designs of threaded parts and calculation methods are given, it is necessary to make some systematization to trace common direction on which goes the evolution of design of threaded parts, threaded joints and their calculation methods. Between different problems, have been considered by researches, the most important problem, in our opinion, are fundamental problems such as development of structures of threaded parts to achieve new useful properties, expansion of use area and development of strength calculation to more exact consideration of a stress condition on different methods of loading. Use new, more exact methods of calculations with use of different materials gives a possibility to increase a reliability of the threaded joints and decrease expenses of their manufacture, assembly, repair and operation as threaded joints as devices and machines as a whole.

In our previous work [1] apart from other problems the fundamental problem of the strength of the threaded joints – how the external axial load is distributed on the threads – in historical aspect was considered and was shown that this distribution in the pair of a bolt and a nut is extremely uneven. This fact leads to large dimensions of the threaded parts and, of course, to large dimensions of the joints as compared with the parts having no threads. In this article the evolution of structures of the threaded joints and methods of strength calculation are considered. It should be noticed that present work is natural sequel of work [1] and may be considered as addition.

KEYWORDS: Machine design, Threaded joints, Strength calculation, Reliability.

INTRODUCTION

To imagine what important a role the different kinds of fasteners have in technique it is necessary to note, for example, that threaded parts are used in telephone up to 75, in lathe more than 1650, in automobiles more than 3500 and in jet aircraft about 1.5 million. In spite of that the number of the threaded parts with taking into consideration of dimensions, permissible load and kinds of the threads reach several thousand, which are presented in standards of different states, new designs of the threaded parts made of different kinds of materials, metals and plastics, having ultimate stress, for example, in aviation up to 1450 – 2100 MPa are elaborated and will be elaborated in the future [11]. From the point of view of economics the meaning of the threaded parts is so large that, for example, on approximate valuing in USA and Great Britain expenses on joints of design elements are within 20 – 40 % of common expenses and though a cost of production of the fastener is usually about 5 % expenses of work time for assembly of the parts are more than half of common expenses for manufacturing of product. Owing to these facts the cost of the fasteners in assembled product is increased up to ten times as compared with their nominal cost.

Manufacture of the fasteners is very important in economics. Thousands firms in different states yearly supply to world market hundreds milliard of such parts and their total cost is very large and, for example, in USA is more than 50 milliard dollars. It gives possibility to suppose that new designs of the fasteners will be created, the area their using will be spread and new more exact methods of the strength calculations will be elaborated with taking into consideration different work conditions, influence of mechanical properties of materials and peculiarity of stress conditions.

EARLY DESIGNS OF THREADED PARTS AND THEIR USE
ANCIENT PERIOD

The first inventor of the screws is apparently the Pithagorean philosopher Archytas of Tarentum (5th century B.C.) when, as a rule, wooden screws were used. Metal screws and nuts only in the 15th century were used.

The first mechanisms having a screw thread, to all appearance were invented 950 years B.C. and used in Egypt for sewage-farm.

In period between two Punic Wars (241 – 218 B.C.) Archimedes was in Egypt in a visit to Canon and Eratosthenes [2] where improved the cochlea design (Fig.1) and then brought this idea to Europe. This mechanism is known as infinite Archimedean screw. The cochlea, as a rule, had the length about 10 – 12 foots and was made of wooden plane bar. On the plane bar surface by spiral a brass ribbon was winded up [3]. All these were placed in a

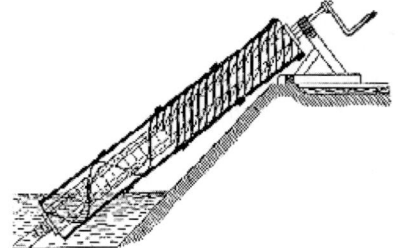

Fig. 1: Cochlea with lever handle [3]

special drum made of wooden strip. A lead angle was from 20 to 45°. The cochlea was brought in operation with one man with help of a level handle fixed on the bar.

Using this idea Archimedes wrapped a pipe around a shaft in a helical pattern to make a crude bilge pump for ships. This mechanism is known as Archimedes' Water Snail Screw Pump.

Being one of the most famous practical scientists of his time Archimedes endeavored the real solutions taken by experience way to describe mathematically and a screw line has been described in treatise on spirals. The family of screws including Acme, Unified Trapezoidal and ISO are known as Archimedean screws because they exhibit straight thread profiles in their axial sections.

Archimedes took the basic inclined plane and wrapped it into a spiral shape. Rotating the spiral in one direction raised the load and rotating it in the other direction lowered the load.

The idea of infinite screw gave Archimedes a shove to create such, simple in present day, well known fastener elements as a bolt and a nut. Approximately in that time first lifting jacks and presses used on manufacturing of wine and oil were used.

As it is well known the helical line is created as a result of uniform axial motion on a surface of uniform rotating cylinder. By analogy the Archimedes' spiral is created but in this case the uniform linear motion takes place by radius of uniform rotating disk. Having forestalled first of all, the idea of the uniform motion, Archimedes geometrically shows a method of creation of the spiral and solves the problem about tangent line to the spiral. This work is very interesting by this that here the motion of a material point is used and this fact, in a sense, anticipates the ideas of present differential calculus. The second problem, which was solved by Archimedes on writing the treatise on spirals, was determination of a circle length. The method of determination of the straight line that equals length of given circle or any it part in the work was given as well. Unfortunately the Archimedes' works were rewritten and translated on other languages with distortions and simplifications and this fact may be explained by scanty competence of rewriters and translators.

But even in works of such scientists as Heron and Pappus, which used the Archimedes' works in their investigations, the helical line is considered rather superficially and descriptively.

Heron (ca.10 – ca.75) in his work "Mechanics" describes five basic mechanisms with help of which the given load set in motion by definite force – a winch, a lever, a pulley, a wedge and a screw [4]. In these works the description of mechanisms essentially simplified and no theoretical reasons are given. So, for example, about lever was said - "When this mechanism was found then have understood that with help of this mechanism it is possible to move the large loads. The piece of wood they called as a lever, if it had circle or quadrangle cross-section. The nearer to a load a stone is placed under the load the more comfortable the motion is done".

Approximately by the same way other mechanisms are described. The action of a wedge was considered as action of blow on the wedge, and work of a screw is considered as wedge acting due to lever but not due to blow. Pappus (ca.290 – ca.350) in eighth book "Mathematical collection" rather in detail but with no

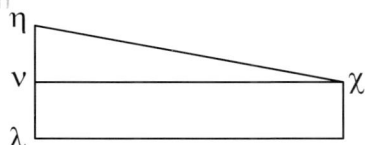

Fig. 2: The plate to create a helical line [4]

theoretical reasons describes a method of creating a screw thread with given pitch on cylinder with given diameter [5]. The creating may be done by wrapping of a brass plate on a cylinder having preliminary marked intervals, which are equal to the thread pitch. The plate (Fig.2) is made of two figures – parallelogram $\chi\lambda$ and triangle $\eta\chi\nu$; plate is equal to the length of the circle of a cylinder, and the height $\eta\nu$ is equal to the value of the thread pitch.

RENAISSANCE PERIOD

Right up to XV century works of ancient mathematician and mechanics were forgotten. Familiarization with Archimedes' works began after the Turks have routed Constantinople in 1453 year. Then Greek refugees brought to Italy remains their manuscripts.

In renaissance period a great interest was to the past, to the ancient art and science. In this period begins mastering of works of Archimedes, Pappus, Heron and others.

Leon Battista Alberti (1404 – 1472) gives description of a screw action in his work "Treatise about Architecture". The screw was considered not as inclined plane as it by ancient scientists was considered, was not shown about law of screw creating, and all thing and problems, which were described, were considered as some simple notes, which had a little sense, however all of these were based on his own observations.

Next, and may be, most famous personality of all renaissance period was Leonardo da Vinci (1452 – 1519). He was interested absolutely in all things what he was surrounded. He had concerned with painting, sculpture, metal casting, weaving, music, anatomy, botany and others.

Working as an engineer Leonardo da Vinci had concerned with organization of ship-canals, sewerages, irrigation from rivers and artesian wells. Not unimportant place in his works chief interest is in mechanics.

He invented a great number of different machines and mechanisms. There were many his works and inventions which on several centuries forestalled his time and found followers only in XVIII – XIX centuries. So Leonardo da Vinci conceived the first flying machines, which used the screw thread principle. Today's propeller driven ships, airplanes and helicopters can be though of as utilizing screws against air or water, which act as the mating nuts.

Considering the screw problem Leonardo does it slightly descriptively and a little theoretically, but very fruitfully for further development of science. In his works, to all appearances, first the screw-cutting lathes for nuts and screws are considered and their technology of manufacturing is described as well [6 - 8].

Considering the manufacturing the nuts (Fig.3) he said, that to make the nut it is

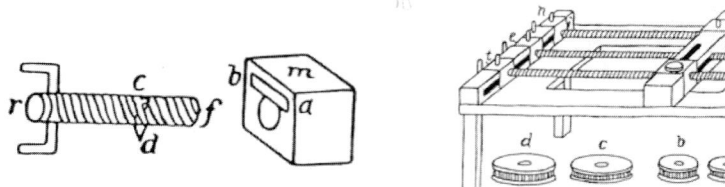

Fig. 3: Nut-cutting device [8]   *Fig. 4: Screw-cutting lathe [8]*

necessary first in wooden bar (m) to make a hole, which must be equal to thickness of the screw blank (fr). Then it should be fastened to the bar an iron plate (ab) having width in two fingers and with thickness which must be equal to the thickness of a string. The metal plate is placed so that the hole was overlapped for half-finger. Then it is necessary to make a canal with shape like the screw has. In this canal the metal plate goes and when the rotating screw will be move forward, then chisel (*cd*) will be cut the nut thread.

In Fig.4 the screw-cutting lathe is presented which may be used for manufacturing of screws. Leonardo da Vinci saying about this lathe writes that rotating the middle toothed wheel, being placed onto the screw blank, it is possible to make the screw thread. If it is necessary to make the screw with more or less of lead of screw thread (thread pitch) then to replace the toothed wheels (*s* and *f*) by wheels (*a* and *b*) or wheels (*c* and *d*). In accordance with these operations it is necessary to move the bearings (*t* and *h*) and also the bearings near the chisel and near the bushing (*g*). The chisel consists of two blades, which moving forward cut the screw thread of the screw.

Considering the principle of the screw act [7] Leonardo da Vinci writes that the nearer to the first degree of facility (i.e. inclination) is the load and the further past one from last degree, the more capably (easily) the load will be to lift. About an eccentrically loaded screw (Fig.5) it is saying that if the center of gravity of the load is not coincide with the screw axis, then this load will be heavier for the nut and screw thread in so much times how much *mn* is in *an*. It should bear in mind that the force leading the screw in action must be increased in the same proportion.

Considering an influence of a thread pitch Leonardo da Vinci notes that to bring into action of the screws having equal thickness the more great force is needed for the screw having lesser thread pitch. When the screws have equal length, thickness and the same thread type the easier will be move the screw having the larger thread pitch. The strength will be higher when the screw will be with lesser thread pitch, however will be more difficult to lead this screw in action.

Considering the different threads profiles he writes that the screw having rectangular thread (Fig.6) can withstand larger load. Usual screws have two surfaces, but rectangular thread has three ones and if in widely spread screws the thread is wide near a low end and thin and weak near head, then here the thread everywhere is equal and strong.

The screws with triangular thread is the widely spread type of the screws. Through the triangular thread is weaker, the screw draws easily, because the nut having rectangular thread profile must come into contact with three surfaces and here only with two ones. Buttress (orthogonal) thread has the same strength as the rectangular thread and is the same in moving, i.e., easy, as triangular equilateral thread, i.e. on the same thread pitch will be better than any other thread.

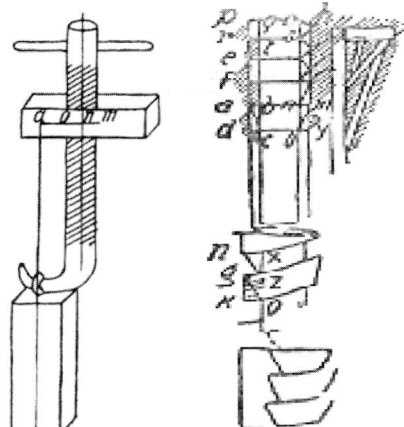

Fig. 5: The screw loaded eccentrically [7]

Fig. 6: The different types of screw threads [7]

Leonardo da Vinci in his works gives specific instructions by formation of helical line, designs the device for test of screws and test method.

Having compared two screw devices (Fig.7) he considers time of moving the nut having one, two, three, four, five and six threads under the constant load on the lever. So were proposed the tests of the screws with different thread pitches.

Leonardo da Vinci also proposed a device for transformation of rotary motion to reciprocating motion. It was in that time newest invention. He elaborated many other mechanisms and devices for practical use, in which as driving mechanism were used a screw drive. So, for example, the device shown in Fig.8 is used for incision of saws.

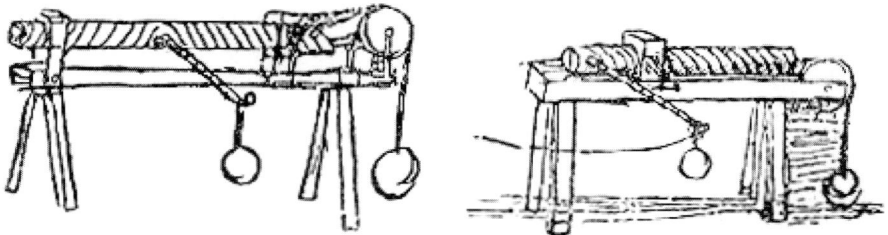

Fig. 7: Device for screw tests [7]

In this mechanism a screw does the feed of a blank. The principle of work of this machine is not difficult and is clear without explanation but it is necessary to note that a handle, visible in front of the device, is not for drive the hammer into action but only to lift the load *A*, hanging on a rope, which is placed on a block. The load putting down drives the machine in action like clock being wound up.

After Leonardo da Vinci, in XVI – XVII centuries, the screw mechanisms was widely used in different machines and devices (Fig.9). The different threads were manufactured in workshops from experience and there were no generally accepted recommendations.

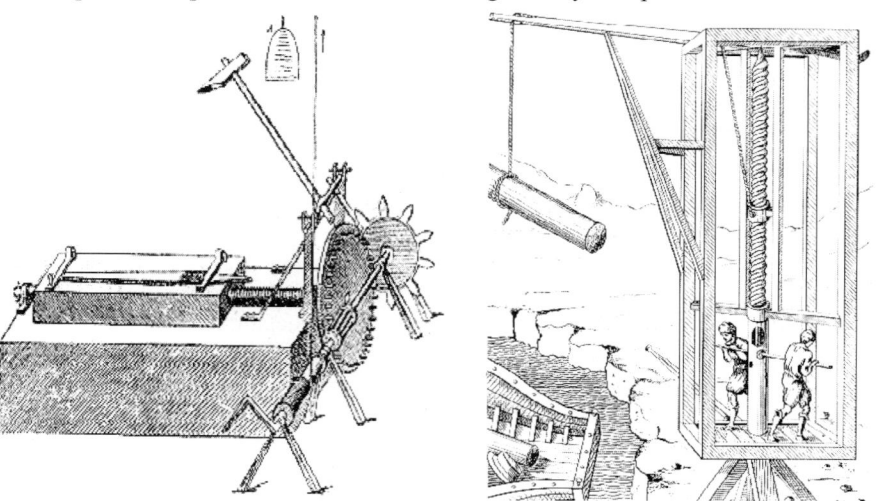

Fig. 8: Device for incision of saws [8]   *Fig. 9: Lifting crane [9]*

First internal threaded parts, now named as a screw, a bolt or a stud, were made with use of a wire. Usually it was a soft metal, for example, copper or brass. The wire was wrapped around plain bar. Nuts also were made of copper. The screw threads in the internal holes of the nut blanks were created by forging them around the wire wrapped rod. Such threaded parts had low accurate and no tolerances. Therefore the pair of a bolt-nut usually was compounded by selection method and as a result the bolts and the nuts were not interchangeable.

Only in 1903 – 1908th years first standard tables are appeared in which the main parameters of the threads are given [10]. The strength of the screws (bolts and nuts) was determined experimentally, and first mathematical methods for strength calculations were elaborated only in beginning of XX century, which are developed up to present.

STRENGTH CALCULATIONS

First strength calculations of the threaded parts and threaded joints were very simple like for machine elements having no threads, i.e., on strain of a bolt shank, on bending, shearing and bearing stresses in the threads.

However, it is well known that to decrease the expenses for material and manufacturing are in use the threaded parts made of material with high strength. These parts are working in conditions when alternate or not completely reversed stresses take place. Use qualitative materials is very actual, especially for threaded parts as different machines and devices have about 70% of all mechanical joints in different main branches of industry: motor-car construction, agricultural equipment, aviation, instrument making and others.

Therefore, use fasteners with high class of strength, for example, 8.8 and higher is constantly increased and, for example, in motor-car constructions of USA, FRG, Japan are about 98% [11]. The imagine about value of economy of material on manufacturing of one bolt having middle length give following data [11]:

Class of strength	4.6	4.8	8.8	10.9
Nominal diameter of Metric thread	M20	M16	M12	M10
Mass of bolt, %	100	58	30	21
Torque on assembly, %	100	82	66	57,7
Limit axial force, kN	32	35	68	87

However the high-grade materials are very sensitive to the stress concentration in a work zone of the thread. The concentration factor K_σ in a threaded zone of bolts and nuts can be equal up to $5.35 - 7.75$.

To take into account the factor K_σ it is necessary to know the most loaded cross-sections in a bolt and a nut. It was experimentally determined that the axial load on thread within a nut height is distributed extremely unevenly.

In fact, on loading the threaded joint when only one, two, three and more threads of a nut subjected to the load the total failure load on the joint is increased so that the proportionality is not observed.

For example, the total failure load acts on three threads considerable less as compared to failure load for one thread being increased threefold. Now this fact is very well explained.

250

For the first time this problem was considered by famous Russian researcher N.E. Joukovsky (1847 – 1921) in 1902, (Fig.10) [12]. More in detail his method of determination of the load distribution on the height of the working threads was described in our work [1].

Joukovsky taking into account a simple discrete scheme shows that if in a joint ten threads are loaded then on first most loaded thread acts about 33 % of the axial load F.

Later this problem was considered by E. Jaquet in 1931 [13], L. Madushka in 1936 [14], B.S. Tsfas in 1961 [15], N.L. Klyachkin in 1972 [16], E.-G. Paland in 1967 [17] and others, which developed the Joukovsky's method. More exactly this problem by I.A. Birger in 1944 was solved [18]. He gives the solution for scheme of loading when all threads were taken into account as continuous. The Birger's solution was given in differential form. Having introduced a new idea about intensity of an axial force $q(z)$ on a unit of the length of the threaded joint on any cross-section z the load on any loaded thread may be determined from following equation

Fig. 10: N. E. Joukovsky
(1847 – 1921)

$$F(i) = \int_{z}^{z+P} q(z)dz \qquad (1)$$

where P is thread pitch; $z = H/i$; i is number of the working thread; H is the nut height.

Birger's method received a wide recognition and have been used by other researches [19, 20, 21, 22 and others]. For more exact results of calculations not only deformations due to thread bending and shearing were taken into account but also radial deformations of bolt and nut bodies in accordance with Poisson's factor and friction forces in the threads. So were added two schemes of loading [22]. The intensity of the axial load distribution for schemes of loading "Bolt-Nut I" when a bolt is stretched and a nut is compressed (see [1]) and "Bolt- Nut II" when a bolt is compressed and a nut is stretched will be as following

Fig. 11: I. A. Birger
(1918 – 1993)

$$q(z) = \frac{Fe^{a(\pm H \mp z)}}{\sinh\ bH}\left[b\cosh\ bz \mp a\sinh\ bz\right] \qquad (2)$$

and for schemes of loading "Tightener" when both a bolt and a nut are stretched and "Post" when both a bolt and a nut are compressed will be as following

$$q(z) = Fe^{\mp az}\left[\frac{b\cosh\ bz \mp a\sinh\ bz}{\beta \cdot e^{\mp aH}\ \sinh\ bH}\left(\frac{1}{E_1 A_1} + \frac{e^{(b \mp a)H}}{E_2 A_2}\right) - \frac{e^{bz}}{\gamma(b \pm a)E_2 A_2}\right] \qquad (3)$$

Here $e \approx 2.71$; a, b, β and γ are coefficients depending on material and type of thread; E_1, E_2, A_1 and A_2 are moduli of elasticity and cross-section areas, accordingly (index 1 for bolt and index 2 for nut). The upper signs correspond to the schemes of loading "Bolt-Nut I" and "Tightener".

The general case of loading when the loads F_1 and F_2 applied to a bolt, the loads F_2 and F_3 applied to a nut was considered also and the following equation was obtained

$$q(z) = \frac{e^{(b \mp a)z}}{\gamma(b+a)}\left[\frac{F_1}{E_1 A_1} - \frac{F_2}{E_2 A_2}\right] + \frac{e^{(\pm H \mp z)a}}{\beta \sinh bH}\left[\frac{F_4 - F_1 e^{(b \mp a)H}}{E_1 A_1} - \frac{F_3 - F_2 e^{(b \mp a)H}}{E_2 A_2}\right](b\cosh bz \mp a \sinh bz)$$

(4)

The lower signs correspond to the schemes of loading when a bolt is compressed.

DISTRIBUTION OF AXIAL LOAD ON BOLT AND NUT THREADS IN REAL THREADED JOINTS

In works have been mentioned above the ideal thread with nominal dimensions were considered. However on manufacturing of the threaded parts and threads the allowances take place. They limit maximum and minimum dimensions of the thread. The values of the allowances depend on classes of accuracy and fits of the threads. Analysis shows that the values of the allowances are commensurable with the dimensions of the thread elements, such as the thread height and height of the working surface of the thread flank. Therefore the deformations of the threads in axial and radial directions to a great extent depend on the overlap t_2 (Fig. 12b,c) of working surfaces of a bolt and a nut. In works [21] and [22] the equations to determine non-dimensional coefficients λ_1 and λ_2 for a nut and a bolt were given as

for bolt
$$\lambda_1 = A_1 + B_1 \tan \rho + \frac{d_2 t_2}{2P^2}\left(\frac{d_2^2 + d_0^2}{d_2^2 - d_0^2} - \mu_1\right)\tan\frac{\alpha}{2}\left(\tan\frac{\alpha}{2} - \tan\rho\right)$$

(5)

and for nut
$$\lambda_2 = A_2 + B_2 \tan \rho + \frac{d_2 t_2}{2P^2}\left(\frac{d_e^2 + d_2^2}{d_e^2 - d_2^2} + \mu_2\right)\tan\frac{\alpha}{2}\left(\tan\frac{\alpha}{2} - \tan\rho\right)$$

where t_2 is thread overlap; d_0 is diameter of a hole in the bolt shank (in most cases $d_0 = 0$); d_e is equivalent external diameter of the nut; A_1, A_2, B_1 and B_2 are coefficients

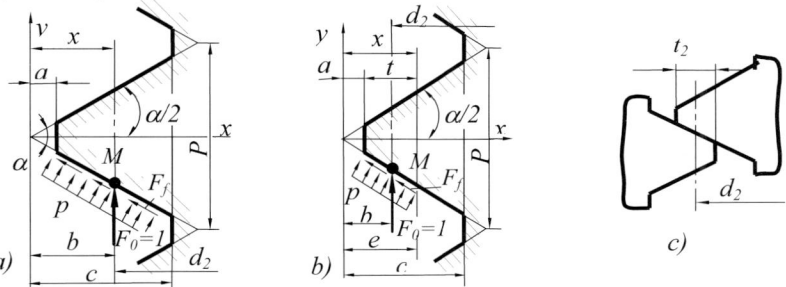

Fig. 12: Calculation scheme for triangular thread profiles

depending on deformations of threads in axial direction due to bending and shearing. Values of the coefficients A_1, A_2, B_1 and B_2 may be determined from calculation scheme, for example, for triangular thread profiles, Fig.12.

So, for example, for metric threads M24...M45 with fit 7G/7e6e for ideal threads the t_2 equal to 0.541 mm, 1.083 mm, 1.623 mm and 2.165 mm for the thread pitches equal to 1 mm, 2 mm, 3 mm and 4 mm, accordingly.

With taking into account the allowances the minimum overlap t_{2min} will be equal to 0.312 mm, 0.718 mm, 1.264 mm and 1.76 mm, accordingly. Then relation t_2/t_{2min} will be equal 1.73, 1.38, 1.28 and 1.23 for the same thread pitches.

Having determined the axial displacements of point M (Fig.12a,b) due to bending δ_b and shearing δ_{sh} by Maxwell-Mohr's reciprocal theorem for flexibilities based on Clapeyron's second theorem it is possible to obtain the values of A_1, A_2, B_1 and B_2. For example, these values for metric thread M24 with fit 7G/7e6e have been calculated and were shown in Table 1.

Table 1. The values of coefficients A_1, A_2, B_1 and B_2 on $\mu_1=\mu_2=0.3$; $f = 0.1$ for nut and bolt threads

Kind of thread		Bolt thread		Nut thread	
		A	B	A	B
All metric threads with any pitch when allowances are equal to zero		0.8328	0.7203	0.8328	0.7203
M24 for joint with 7G/7e6e fit with pitch P, mm	3	1.2203	0.7565	0.8546	0.5854
	2	1.3598	0.9124	0.8969	0.6474
	1	1.5350	0.9830	1.1185	0.6546

Here μ is Poisson's factor and f is friction factor.

These values of A and B illustrate that the less the thread pitch for the same nominal thread dimension the greater the influence of the allowances on the axial thread displacements.

The loads on the most loaded thread for ideal and real threads are shown in Table 2.

Table 2 illustrates that the less the thread pitch the less the load on the most loaded thread and when the allowances are taken into account the load on the thread may be considerably decreased. For *Tightener* and *Post* schemes of loading the axial load F is distributed more uniformly.

Table 2. The loads $F(i)$ on the most loaded thread in percentage of axial load F for metric thread M24 7G/7e6e: d_e=36 mm, H=18 mm, μ=0.3, f=0.1

Thread pitch, P, mm; number of threads in joint, n	Scheme of loading			
	Bolt-Nut I	*Bolt-Nut II*	*Tightener*	*Post*
$P=1^*$, $n = 18$	12.24	16.15	7.65	10.57
$P=1$, $n = 18$	11.19	14.30	7.27	9.71
$P=2^*$, $n = 9$	22.42	28.47	14.74	19.52
$P=2$, $n = 9$	20.32	25.00	14.00	17.88
$P=3^*$, $n = 6$	31.19	38.39	21.50	27.48
$P=3$, $n = 6$	28.35	33.94	20.51	25.35

*Allowances are equal to zero. Here μ is Poisson factor, f is friction coefficient.

Solutions of this problem for rectangular, trapezoidal, buttress, round-profiled and any asymmetric triangular thread profiles in work [23] were obtained as well

So it is possible to receive with high accuracy the loads on any threads in any thread joints with any thread profiles with taking into account the classes of accuracy and the fits for the threaded parts made of metals or plastics.

SPECIAL DESIGNS OF THREADED JOINTS

USE OF DIFFERENT MATERIALS FOR BOLTS AND NUTS

In different machines and devices often with steel bolts the brass, bronze or plastic nuts are used if it is necessary to obtain special useful properties, for example, low friction coefficient, wear, weight or corrosion resistance. Especially these properties are useful in bolt-nut drives, which often called as screw drive. In these materials the moduli of elasticity differ from each other and usually for a nut E_2 is less than for a bolt E_1.

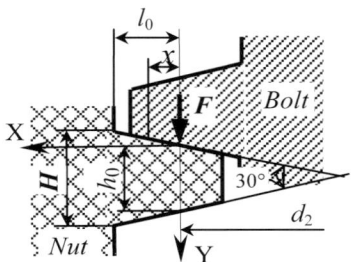

Fig. 13: Calculation scheme to determine the displacement of axial force

In this case when the engaged threads are loaded by force F (Fig.13), the line of the force acting, due to the bending of the threads, is displaced in direction to the nut which has lesser value of the E_2.

The solution for determining of displacement x in work [24] is given. Taking into consideration that angles of turn for the nut and bolt threads are equal the value of displacement x can be determined as

$$x = \frac{\sqrt{k}-1}{\sqrt{k}+1}l_0 \qquad (6)$$

Here $k=E_1d_0/E_2d_1$; d_0 is internal diameter of the bolt thread; d_1 is external diameter of the nut thread; l_0 is

the distance from the middle line to the root of the nut thread. It is shown that the less the E_2 as compared with E_1 the more will be displacement of the force F. This displacement is rather great and, for example, for the polycarbonate nut past one will be up to the nut thread root. This fact is very useful as the bending stresses decreased, however will be increased the shearing stresses. Therefore when the rectangular and buttress thread profiles are used for the plastic nuts working with steel bolts the main crushing stresses will be the shearing stresses and this fact was proved.

Taking into account this displacement x more exactly will be determined the load distribution on threads and found most stresses cross-section. The calculations shown that the more the moduli of elasticity of nut and bolt material differ from each other the more evenly the axial load will be distributed on the threads. It is very useful from point of view of wear for screw drives.

SCREW INSERTS

In consequence of that axial load is distributed extremely unevenly the researches suggested different method to even this distribution. The nuts having variable thread pitch or pitch diameter were suggested [19]. However it is effective only when the axial load is determined very exactly and do not change in operation as geometrical parameters for nut thread must be chosen in accordance with this load. More effective method to even the axial load distribution is use a spiral screw insert [19] (Fig.14) or use the special flange nuts [25].

The screw spiral insert is like usual spiral spring having diameter-shaped cross-section and made of chrome-nickel steel. The calculations show that the screw insert being introduced between the bolt and nut threads leads to more even axial load distribution on the threads. In this case the number of working threads will be considerably less as compared with the joint having no insert. Besides the load evening the inserts protected thread against wear and corrosion and often can be used when it is necessary to repair the broken threads and owing to these facts the life of machines will be increased.

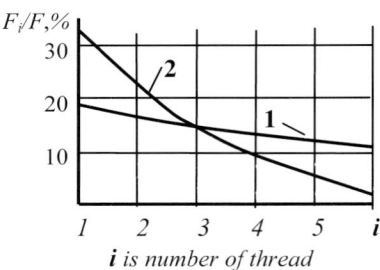 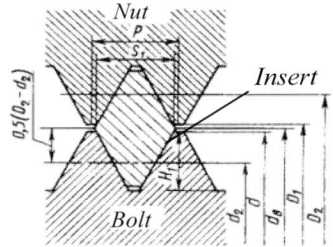

Fig. 14: Screw insert and load distribution on thread
1 – for joint with screw insertion   2 – for usual threaded joint

Calculation of such joints is given in work [19]. The most loaded cross-section of the threaded joints and loads on the thread can be found using Birger's method on the base of the problem solution by considering the rotation of the ring cross-section with respect to each other by the torque when the ring has a diamond-shaped cross-section.

SCREW DRIVES

Ball screws, first invented in the end of 18th century, did not come into widespread use until the 1940s when they were adapted in automobile steering gears. Since that time in different machines and devices ball screw mechanisms, called as screw drive, are

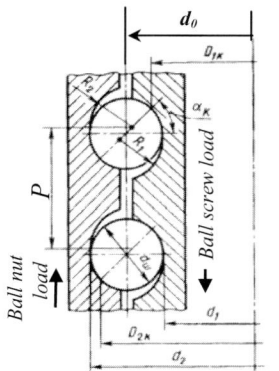

Fig. 15: Main profiles of screw and nut in normal cross-section [19]

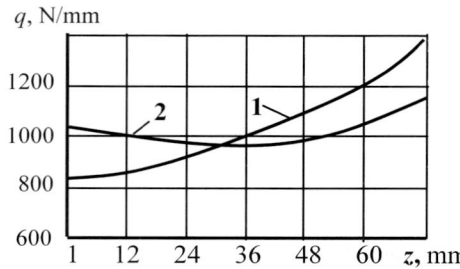

Fig. 16: Intensity of load distribution in screw drive [19]. 1 – joint of "Bolt-Nut I" type; 2 – joint of "Tightener" type
$d_0 = 100$ mm, $d_b = 7$ mm, $F = 72$ kN,
z is height from nut face

widely used due to the main advantages: high efficiency η up to 90 %, for example, the usual drives of bolt-nut type have η = 0.2 – 0.4; high evenness of motion; low static friction; possibility of full removal of clearance in thread, ensuring high axial rigidity [19]. Ball screws offer an efficient means of converting rotary motion to linear motion. A ball screw is an improvement over a trapezoidal screw just an antifriction ball bearing is an improvement over a plain bushing. The load distribution on the ball nut height in Fig.16 is shown. Ball screw assemblies have a number of bearing balls that transfer the load between the nut and screw (Fig.15) [19].

The thread form in which the bearing balls ride may be different: ogival shape formed from two arcs of the same radius with offset centers; semicircular (see Fig.15); rectangular; trapezoidal; triangular and combined.

In operation bearing balls circulate in a ball nut. Balls enter to the ball path between the nut and screw carrying the load one or more turns around the screw.

The bearing ball is then picked up and returned to the beginning of the circuit through the return guide.

SPECIAL FLANGE NUTS

To increase of fatigue strength of threaded joints the special flanged nuts were elaborated, Fig. 17, [25]. Creating of conditions for the opposite law of load distribution on the threads for leveling the stress concentration in the thread roots of working and nonworking threads of the bolt is due to turning deformation of the nut body on circular bending of the supporting flange of the nut. The required value of the external flange diameter D_2 can be determined from the Eq. (5) connecting all the main dimensions of the flange nut.

$$\delta \geq 1.2H^2 \left[(D_1/D_2)^2 \ln(D_1/D_2) \right] / (D_1^2 - d_2^2) \qquad (7)$$

where $H = (D_1 + d_2)/2$.

When H=6P the condition that the K_σ=constant take place. The angle γ in the Fig. 17 is equal to 3°- 5°.

Tests of flanged nuts with bolts made of steel 40X2HMA having R_m=1,320 MPa with metric thread M16X1.5 with class of accuracy of nut thread 6H and bolt thread 6h on frequency of loading 1.15Hz show that the fatigue life is increased up to 2.08 times as compared with the usual nuts without flanges. On the base of the nut shown in Fig.17 a lot of different flanged and small-dimensioned nut forms were designed and tested. Some of them in Fig. 18 are shown. Tsese nuts providing even stress distribution can also have other special useful functions. The clearance h between the nut central circular ldge and the supporting surface of the joined

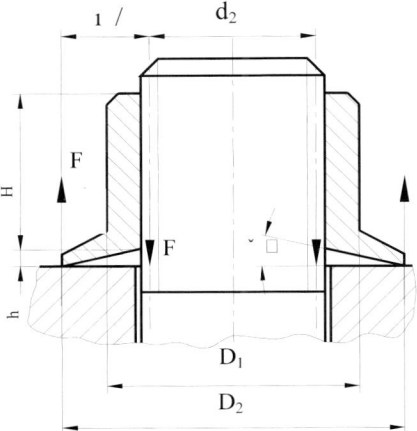

Fig. 17: Special flanged nut

256

parts gives the possibility to check the tightening torque (Figs 18.1 and 18.2). The small-dimensioned nuts can be made of standard nuts if the internal hole of the supporting face will be machined as a cone. The circular cut hollow with the width *t* can be used to check the tightening force (Fig. 18.3). The self-adjusting nuts with spherical washers are shown in Figs 18.4, 18.5, and 18.6. Self-adjusting nuts having spherical convex supporting surfaces are shown in Figs18.7-18.11. In these nuts the clearance between the smooth parts of the bolt and the nut gives a possibility to check the tightening torque. Besides these nuts have increased self-braking and partly unload the bolt of axial force.

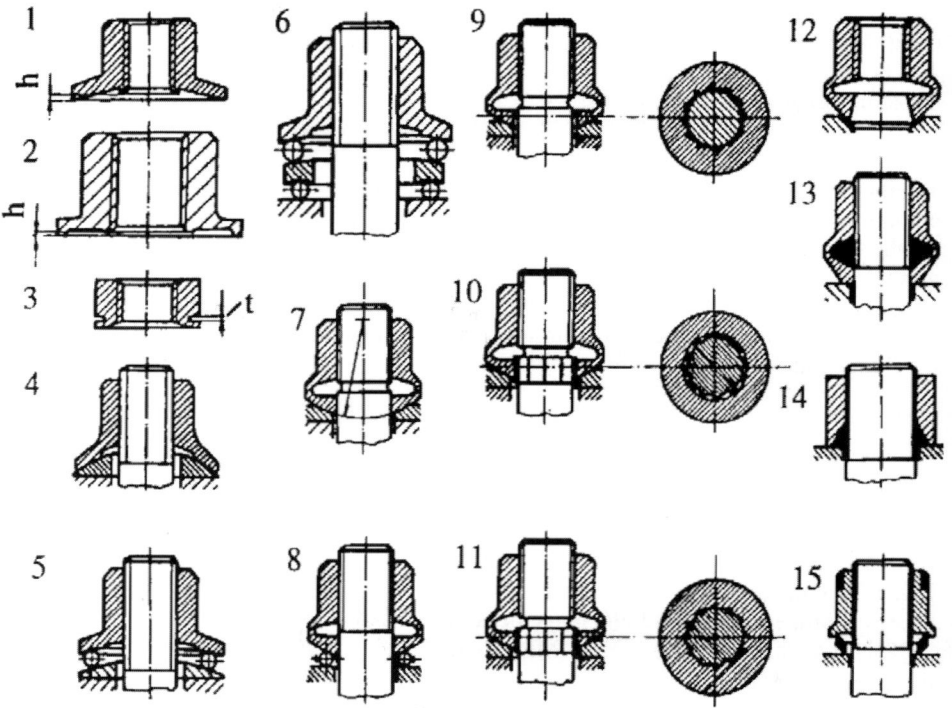

Fig.18: The special flanged and small-dimensioned nuts

The joints shown in Figs 18.9, 18.10, and 18.11 have ratchet teeth on the nut body (Fig. 18.9), on the bolt body (Fig. 18.10), and on the both bolt and nut bodies for the self-closing joint (Fig 18.11). The self-centering nut with external conical surface is shown in Fig. 18.12. Self-tightening nuts are shown in Figs 18.13, 18.14, and18.15. However it should be noted that the flanged nuts increase the axial pliability of the bolt system and this fact leads to decrease of the part of the external force exerted on the bolt in preliminary tightened bolted joints. This fact increases the fatigue life of bolted joints.

CONCLUSION

A brief historical survey shows that in spite of that a screw, screw thread and threaded parts is well known and used during more that 2000 years exact technical strength calculations and designing of different threads and threaded parts only were fulfilled in the course of the last 100 years.

REFERENCES

1   Strizhak, V. and Penkov, I. Distribution of Axial Load on Bolt and Nut Threads. Proceedings of International Symposium on History of Machines and Mechanisms, Cassino, Italy, 2000, pp. 101 - 110.
2   Archimedes. Works. Translated by Veselovsky I. N., Physic-Mathematical Literature, Moscow, 1962, (in Russian).
3   Kirby, R. S., Withington, S., Darling, A. B. and Kilgour, F. G. Engineering in History. McGraw-Hill Book Company, Inc., New York, Toronto, London, 1956.
4   Herons von Alexandria Mechanik. Hrsg. und überzetzt von L. Nux und W. Schmidt, Leipzig, B. C. Teubner, 1900.
5   Pappi Alexandrini. Collectiones quae supersunt e libris manu scriptis. Edidit latina interpretatione et commentaries instruxit Fr. Hultsch. Vol. I-III, Belorini apud Weidmannos, 1898.
6   Gukovsky, M. A. Mechanics of Leonardo da Vinci. Academy of Sciences, Moscow-Leningrad, 1947, (in Russian).
7   Leonardo da Vinci. Selected Natural-Science Works. Translated and Edited by Zubov V.P., Academy of Sciences, Moscow, 1947, (in Russian).
8   Sidorov, A. I. Essays about Engineering History. State Technical Publishing House, Moscow, 1928, (in Russian).
9   Bogolubov, A. N. Mechanics in Humanity History. Science, Moscow, 1978, (in Russian).
10 Leinweber, P. Gewinde. Springer-Verlag, Berlin, 1951.
11 Petrikov, V. G, Vlasov A. P. Progressive Fasteners. Mechanical Engineering, Moscow, 1991 (in Russian).
12 Joukovsky, N. E. Collected Papers. Vol. VIII, Theory of Elasticity, Railways, Automobiles, ONTI, Moscow, 1937, pp 48-56 (in Russian).
13 Jaquet, E. Über eine newartige Schraubenverbindungen. Ingenieur-Archiv, Bd.2, 1931.
14 Maduschka, L. Beansprughung von Schraubenverbindungen und zweckmässige Gestaltung der Gewindeträger. Forsch. Geb. Ingenierwessens, 1936, B.7.
15 Tsfas, B. S. Solution of Joukovsky's Problem of Distribution of Pressures on the Bolt and Nut Threads Given in Closed Form. Proceedings of Educational Institutions, Mechanical Engineering, No.9, 1961, pp. 38-49. (in Russian).
16 Klyachkin, N. L. Calculation of Group Threaded Joints. Uljanovsk, 1972. (in Russian).
17 Paland, E.-G. Gewindelastverteilung in der Schrauben-Muttern-Verbindung. Konstruktion, 19, 1967, Heft 9.
18 Birger, I. A. Calculation of Threaded Joints, Oborongiz, Moscow, 1959. (in Russian).
19 Birger, I. A. and Iosilevich, G.B. Threaded and Flanged Joints. Mechanical Engineering, Moscow, 1990. (in Russian).
20 Glukharev, E. G. To Calculation of Joints of Paddles of Gas Turbines. Power-Machine-Engineering, No.7, Moscow, 1958. (in Russian).
21 Meng, V. V. and Strizhak, V. I. Influence of Scheme of Loading upon Load Distribution on Threads. Proceedings of Educational Institutions, Mechanical Engineerings, No.10, 1976, pp. 37-40. (in Russian).
22 Strizhak, V. and Penkov, I. Dependence of Load Capacity of Threaded Joints on Design Parameters. Proceedings of OST-95 Symposium on Machine Design, Oulu, Finland, 1995, pp. 66-75.
23 Strizhak, V. and Penkov, I. Effect of Thread Profiles on Load Distribution on Threads. Proceedings of ROM2000, International Conference on Reliability of Machines and Their Life Prediction. Ivano-Frankivsk-Jaremcha, Ukraine, 2000, pp. 235 – 244.
24 Zemlyakov, I. P. Strength of Threaded Parts Made of Plastics. Mechanical Engineering, Moscow, 1972.(in Russian).
25 V. Strizhak and V. Meng. Increase of Strength of Threaded Joints. Proceedings of Symposium on Fatique Design 1998.182, Vol. II,VTT Technical Research Centre of Finland, Espoo 1998,pp. 625-636.

INDUSTRIAL ARCHAEOLOGY: FROM THE 17th TO THE 21st CENTURY. RECONSTRUCTION OF HERSCHEL'S TELESCOPE

Emilio Bautista Paz, José Luis Muñoz Sanz, Pilar Leal Wiña, Javier Echávarri Otero
Machine and Mechanisms Laboratory. Dpt: Mechanical Engineering and Manufacturing
E.T.S. Ingenieros Industriales. Universidad Politécnica de Madrid
José Gutiérrez Abascal 2, 28006 Madrid, Spain
e-mail: ebautista@etsii.upm.es

ABSTRACT- Towards the end of 1997, the director of the National Astronomical Observatory, Jesús Gómez, commissioned the Machines and Mechanisms Laboratory of this School, to produce a viability study for the reconstruction of "Herschel's Great Telescope". This device was built in the 18th century as part of the fine set of instruments that on the orders of Charles IV, were to equip the Madrid Royal Observatory, which had been created by Charles III in 1790. Unfortunately, this great telescope was destroyed in the War of Independence. All that survived were a few remains, and the image of its structure, thanks to the magnificent illustrations executed by the hand of Mendoza (Fig. 1), the naval officer to whom the King had entrusted its construction.

The project to be undertaken by the Engineering School group, not only had to embrace reconstructing the telescope and how it worked, but also quantify the costs of the undertaking and take other aspects into account, such as a land study, transport, installation and assembly.

There were also other details which were not part of the study, such as, location, the foundations, and landscaping. All this would involve analysis and decision making that could only be done by an architect.

Now, 5 years later, the construction of a replica of the Great Telescope is about to become reality in *the Bermeo Shipyards, a shipbuilding company located on the Vizcaya coast. It will be housed in a transparent glass building designed by the architect Antonio Fernández Alba, and stand on supports made of granite from the Madrid mountain quarries,. According to astronomer Pere Planesas, Jesús Gómez was the driving force behind the reconstruction project, which was drawn up by a team of industrial engineers from the E.T.S. de Ingenieros Industriales de Madrid (Madrid Higher Technical School of Industrial Engineers), headed by Emilio Bautista and made up of Pilar Leal, José Luis Muñoz and the late Ignacio Medina. (EL PAIS [1], "Del Retiro a las estrellas", (From the Retiro to the Stars), Wednesday 20 February, 2002).*

We believe it is worthwhile here, to emphasise that an 18th century device has been recovered two centuries after its destruction, due in part, to the participation of our school.

HISTORICAL DATA

There are several obscure periods in the history of the 25 foot telescope. The data set out below will attempt to summarise the facts obtained by studying and compiling information from archives in Madrid, such as the references [2] and [3].

The date the telescope was ordered to be built, the circumstances surrounding its construction, and the price paid are all unknown. Neither is anything known of the vicissitudes surrounding its early demise. There is a letter dated January 1796, where Herschel accepts an assignment to build two telescopes for the King of Spain [4]. The letter states that the largest would be 25 feet long with a 24 inch diameter mirror. This document also details the price, the method of payment, etc ....

Fig.1

Herschel was a pioneer among astronomers of the time. He substituted position astronomy, used by almost everyone, for observation and built telescopes with ever greater apertures. The technical difficulties encountered when casting and cutting the mirrors, meant that the quality diminished when Herschel exceeded a 2 foot diameter.

No-one could surpass Herschel in his dual role of observer and telescope builder. Half a century was to pass by before Lord Posse was successfully able to complete a telescope with a six foot aperture.

Neither the builder, Herschel, nor the size chosen for the telescope, a 25 foot focal length with a two foot diameter, could have been bettered at that time. The Madrid Observatory could be proud that it was going to be introduced into the science of Cosmology, an up and coming branch of Astronomy, equipped with one of the best telescopes in the world: in Herschel's opinion, it was number one in optical quality and second only in size.

Someone had to be in charge of all the formalities necessary for getting the telescope built. All the stages of its building had to be closely monitored and clear instructions had to be written regarding its assembly and use, as well as its shipment to Spain. The person chosen was José de Mendoza Ríos ([5] and [6]).

Thanks to Herschel's manuscripts, the exact chronology of the casting and optical treatment of the mirrors is known. At that period, the mirrors, which were a fundamental part of the telescope, were made of bronze, with alloys that were usually kept secret by the makers. However, as for the other components of the telescope, such as the tube, the structure and ancillary instruments, all that can be established with any certainty is the date of completion: the date recording when Herschel made observations with the completed telescope.

In January, 1797, Herschel had completed the telescope [7], for that is when he recorded his first observation with it. However, he cannot have been satisfied with its performance, since he returned the mirrors to the works to perfect their polishing. In March, 1798, after an intense work on the mirrors, Herschel again used the telescope to observe the planet Uranus, which he had discovered a few years previously. It was then that he commented that he had never seen the planet with such definition and that the quality of this telescope was far superior to that of the 20 foot one.

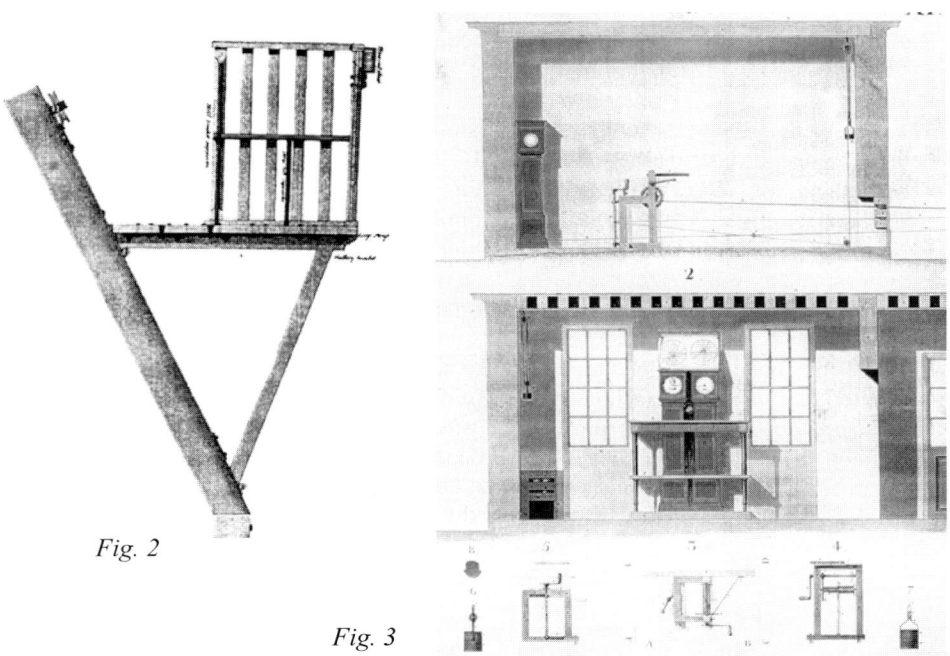

Fig. 2

Fig. 3

The mechanical part of the telescope was similar to that of the 40 foot one, but Herschel added interesting innovations; for example, a chair to go up to the observation gallery (Fig. 2), which, although it had been planned for the 40 foot telescope, was never actually incorporated, and in addition, a more complete outhouse for the ancillary observation instruments (Fig. 3).

Once building had finished, preparations were made to ship it to Madrid (a total of 52 cases). The ship that was to transport it, set sail from the port of London on January 7 1802. When it arrived in Bilbao, four wagons were purpose-built for the largest cases and the tube. The rest were loaded on to carts until all 500 *arrobas* of equipment had been moved (arroba = a unit of weight of between 11 and 16 kilos according to region).

Having reached Madrid, the instrument, *"gave rise to a fierce argument as to where it should be located. The cemetery of San Blas, recently established opposite the Observatory was chosen, with the idea of moving the former to another site. The Retiro priests refused, so Villanueva, the builder of the edifice, in deference to the aforementioned protestors, opined that it should be placed in the dome of the edifice, the necessary modifications being made to house the rotary tower, but since the edifice would become disfigured, in this manner, fear was expressed as to the action of the winds at such a height"*. Finally, the first option triumphed., *"the cemetery was removed to the place it presently occupies and the telescope located in its place; this is slightly more to the south of where the inclinometer building is presently located."* [8].

DESCRIPTION OF THE DEVICE

The 25 foot Herschel telescope is supported on a special structure with a circular base consisting of oak pillars and beams. The tube is suspended on one of its sides by a block and tackle supported on a crossbeam situated at the highest part, while the other end leans on the base. The whole structure can be rotated at the observer's will, with respect to the base, and the inclination of the tube varied, these being the main movements of the telescope.

The size of the different elements comprising the structure was decided by taking the existing illustrations of the telescope as a reference. These pictures provide an extremely faithful description, but in certain cases there is insufficient detail. As for size, there is only a reference in the bottom part of the pictures which allows for the scale of the drawings to be deduced. In order to obtain a concrete measure, a length, a thickness, or any other size, measuring was done directly on top of the illustrations. A model of the telescope owned by the National Geographical Institute, was also of great help (Fig. 4).

In Fig. 5 the above-mentioned model can be seen together with Ignacio Medina Cubillo, a teacher at this school, who sadly passed away in September, 2000. He was an enthusiastic advocate of this project and steered the work of Alejandro I. Grande, a student of the school at the time, who chose this task as his end-of-course project [9].

Fig. 4

Fig. 5

As the procedure described can lead to errors due to measurement or lack of precision in the illustration from which the data was taken, a preliminary study was carried out to discover what the most sought after elements would be, and if the calculation of their size was correct. The appropriate sizes were deduced from these prior calculations and the next step was to design the mechanical metal parts and estimate the manufacturing costs in wood and metal for the construction of all the mechanisms, both for those on the external part of the telescope as well as for those in the control room.

Estimate for the work to be done in wood

An approximate estimate of the weight of the structure to be made out of wood was 5.5 tons, bearing in mind that the wood would be oak (as it was in the original) and the longitudinal beams would be made from a single piece (the price for oaks of this height have been based on British forests). The beams would be complete tree trunks stripped of bark and cut to three or four faces, meaning that 30% of the raw material would be lost.

Design of metal parts

Costs were estimated and the manufacturing methods defined for a total of 80 different metallic parts. It is worth pointing out that 60% of these parts would need a forge workshop to manufacture them, as these were component parts that needed to be an exact replica of those in the original telescope.

This phase represents approximately 20% of the cost of the work in wood.

Of all these elements, the main ones would be:

- On the fixed base, 29 flatiron plates to join the wooden segments of the base; the carriage rotary track, in 16 metallic segments, in order to avoid wear of the wooden base; the central axis, which would allow the carriage to rotate and keep it centred in position on its base.
- In the mobile base, the central flatiron plate, whose purpose is to join the horizontal beams to the base of the mobile carriage.
- On the elevator, the iron frame, which serves as a base for the wooden handrail; on the balcony, the seat column to reinforce the resistance of the wooden seat, flatiron crossplates, to reinforce the floor and allow the sections of the balcony floor to be articulated; a pulley to raise the balcony allows the rope that raises it to pass through, and serves as a an axis for the longitudinal guide wheels,...

Control room mechanisms

These consist of a crank-handle (Fig. 6) for manually rotating the carriage; pulleys to support the counterweight, and indicators that show the different angular magnitudes that determine the position of the tube on graduated scales, etc.

265

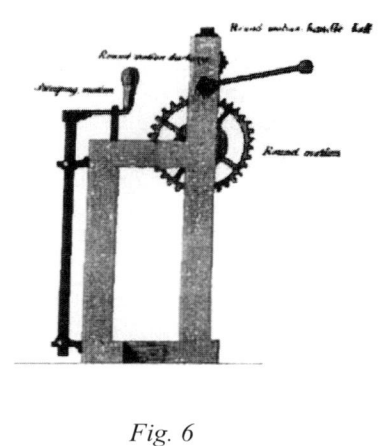

Fig. 6

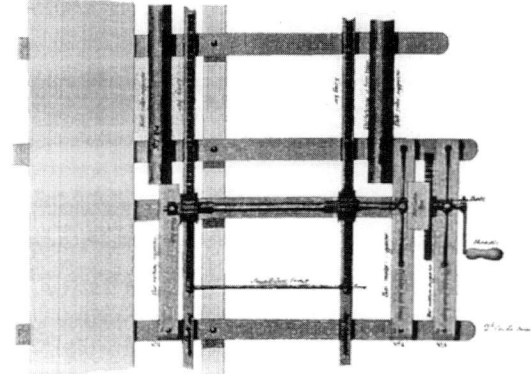

Fig. 7

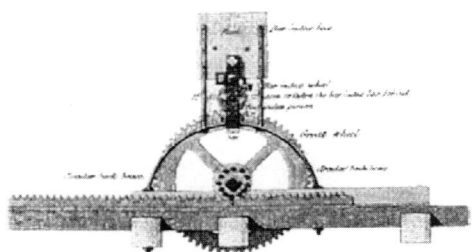

Fig. 8

Fig. 9

4.

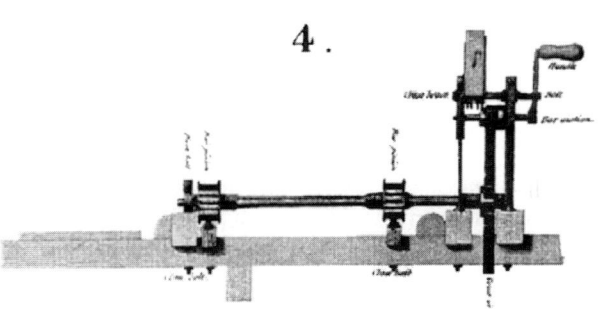

Fig. 10

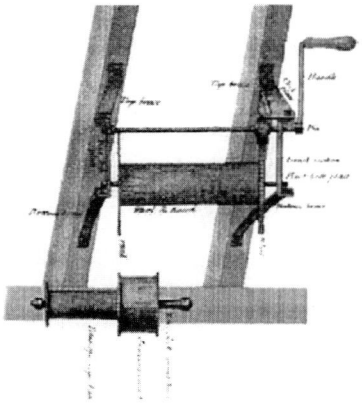

Fig. 11

Outside mechanisms

There are also a set of mechanisms on the outside. Amongst others, a double rack to transform the rotation of a pinion into linear movement in order to vary the inclination of the telescopic (Fig. 7, 8, 9 and 10); the axis on which the tube is balanced supports its weight and allows it to rotate; the hinge joining the tube to its axis which allows the inclination of the tube to be changed during rotation; the pinion of the drive chain from the crank-handle to the rack; the crank-handle on the outer part of the ladder used to raise or lower the elevator, balcony and mirror (Fig. 11, 12 and 13).

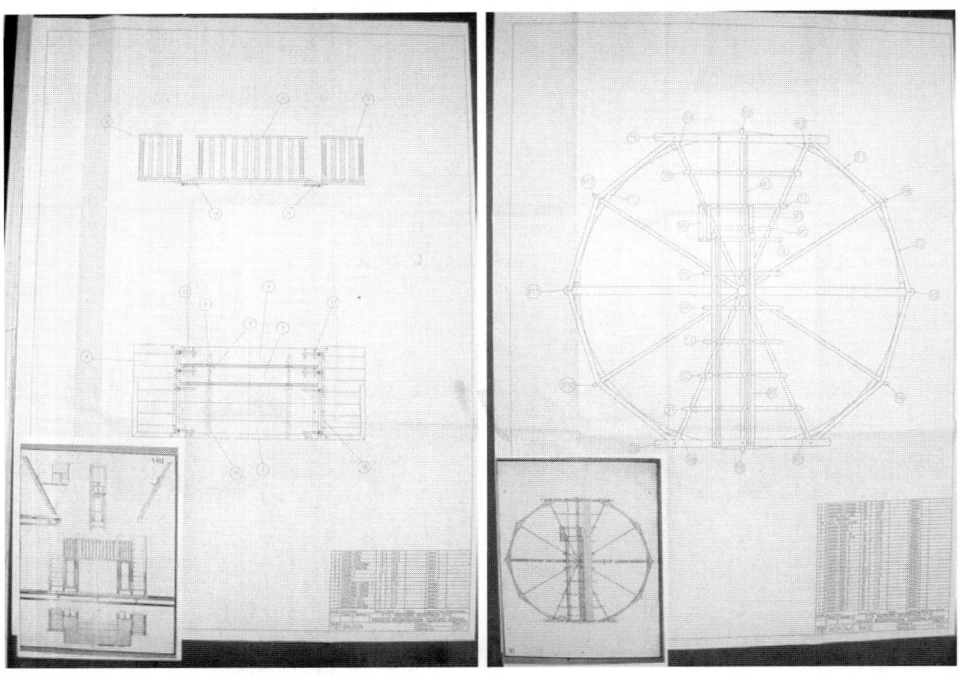

Fig. 12   Fig. 13

TRANSPORT

Transporting the telescope's support-structure was specifically analysed since it is approximately 10 m. Long. This means that any incidents that might arise out of moving the four "stairs" from their place of manufacture to their final location need to be foreseen.

One aspect to be looked at in detail is how the vehicle transporting the four "ladders" will move in the Observatory grounds given the layout of the roads.

Also, the initially planned site for the telescope, and the fact that there are buildings in the grounds must be taken into account. Therefore, the unloading area will need to be accessible to the transport vehicle, and any additional means needed to move the "ladders" from there to the their final assembly site, must be examined.

The remaining parts of the telescope are much smaller, which means that no detailed analysis is needed for their route to the assembly point. In order to produce a detailed study for the access routes for the "stairs" transport, a scale plan of the road layout and the position of the buildings was needed. The possible turning circles and any obstacles that might get in the way of the transport vehicle had to be calculated.

Considerations similar to those above had to be made concerning access for the works equipment, particularly the crane needed to unload the "ladders" and other equipment needed to erect the building housing the telescope.

ASSEMBLY

In order to assemble the telescope, it is essential that the building housing it is equipped with all the facilities needed. The architect who designed the project will also need to bear in mind all the special features of the telescope in order to plan everything required for its assembly.

LOCATION

In accordance with what was initially agreed, it appears that it will be located in the Observatory grounds, because of the ground conditions and the availability of space.

This area is higher than the buildings surrounding the Observatory and faces south. This is what the builder of the telescope initially planned for its installation, and the area is even terrain, which makes installation simpler.

Fig. 14

Moreover, there are no trees or landscaping in this area which might prevent the site from being made to match other wooded or landscaped areas later on.

So, we have passed from the 18th to the 21st century [10], by reconstructing a device, that in its time, was the talk of the entire world, and which, the same as two centuries before, would end up being taken from the Basque Country to Madrid.. *"59,373 reales were paid to Messrs. Arzula of Bilbao, for carts, wagons and transportation and on April 17, they set out from the Basque capital for Madrid."*, for it to be then installed and brought into service, ...*"on August 18, 1804, it was now possible to look through the Telescope" (Fig. 14).*

REFERENCES

1. El País. 2002. "Del Retiro a las Estrellas".
2. Juan de Villanueva. 1979. "El observatorio Astronómico de Madrid". Editor: Xarrait.
3. P. Carrasco Garrorena. 1931. "El Gran Telescopio Herschel-Mendoza, adquirido para Madrid a fines del siglo XVIII".
4. Herschel 1796. "Carta de aceptación de construir dos telescopios para el Rey de España".
5. José de Mendoza y Ríos. 1801. "Explicación sucinta de la vista del Telescopio de 25 pies ingleses de largo, para dar una idea general de este intrumento".
6. José de Mendoza y Ríos. 1801. "Iron, brass and other works belonging to the Telescope".
7. Hartman. "History of Astronomy from Herschel to Hertzsprung".
8. King. "History of the Telescope".
9. A. Grande. 1998. "Estudio de Viabilidad para la reconstrucción del Telescopio de Herschel de 25 pies del Observatorio Astronómico de Madrid".
10. E. Bautista, J.L. Muñoz , P. Leal, A. Grande. 2003. "Arqueología Industrial: del siglo XVIII al XXI. Reconstrucción del Telescopio de Herschel".

PISTON ENGINE VERSUS ROTARY ENGINE - POSSIBLE STRUCTURAL SOLUTIONS IN THE INVENTION AND DEVELOPMENT OF THE STEAM ENGINE BEFORE 1900

Klaus Mauersberger
Technische Universität Dresden
Kustodie
D-01062 Dresden, Germany
e-mail: Klaus.Mauersberger@mailbox.tu-dresden.de

ABSTRACT - Besides the so-called positive results of engineering creativity, there are also lost inventions, failed innovations, erroneous paths and obsolete functional principles which may reveal the complex and complicated development of technology. The historical facts seem to confirm a general principle according to which technical innovations are normally made in a potential field of morphological and functional variants. Depending on the prevailing social conditions, the technical demand and economic situation, basic structural ideas are formed which are subject to modification, further development or elimination and replacement when practical applications are opened up.

KEYWORDS: Kinematics, History of TMM, Mechanism Design, Rotary Steam Engines, Chamber Trains

INTTRODUCTION

Science and technology historians have recently started intense research of the issues of lost inventions and the even more extensive area of failed innovations. Its relevance becomes apparent when the question is posed if and how also those intellectual and practical achievements - which did not become promising or at least practical technical results, which did not contribute to economic gain or which even ended up in a complete fiasco - have effects in the longer run and can be historically assessed by later researchers. Because - as a rule - the technological developments that we still know today and that are a testimony of earlier engineering efforts are those which had the power to overcome critical development processes, which entered the production-line status or which became widespread thanks to favourable conditions that were not at all accidental then. Are, however, the sidelines and the sometimes painful mistakes of the technological past not worth the same attention? Dealing also with the errors, setbacks and failures finally means that we become more aware of the lively interrelation between invention and innovation.

Thus it is possible that with the creation of any engineering object alternatives may be found. It is already at the stage of mental anticipation, and even more in the design phase or the experimental stage that the variants are weighed against each other, evaluated and selected. The criteria that are applied in these processes cannot just be found in the experience and subjective knowledge and skills of the individual inventor or the team. The favourite solution expresses the time spirit as well: each engineering effort includes and expresses the actual kind and the degree of maturity of the technological knowhow, the specific mental approaches of engineers and technicians, their relation to science and practice.

PISTON-TYPE ENGINE VERSUS ROTARY ENGINE

A prototype of the reciprocating steam engine of the early industrialisation process is the James Watt design: a double-acting rotative steam engine with working beam and crankshaft. Even after further mechanical and thermal advancement the basic design of the **reciprocating steam engine** has become state of the art - at least until the general use of the steam turbine (Fig. 1).

MACHINE DE WM. POWEL, de Rouen. Médaille d'or.] — Gravure de M. Barradet.

Fig.1: Woolf's Steam Engine at the Paris World Exhibition in 1867.
The basic design of James Watt's steam engine with working beam can still be found in this efficient compound machine.

This fact makes us sometimes forget that there were alternatives to the steam drives developed. A development in the 19th century is particularly interesting: the advent of rotary steam engines, mostly in the form of rotary piston engines. American patent statistics show that there were more applications for this particular type than for the traditional reciprocating steam engines for the years following 1850. An excursion into history shall illustrate what might have motivated inventors to choose exactly this alternative, which ignored the change-of-motion mechanisms, when trying to perfect the operating steam engine and which were the prototypes they found in the old literature on mechanical engineering.

The change of motion in mechanisms and machines is a central issue of mechanisation. This is equally true for industrial production, for the manufacturing of instruments for scientific purposes and also for technical products for entertaining purposes. It is the purpose of changing motions to harmonise different types of motion (e.g., rotation and translation) between the driving source, power transmission and the place of operation. The classical mechanisms to convert motion are cranks, camshafts and mangle wheels. The far-reaching mechanisation of production, which began shortly after the arrival of the industrial revolution, produced a wide range of gear types whose basic elements had largely been available or conceived long before.

The following structural alternatives with regard to the conversion of motion or its relinquishment developed as a result of the advent of the modern mechanical engineering:

1. **Conversion of motion becomes necessary if the drive energy is provided by a waterwheel as a circular motion, for example, but the operational machine requires a different type of motion.**

   The waterwheel, which had been the most widespread mechanism of exploiting natural powers until the 19th century, was largely used to drive piston pumps or stamp mills. For this purpose, it necessitated suitable motion conversion mechansims. Although the classic designs described earlier in this paper had proved useful for certain applications, the search for alternatives went on and very often resulted in complicated or odd solutions. As early as in the beginning of the 16th century, Agostino Ramelli, for example, popularised the half-thoothed wheel in an internal rack, a modification of the mangle wheel, as an alternative to the overwhelmingly simple crank. Such bizarre mechanisms were still presented as innovative solutions in the machinery handbooks of the 17th century (Fig. 2).

   The search for practical solutions of such alternative structural designs was the result of actual and supposed drawbacks of the crank mechanisms (mass action, nonuniform motion). In 1763, Henning Calvör made the first change gear with ratchet wheel and internal racks in a quarter (Fig. 3) which was not workable under the rough mining conditions. He failed to harmonise material issues, accuracy in fitting, low friction and tooth strength at that time.

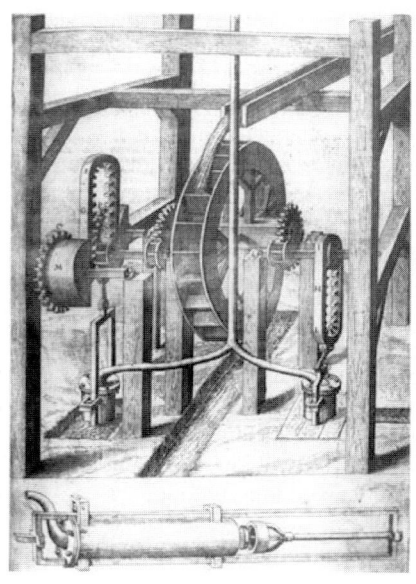

Fig. 2: Old Pump Works by Salomon De Caus (1615)
The water wheel drives two piston pumps by a complicated mangle-wheel mechanism
via a toothed-wheel gearing.

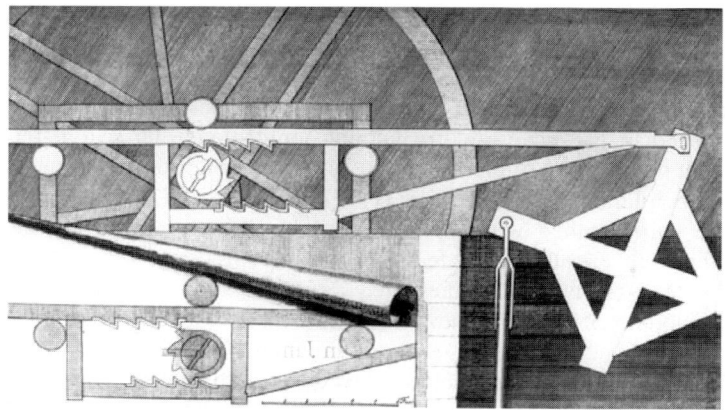

Fig. 3: Drive Mechanism for a Water Raising Structure for Mining Purposes according
to Henning Calvör (1763)
The circular motion of the water wheel in the background is transferred to the half-
toothed ratchet wheel. The latter drives an internal rack in the swaying quarter on
roller bearings. The motion is transferred by a linkwork to the pump rod.

However, the mechanical manuals of the preindustrial age offer a good supply of structural ideas even for the engineer of the 19th century. The so called "Spiegelthaler Wasserkunst" for pumping water is an example of this (Fig. 4).

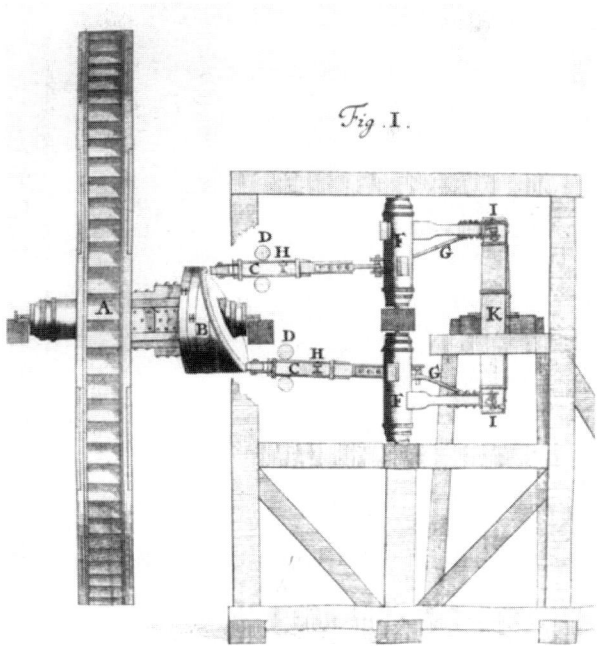

Fig. 4: The „Spiegelthaler Wasserkunst" with spatial cam (1733)
The water wheel (A) is connected to the rotating cylinder (B) with follower contacting
by friction rolls (C). Their repetitive translational motion is transferred by the
linkworks (F, G) to the pumps.

It had been built in 1733 on the inventors' costs and later destroyed because it suffered a failure. However, it illustrates the amazing attempt to make a real cylindrical cam with followers and rolls to reduce friction. The structure, which was similar to the later inclined wheel cam, was taken as a basis for new developments by no less a person than James Watt. A very elegant application of a scalloped cam for the conversion of motions was developed by the French scholars Desargues and De La Hire (Fig. 5).

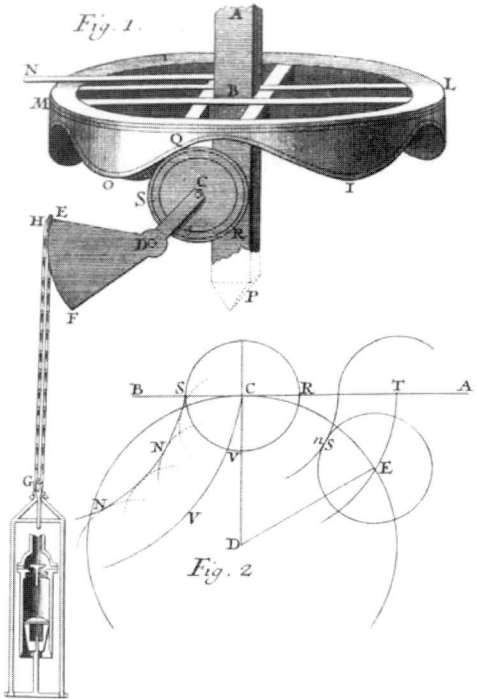

Fig. 5: Scalloped cam in a horse driven pump of Beaulieu Castle (1695)
The flat chain HG wraps around the sector arm DEF and moves the piston rod of the
water pump, attached to the yoke near G, up and down. The scalloped cam, which
pivots at F, is turmed about ist vertical axis by a horse pulling the arm extending to the
left at N.

However, crank mechanisms in connection with flywheels remained unsurpassed with regard to their reliability and functionality in practical continuous service.

2. **The steam engine has originally been designed as a pump engine for mining operation. The structural models for this drive system from the mechanical point of view had been piston pumps, linkwork and "waterpillarmachines". Watt's classic steam engine was a piston steam engine with working beam for pumping purposes.**
   However, Thomas Savery developed an early alternative design to the piston steam engine (by Denis Papin and Thomas Newcomen) – a steam-powered rotary water pump with the steam acting directly on the water.

3. **It was only when the reciprocating steam engine became a driving engine**

that could be used independently of a certain location that the use of mechanisms to convert motion became necessary. A prototype of such mechanisms, Watt's steam engine, was again a double-action piston steam engine with working beam and sun-and-planet gearing or crankshaft, respectively. The search for alternative motion converting mechanisms has always been dominating structural development efforts.

On the way towards the "Agent of Industry", the initial effort to avoid the conversion of motion had lead to a remarkable parallel development, the so-called 'intermediate machine', where the steam engine was exclusively used to drive a pump system. These pumps lifted water to a tank on a higher level from where the water directed to the top of a water wheel. The circular motion of the water wheel was suitable to drive the downline working machines (e.g. of a spinning mill).

The double-action beam engine (forward and backward motion under load) initially required a fixed connection (straight-line linkage) of the piston rod to the swaying end of the working beam. An excellent linkage was realised by Watt's parallel motion design (Fig. 6).

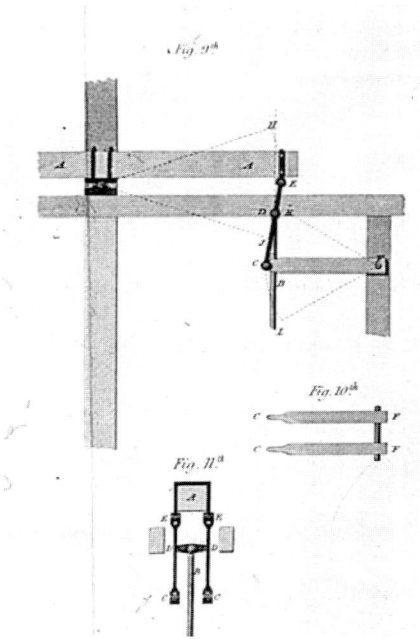

Fig. 6: James Watt's 'Parallel Motion' (1782)
One of Watt's patents for guiding the upper end of the piston rod of a double-acting engine (according to Muirhead, 1854)

Due to patent disputes, it was much more difficult to convert the translational motion of the piston into a circular motion. The necessity to do without the simplest solution - the 'classic' crankshaft mechanism - resulted in numerous alternatives. Alone the patents Watt took out between 1781 and 1784 demonstrate the large variety of inventions. The fact that James Watt used earlier models proves his good knowledge of the technical literature. This is, for example, true for the forgotten cylindrical cam used in the *Spiegelthaler Wasserkunst*, which had already been described - in modified ways - in numerous machine books by Ramelli, Jacob Leupold et al., however, in the far more complicated reverse direction of motion conversion (Fig. 7).

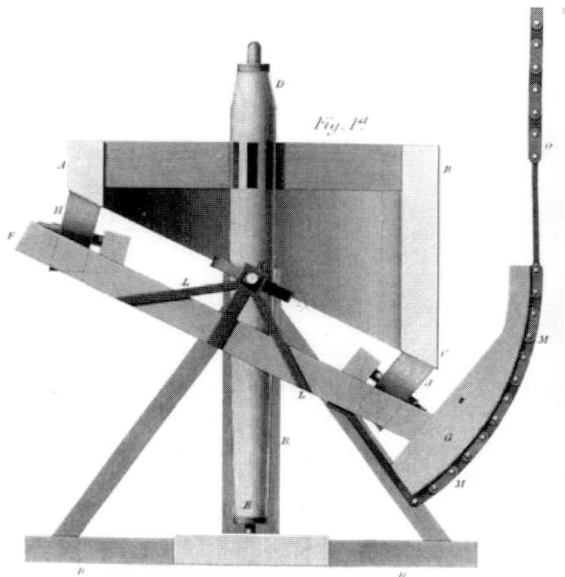

Fig. 7: "Inclined Wheel" by James Watt (patent in 1781)
The vertical shaft at D should be rotate driving by action of wheels H and J of cam, or swash plate ABC. Watt tried this device but discarded it.

The efforts made to find a structural alternative to the crankshaft soon lead to sometimes strange solutions and to very complicated drives and intricate mechanisms (Fig. 8).
Watt's well known sun-and-planet gearing had obviously been of practical relevance for a longer period of time. It was fully functional for the first time in 1786.

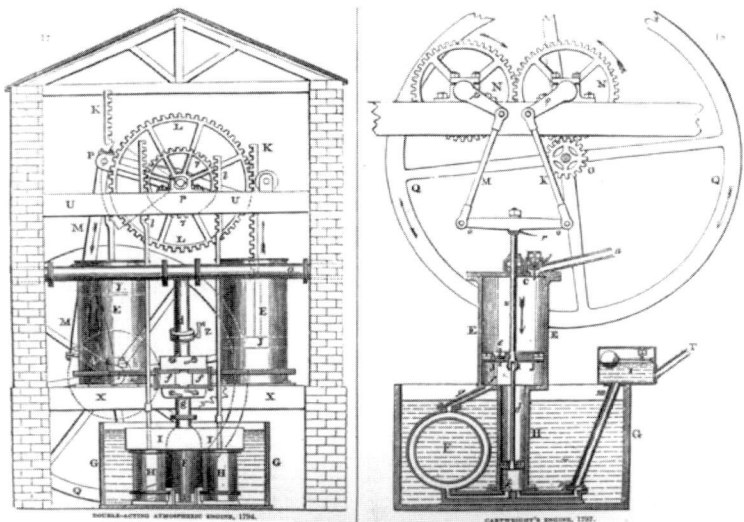

Fig. 8: Alternative devices for the conversion of reciprocating motion to rotary motion left: The idea of a two-cylinder steam engine (1794) with rack-driving is attributed to James Watt.
right: Edmund Cartwright's geared straight-line mechanism with double-crank (1797).

4. **To avoid the use of converting mechanisms which cause energy losses, direct (rotary) pumps are being developed. At the same time, efforts are put forth to advance the rotary steam engine by providing the circular drive motion for the machine not with the help of the piston engine but with rotary steam engines.**

The first known basic plans of rotary piston pumps can be attributed to Ramelli (1588) as well (Fig. 9).

Such pumps (in german "Kapselkünste") were considered innovative designs and could be used for fairly small machines because of their limited head and throughput (no suction). However, the manufacturing process, the accuracy in fitting (tightness) and efficiency were rather unsatisfying at that time so that these designs did not become true alternatives to the piston pumps. The 'Machina Pappenheimiana' described by Leupold in 1724 was also well-known.

It served as a design model for a large number of gear-type pumps and rotary steam engines. Its geometrical design had been influenced by such scholars like Kepler and Schickhard. The analogous basic idea – a direct-action rotary fire machine – was born in 1699 when Guilleaume Amonton invented the 'fire wheel'. This intricate invention, which resembles a multi-chamber water wheel that is moved by displacement and gravitation, was hardly ever put into practice and remained a design idea that is exhibited in Amonton's memoires in the Paris Academy.

Fig. 9: Rotary positive-displacement pump by Agostino Ramelli for raising water
(1588)
The functional structure of this chamber train shows a striking similarity with those of
later rotary steam engines (cf. Fig. 10).

It was only by the end of the 18th century that rotary steam engines became fully functional. Here again, James Watt had actually prepared this development by his inventions. His 1782 patent of a rotary engine with rotating piston is based on traditional designs. In 1797, Edmund Cartwright designed a refined form of this prototype with three nose-like projections on the rotary piston and used the basic idea of Ramelli's chamber pump even more clearly (Fig. 10).

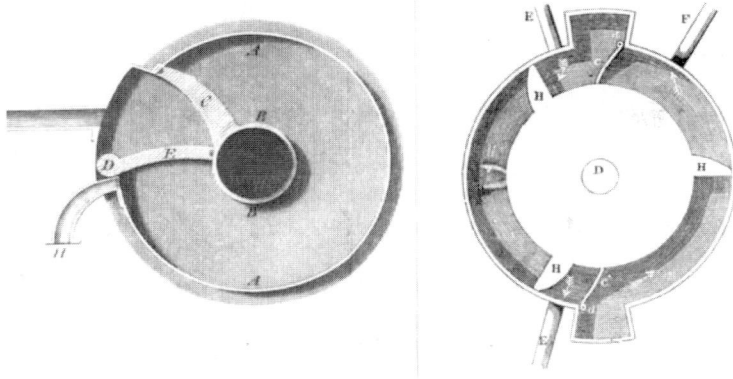

Fig. 10: Early designs of rotary steam engines
left: Watt's rotary engine with rotating completely piston and valve flap (Patent 1782)
right: Modified variant with three noses according to Cartwright (1797)

And there is also an unmistakable similarity between the chamber wheel-rotary engine, which William Murdock designed two years later, and the basic form of the Pappenheim structure. However, these interesting variants were not very promising from a practical point of view at the time they were designed mainly because of insufficient tightness. Only later, some effective improvements were made by Galloway (1834), Hall (1869) and Fabry which allowed production at a rather large scale (Fig. 11).

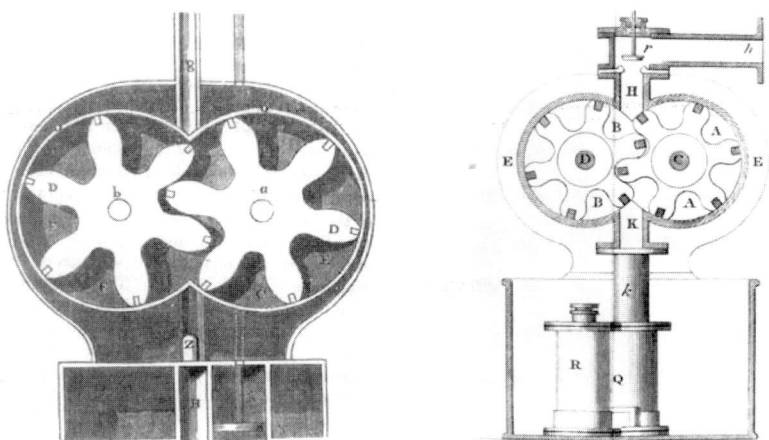

Fig. 11: Rotary engines following the chamber-wheel train principle
left: following a design by Murdock (1799)
right: operative machine by Galloway (1834)

280

There is also an apparent similarity between the basic structure and another chamber pump with sliding gates by Ramelli.

In 1875, Franz Reuleaux also used the basic structure of the said Ramelli's pump and the corresponding steam engine by Davies (Fig. 12) in his basic works on kinematics to illustrate the reversal of the functional principle of chamber-crank trains and chamber-wheel trains thereby using a distinctive symbolic language that was typical for him.

5. **The maturation of water turbines brought the basic structural form of steam turbines. Traditional operating machines initially required slow-running drives. Only later when electrical power was generated, steam turbine designs became popular.**

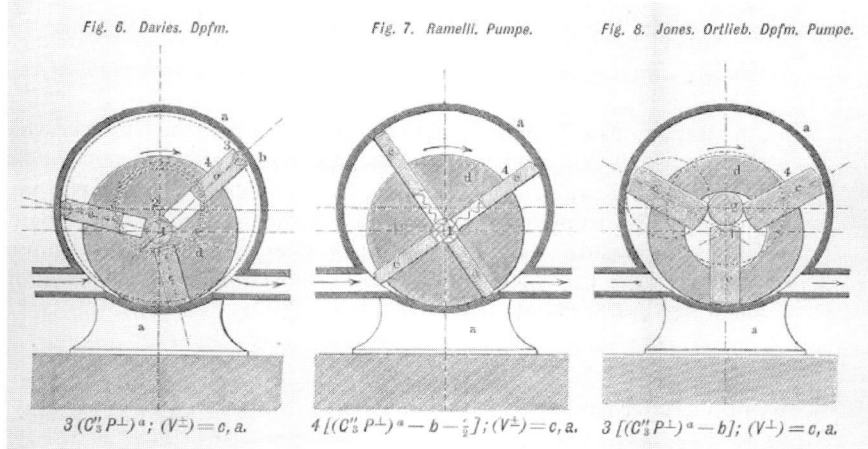

Fig. 12: Kinematic Analysis of Chamber Crank Trains by Reuleaux (1875)
This figure again shows analogies of structural characteristics between rotary engine, rotary pump and historic chamber pump ("Kapselkunst").

It remains open which of the two alternatives could prove feasible for commercial operation under the technical conditions of the industrial age. It is well known that the piston steam engines made triumphant progress in the early 19th century and were increasingly being replaced in their most prominent fields of application (shipbuilding, power generation, mining etc.) in the beginning of the 20th century. Although their efficiency could be increased (higher speed, vapour pressure and temperature), the comparatively slow-running steam engines were substituted by electric drives. Since the turn of the century, electric power has mainly been generated to drive fast-running and efficient steam turbines. Although one may see a simple structural principle in the (comparatively slow-running) rotary engines, they had never been a serious alternative to piston engines. Most of the rotary engines lead a shadowy existence in the dusty drawers of patent offices. According to experts, only the rotary steam engine developed by the Swedish Hult in 1899 was economically successful to some extent. There have not been any other competitive machines for the piston engine far and wide until recently. Thus the years around 1900 marked the end of a chapter in the history of engineering which has almost fallen into oblivion. What were the reasons why this simple and therefore promising design had not been successful from a technical and also economic point of view?

In some respect, the feverish search for the rotary steam engine was in accordance with the ancient myth of the perfect (harmonious) circular movement which engineers and technicians also tried to perpetuate. Moreover, the experts of that time believed - and this is a historic aspect - that they could make use of the advantages of the waterwheels. Unfortunately, they ignored the fact that the medium steam had completely different characteristics with regard to fluid mechanics and thermodynamics when compared with incompressible water and that it therefore required other technical solutions. The analyst Reuleaux described the mechanical problem thus that the engineers of the 19th century - following a supposed principle of simplicity - tried to operate the steam engine as a continuous transmitting system and not as a step motion system (in the terminology of Reuleaux). Many engineers even believed that motion converting mechanisms, e.g. the crank, would absorb much of the power. The engineering sciences could, however, prove that no mechanical work was lost in piston engines except for negligible friction losses. On the other hand, it was extremely difficult to put the ideal mechanical principle of 'rotation' into practice. Moreover, chamber trains are least suitable for steam operation because of their lacking ability to exploit the expansion effect. An additional critical issue was the fact that it was hardly possible to provide a continuous and smooth steam supply. The major problem was, however, unsatisfying tightness during continuous operation; the simple prism pair 'piston and cylinder' was far more advantageous in this respect and was tested in mature structural designs.

Nevertheless, an untold number of inventors were attracted by the idea of the rotary steam engine. A look into patent statistics evoke the idea of a mania. Before 1859, approximately 210 patents were granted for such machines in England; the Washington patent office received approximately 500 drawings and models until the same year of which only a small fraction was put into practice. Even as late as 1900, when the mechanically and thermodynamically ripe piston engines had reached their technical and economic peak, there were still patent applications for rotary engines in the Anglo-Saxon area. In the beginning, some inventors might have intended to get round patents on motion conversion by searching for original alternatives. For James Watt, it was the critical issue of limited cylinder travel of the then still heavy piston engine monsters which prevented him from adding a circular motion to the pumping machine and which made him apply the innovative solution of a direct rotational effect. The impulses for engineering efforts may have been diverse including psychological aspects, but the actual technological and economic situation soon made the inventors of rotary engines realise the limitations of their ideas. The developing engineering sciences contributed to the creation of the theoretical background which enabled engineers to critically examine the often unsuccessful designs of rotary steam engines.

CONCLUSION

As a conclusion it can be said that the competition between piston and rotary engines proves the readiness of inventors and design engineers to follow the principle of simplicity in technical designs. However, such criteria like efficiency and reliability in continuous operation were in favour of the piston engine. All the more, the multifaceted history of the development of rotary engines provides essential evidence of the struggling for alternative designs and functional principles in mechanical engineering. We discover the historic chamber trains that are described in the early machine literature again in the rotary engines (drive engines and pumps), which provides evidence of the continuity of the flows of empirical engineering know-how from the Renaissance to the industrial revolution. It is also interesting to know that the fast-running steam turbines, which became popular at the turn of the century mainly for the generation of electric power, rang in a new era of promising rotary engines; and in the 1960s chamber trains and rotary steam engines enjoyed another strange revival in the Wankel engine – as we can see, the history of technology is not free from repetition.

Of course the present discussion is determined by an another technological standard. But she is controversial like in the 19th century (see References). The topic of this paper is the historical background of the development of automotive engine alternatives. The historian may possibly give some inspiration for the actual debate about the advantage and disadvantage of rotary engines.

REFERENCES

Ferguson E. S., Kinematics of Mechanisms from the Time of Watt: United States National Museum, Bulletin 228, Washington D.C. 1962,

Mauersberger K., Bewegungswandlung als wissenschaftliches Maschinenproblem im Vorfeld der industriellen Revolution: NTM Schriftenreihe zur Geschichte der Naturwissenschaft, Technik und Medizin, Vol. 26 (1989)1, pp. 91-107,

Mauersberger K., Bewegungswandlung im Entwicklungsprozess der Betriebsdampfmaschine: NTM Schriftenreihe zur Geschichte der Naturwissenschaft, Technik und Medizin, Vol. 27(1990)2, pp. 57-79,

Korp, D., Der Wankelmotor. Protokoll einer Erfindung, Stuttgart 1975,

Evans; R. L., Automotive Engine Alternatives, New York 1987,

Knie, A., Wankel-Mut in der Autoindustrie. Anfang und Ende einer Antriebsalternative, Berlin 1994.

7. Mechanism Analysis and Design

Ceccarelli M.: *Historical Evolution of the Classification of Mechanisms*

Dijksman E.: *On the History of Focal Mechanisms and Their Derivatives*

Luck K., Rehwald W.: *Historical Evolution of the Pole-Theory*

Moon F. C.: *The Reuleaux Models: Creating an International Digital Library of Kinematics History*

HISTORICAL EVOLUTION OF THE CLASSIFICATION OF MECHANISMS

Marco Ceccarelli

DiMSAT, University of Cassino, Via Di Biasio 43, 03043 Cassino (Fr), Italy

ceccarelli@unicas.it

Abstract – The variety of mechanism classifications is reviewed to give a survey of the evolution of this basic analysis of mechanisms over the time. Classifications, catalogues, and basic elements of mechanisms are illustrated as the means that have been used to overview mechanism design and give unifying principles for existing mechanisms.

Introduction

Classification is a need when an object or a concept has been evolved with a great variety. Thus, a classification means is required for orienting choices but also for guiding further evolutions in the acquired knowledge and for suggesting new additional entries of the variety. In many disciplines of Science, classification of subjects and objects is obtained by formulating procedures that can be summarized basically in two approaches, namely topological views and mathematical algorithms. In practice the two approaches give tables for catalogue purposes and formulation for design aims, as final results, respectively.

In MMS (Machine and Mechanism Science) Classification of Mechanisms has been recognized as a need since the Antiquity when a multitude of mechanical designs were available and a designer/user needed to be oriented for a proper choice in practical applications. But at the same time, since the beginning of engineering activity, a deep knowledge of mechanism design was recognized as necessary in order to discover and/or formulate algorithms and procedures that could give a unified view of the great variety of existing mechanisms.

TMM has evolved to MMS in the last decades by enlarging the technical contents with other disciplines like Robotics and Mechatronics, but mainly by looking at Mechanism Design with a broader view for theory and applications.

In this paper we have attempted to overview the evolution of mechanism classifications that have been proposed over the time.

The History of MMS has been overviewed by several authors, mainly since the beginning of TMM (Theory of Machines and Mechanisms) in 19-th century.

Many authors have attached the problem of outlining the History of MMS at different level of content, in the past like for example Chasles (1837) and Reuleaux (1875), and recently like for example Hartenberg and Denavit (1956), Ferguson (1962), Hain (1967), Nolle (1974), Dimarogonas (1993), Angeles (1997), Ceccarelli (1998, 2001). Those authors have addressed to mechanism classification as a part of TMM evolution giving it a marginal attention, like for example in (Hartenberg and Denavit 1956), or a certain importance, like for example in (Ceccarelli 2001).

The subject of mechanism classification has become of primary importance since TMM has evolved to a mature discipline and it has been considered as a basic means for technical development during the Industrial Revolution in 19-th century. Thus, new ideas for mechanism classifications have been proposed as a design tool together with a review of previously adopted schemes. The review of these classifications developed in 19-th century is the core of this paper, since still nowadays we use those concepts and approaches, but with the modern Computer Technology that seems not to have stimulated further conceptual developments. This renewed use of mechanism classifications via computer for type synthesis of mechanisms has stimulated a re-examination of past procedures in order to develop computer-oriented classifications (expert systems on mechanisms). Still nowadays the classification of mechanisms is an open problem concerning with aspects for exhaustive expert systems for computer use in design procedures for type synthesis.

In this paper, the proposed overview gives a historical panorama by using significant pages of fundamental works. Thus, the historical developments are illustrated by discussing mainly the reported figures covering the main aspects of the historical evolution of mechanism classifications.

Early Classifications in the Antiquity

First studies on mechanical systems can be ascribed to Aristotele who in 3-rd century B.C. in his work "Mechanical Problems" attached problems of Mechanics with application also to mechanical design of machines. Indeed, the machines of the time were based on elementary mechanisms, although some elaborated mechanisms were built like those for theatre machines.

In Antiquity great attention was addressed to elementary machines like the lever, wheel, inclined plane and screw. Unfortunately, no original documents have passed the time but they have been reproduced and interpreted during Renaissance.

Fundamental contributions can be recognized in the works by Ctesibius, Filon, and Heron who were first mechanical engineers and investigators at the School of Alexandria of Egypt during the 3rd and 2nd centuries B.C.. Following their works Archimedes seems to be the first who attached the Mechanics of elementary machines by looking at the Kinematics and Statics of a lever and screw as fundamental machine components. This can be also considered a first attempt of cataloguing the existing machines under a unified view.

Roman engineers improved and designed machinery and they elaborated first catalogues of machines to give an overview of the variety but possibility of mechanical systems. In 1-st century B.C. relevant are the work (Frontini 1930) in the field of hydraulic mechanical systems by Sextus Julius Frontinus, and the work (Vitruvius 1511) in the analysis of machines by Vitruvius. Besides a description of the mechanical design, Marcus Polione Vitruvius explained the operation of a system with technical details that makes very clear also the teaching purpose of the work.

Early Collections of Mechanism Designs

In the Renaissance the growth of Society and production required enhancements in the machinery. Thus, new designs were conceived and used. Collections of these machines were made mainly as handbooks for designers themselves, as personal notes or preliminary explanatory sketches for construction purposes.

Examples are shown in Fig.1 to stress basic views in the form of a limited selection. In Fig.1 a), (Bechmann 1991), sketches are shown as due to Villard de Honnecourt, who in 13-th century drew several automatic devices by a primitive description by using a variety of mechanisms. The purpose of the sketches seems to be the illustration of overall machines with hints on their operation, but with deficiency of explanation that requires the designer consulting, as pointed out in (Ceccarelli 1998).

In Fig.1b), (Beck ed. 1969) a study shows different gripping devices that were designed by Mariano di Jacopo, il Taccola (1382-1458?) for a specific application. The variety of grippers can be thought as a result of both studies on the grasp and mechanism design.

In Fig.1c), (Brunelleschi 1420), the work by Filippo Brunelleschi (1377-1446) is reported as referring to crane machines that he used for his architectural goals. Different crane structures and mechanisms are shown for different construction situations.

These early collections of mechanical devices do not show a clear order in cataloguing machines and mechanisms, but they seem to be handbook notes for personal use of the designers and their co-workers. Catalogue purposes became more evident once the wide use of machines made the machines of common practice. Thus, publications of collections of machines were proposed as early handbooks at different levels of mechanism study, but to show the practical possibility and use of well-established mechanical designs. Examples of these collections are shown in Fig.2 from the time of Renaissance.

Specific attention to mechanisms that are the core of a machine was also addressed as the case of Fig.2 a), (Galluzzi 1991), illustrates in which the pump mechanisms are listed and discussed. The work by Francesco Di Giorgio is also an important indication of the maturity of a community of machine designers who were formed mainly in Tuscany and particularly in Siena as stressed in (Galluzzi 1991).

Figure 2b), (Cianchi 1984), shows notes by Leonardo da Vinci studying alternative solutions for a mechanical transmission. The ingenuity of Leonardo is great from engineering viewpoint even when he studied and re-designed existent systems with the aim to improve the clarity and effectiveness of operation and mechanical design. This gives a sure impression of the high skill of Leonard as practical mechanism designer.

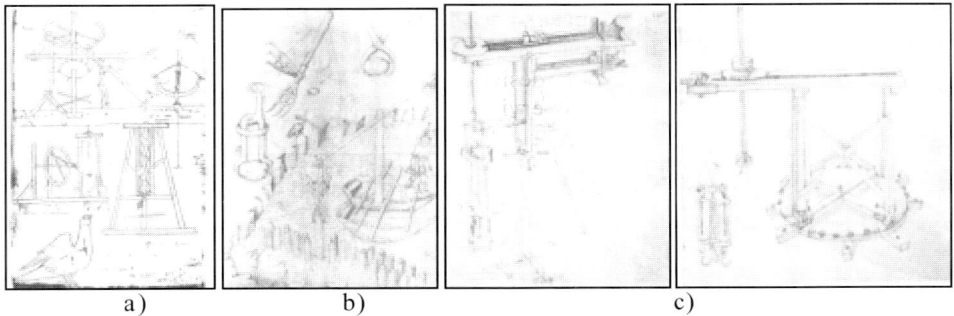

a)   b)   c)

Fig. 1: Examples of early collections of mechanisms by: a) Villard de Honnecourt in 13th century; b) Mariano di Jacopo (il Taccola)(1382-1458?); c) Filippo Brunelleschi(1377-1446).

288

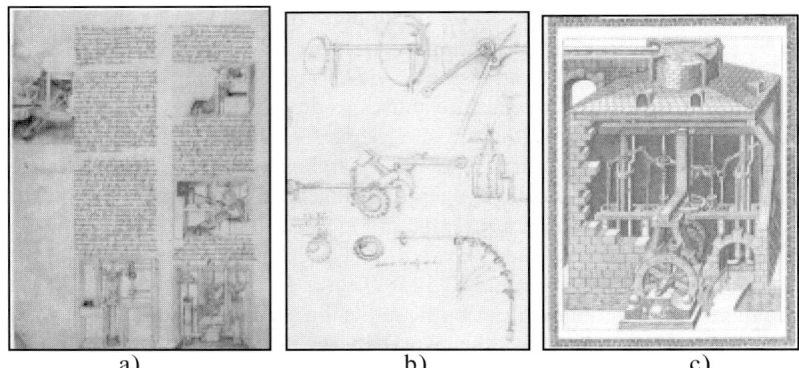

Fig. 2: Examples of early collections of mechanisms by: a) Francesco Di Giorgio (1439-1501); b) Leonardo da Vinci (1452-1519); c) Agostino Ramelli(1531?-1608?).

In (1588) Agostino Ramelli published a first very successful collection of machines with explanatory notes for a large public. In Fig.2c), (Ramelli 1588), an example is shown regarding a complete machinery that is also explained with a specific text. The machinery is completely illustrated and the explaining text is specifically addressed to the machine operation. This kind of approach will be common to machine collections as Theatrum Machinarum in the 17-th and 18-th centuries.

The use of word Theatrum (theatre) in 17-th and 18-th centuries refers to the aim to show machinery to a wide public with both illustrations and explanations, which can be helpful for a general understanding but indicating the expertise of the author who is available for consulting. Examples are shown in Fig.3.

The collections of mechanisms were more and more riches of used machinery and the mechanisms also evolved with more elaborated solutions. Thus, the schemes became more detailed and presentation became more attractive, like in the examples in Fig.3a), (Besson 1578), from France and Fig.3b), (Branca 1629), from Italy. It is to note from Fig.3a) the machine power is indicated by showing a man operating the system; from Fig.3b) the machine system is illustrated emphasizing its automatic operation.

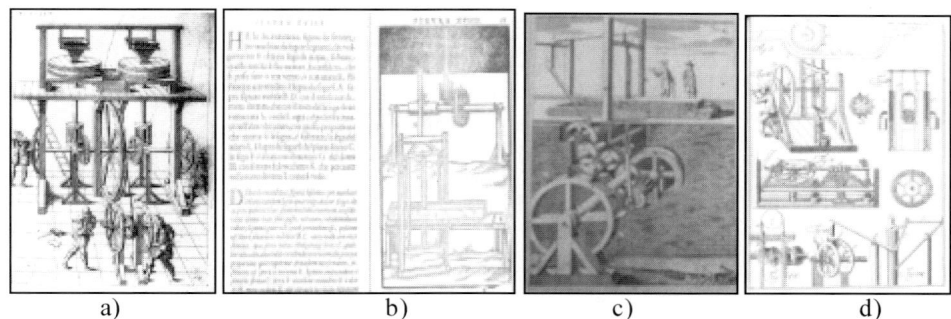

Fig.3: Examples of collections of mechanisms in the form of Theatrum Machinarum by: a) Jacques Besson in (1578); b) Giovanni Branca in (1629); c) Georg A. Boeckler in (1661); d) Jacob Leupold in (1724).

Both cases follow the tradition of machine collection with explanatory notes that was established by Ramelli, but they added a better order in the presentation of machines by grouping the machines as functions of the applications and the mechanism types. This last aspect has become a technical common view, even with a more technical representation of machinery, as observable in the examples of Fig.3.

In 17-th century the mechanism collections have been enriched and the size of the mechanical designs were made for larger and larger power, but preserving somehow the old mechanism designs with practical enhancements for productivity aims, like in the example of Fig.3c), (Boeckler 1661).

The collections of mechanisms became also a sort of Encyclopaedia for the wide variety of illustrated devices, like the case of the work (Leupold 1724). In addition, this kind of survey gave also the possibility to show unusual mechanism for that time, like the case of the spatial 4-bar linkage in the right bottom corner of Fig.3d), (Leupold 1724).

Thus, mechanism collections evolved clearly to encyclopedic surveys that show technical details also for a large public, like the case referring to the Eciclopediè by D'Alembert and Diderot (1774). Moreover in Fig.4, (D'Alembert and Diderot 1774), a clock mechanism is shown in an early modern drawing for assembling that is used to show both basic components and mechanism complexity. This shows also the application of a variety of mechanism solutions as well-established practice. However, the encyclopedic view for mechanism design, that started during the Illuminism, was persistent along the 19-th century, with an evolution to more technical contents, like for example in (Poppe 1803) and (Rankine 1887).

Both in the Theatrum Machinarum and early Encyclopaedias there are not clear attempts to classify the variety of mechanisms under few principles that could help the choice of a solution among the many available ones. But the aim seems to astonish a reader with the variety and complexity of the illustrated mechanisms and machines. From technical viewpoint the Theatrum Machinarum and early Encyclopaedias seem to be a ordered collection of available machines that contain also explanatory indications for successful applications. Thus, catalogue purposes can be considered well established at the end of 18-th century but classification aims with unifying views or principles are still at the beginning also because the mathematization of mechanism design has no yet reached the necessary algorithm formulation.

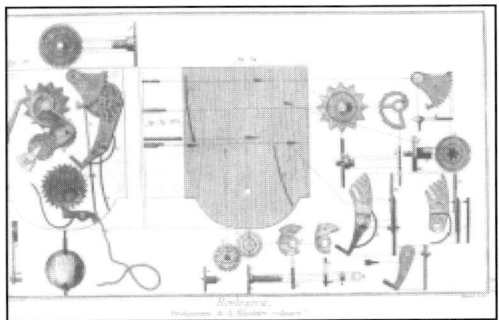

Fig.4: An example of mechanism design from the Encyclopediè by J.B. D'Alembert and D. Diderot in (1774).

Basic Elementary Mechanisms over the time

Since the beginning of an Engineering Technology an alternative way to classify mechanisms has been considered the identification and study of basic elements that can be used to assemble a mechanism or a machine. Indeed, the analysis of basic components has been understood as a principle under which recognizes common nature and design rules for the variety of mechanisms.

In 3-rd century B.C. Archimedes was the first who attached the problem for an overview of mechanisms from engineering viewpoint by studying the Mechanics of the so-called basic machines. Particularly he analyzed the lever and screw, and he used the results also to design new machines, like for example the so-called Archimedes screw pump, as pointed out in (Koestier 1999). His approach was considered as a milestone reference for long time, through the Roman engineers up to the Renaissance designers.

In the Renaissance, with the growth of interest and use of machines, basic elements address great attention mainly for mechanical design purposes. Thus, the examples in Fig.5 show studies of basic elements of mechanisms in terms of joints but the classification aim is not yet evident. In Fig.5a), (Cianchi 1984), Leonardo studied both operation and construction of basic mechanism joints and he reported those possibilities as an early catalogue variety. In Fig.5b), (Ramelli 1588), Ramelli studied the practical construction and operation of gear systems as fundamental parts of power transmissions.

A fundamental work that can be considered a milestone for TMM, is the book by Guidobaldo Del Monte in (1577). This is a first rigorous treatise that is fully dedicated to the analysis of Mechanics of machines and mechanisms by looking at elementary machines: lever, pulley, wheel and axle, wedge (inclined plane), and screw as basic principles for analysis and design of mechanical systems. Each basic machine is studied in a chapter with great details, even by using very early kinematic schemes, as shown in Fig.6, (Del Monte 1577). The approach refers to Archimedes's view by which a fundamental system is the screw that can be studied by referring to the Mechanics of inclined plane that finally can be modeled as a lever.

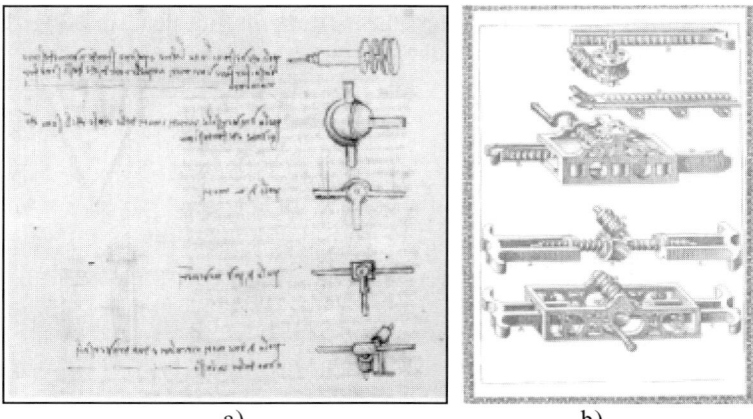

a)   b)

Fig.5: Early studies of elementary components for mechanisms: a) joints by Leonardo Da Vinci in 14-th century; b) gear connections by Agostino Ramelli in (1588).

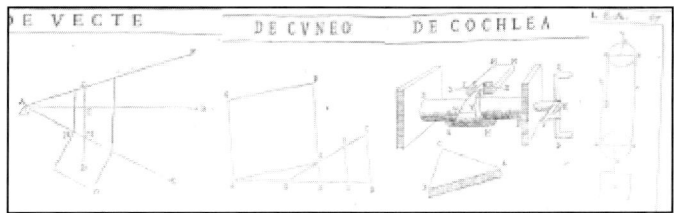

Fig.6: Early studies of elementary mechanisms by Guidobaldo Del Monte in (1577).

Thus, once a system is sketched, the basic behavior is formulated as an equivalent inclined plane by using mechanical principles for levers. This can be considered a first unifying principle for mechanism classification with a mathematical formulation.

A further improvement of the study of Del Monte's approach can be found in the early work by Galilei in (1593) in which the description and discussion have been deepened. This brought the attention of the Academic world to the mechanical practical systems. The subject of Mechanics of machines and mechanisms addressed attention at the end of Renaissance for mathematization of mechanism Mechanics and several studies were published that used approaches similar or deduced from Del Monte's study. A relevant example is reported in (Stigliola 1597) from Naples with the aim to stress that such an interest on mechanism design was well established even with very early standardized kinematic schemes that were used also for classification purposes.

The interest on basic elements of machines was considered along the 17-th and 18-th centuries by following Del Monte's approach and no relevant novelties have been proposed.

A renewed interest on basic mechanism elements arose when a modern view of mechanisms was conceived with milestones works at Ecole Polytechnique not only in TMM, but also in Geometry and Mechanics. A significant example is the work (Coulomb 1821) in which the basic mechanisms are treated with a modern approach and detailed schemes as shown in the examples in Fig.7a), but mainly by examining their operation through results of early modern experimental Mechanics.

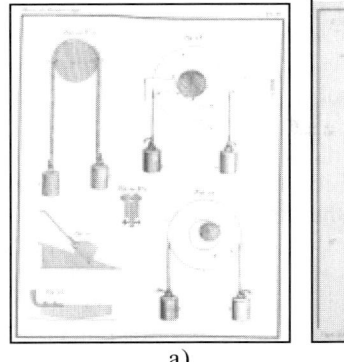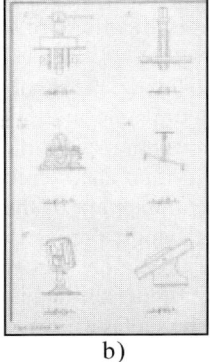

a)   b)

Fig.7: Examples of schematic design for elementary kinematic elements by:
a) C.A. Coulomb in (1821); b) F. Masi in (1883) according to (Reuleaux 1875).

292

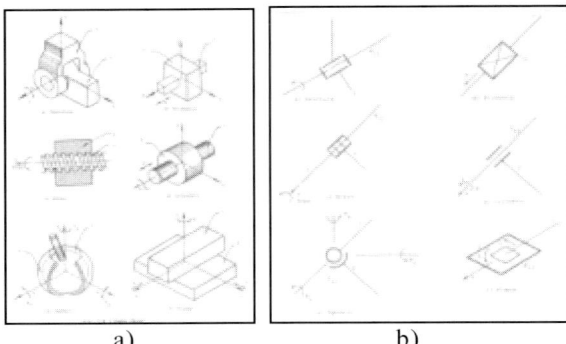

a)   b)

Fig.8: Basic joints of mechanisms as studied at present time: a) mechanical designs; b) kinematic schemes.

In addition, milestone works on mechanism classifications by Robert Willis in (1841) and Franz Reuleaux in (1875) gave a further attention to catalogue and classification of basic components of mechanisms. Reuleaux considered the elementary components in term of joints as fundamental for mechanism architectures, even for a mechanism notation. The joints were analyzed in great details and suitable schemes were elaborated for catalogue purposes, like in the example in Fig.7b), (Masi 1883). Those schemes of joints were the bases for mechanism classification and notation and still nowadays they are used with almost the same mechanical design and kinematic schemes as shown in Fig.8. Indeed, this approach has still the basic features of the view by Del Monte.

Early Modern Classifications with the Establishment of TMM
Modern TMM was established at Ecole Polytechnique at the beginning of XIX-th century, when industrial engineering was recognized as independent of a military formation. Indeed, this need was felt and gave start of technical Universities in many countries all around the world.
At Ecole Polytechnique in Paris, Gaspard Monge (1746-1818) and Jean Nicolas Pierre Hachette (1769-1834) classified mechanisms as a function of the type of motion transformation that can be obtained from input link to output link in a mechanism. Tables of a general classification are shown in Fig. 9, (Lanz and Betancourt 1808) and in Fig. 10, (Hachette 1811).
They classified all existing mechanisms in 21 types of motion transformation by considering the change of one motion into another motion in the categories of linear, circular, or curvilinear motions with the possibility of continuos or alternative operation. The classification algorithm can be recognized in the indices of rows and columns that in general tables give a mapping of the above-mentioned principle of motion catalogue. J.M. Lanz and A Betancourt prepared the book (Lanz and Betancourt 1808) by using their previous experiences on mechanism design and first classes on mechanism Kinematics by Monge, under the supervision of Monge and Hachette.
The novelty of Monge's classification with respect to previous machine catalogues consists in the fact that Monge addressed attention to capability and architecture of mechanisms in order to illustrate design possibility of mechanisms and machines but without referring to specific applications. In fact, previous machine collections, both

Theatrum Machinarum and early Encyclopaedias, had the aim to show and explain general behavior of machinery as a whole for well-defined applications.

Gian Antonio Borgnis extended this view by including a description of machine components in terms of "receptors, modificators, frames, regulators, and operators"; but then he catalogued the existing machines also by referring to practical uses, as pointed out in (Ceccarelli 2000). His Encyclopedic work in 9 volumes (Borgnis 1818-21) was used as reference handbook by practicing engineers along the whole 19-th century, as a first modern technical handbook. A specific volume on technical terminology on machines and mechanisms completes the catalogue and classification purposes of Borgnis's work.

In the tables by Lanz and Betancourt in Fig.9, and by Hachette in Fig.10 there is also a first attempt of notation for catalogue purposes, if one considers the names and number codes on rows and columns.

Since then, notation was recognized fundamental for mechanism catalogues and analysis. Charles Babbage presented a relevant attempt of a detailed notation in Babbage (1826). But, as one can understand from the example of Fig.11 a), the Babbage's notation was considered too cumbersome and unpractical and thus it was quickly forgotten. The complexity of a graphical representation for a table codification was not considered useful and even difficult to understand.

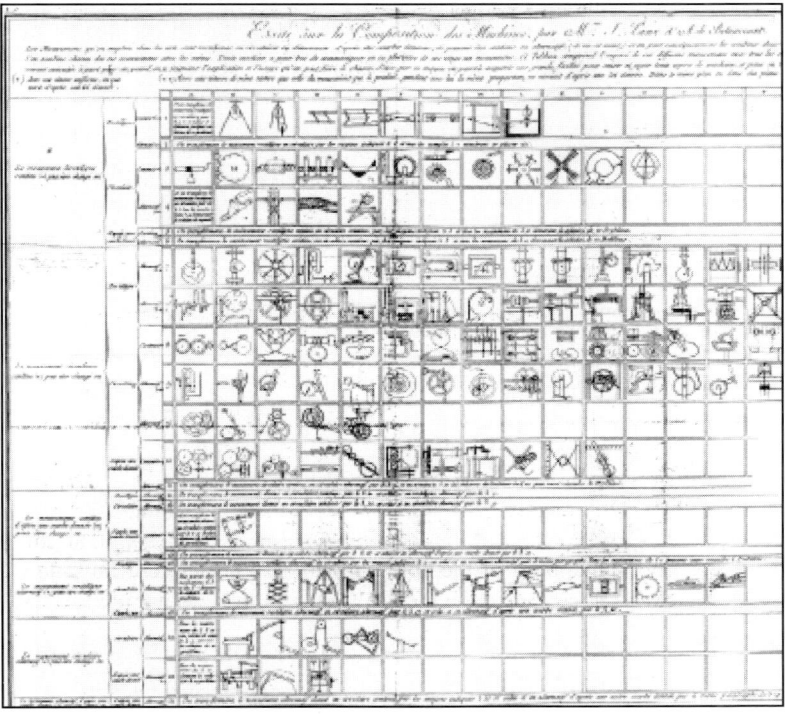

Fig.9: The table for classification of mechanisms in the "Essai sur la Composition des Machines" by JoseMaria Lanz and Agustin de Betancourt published in 1808.

294

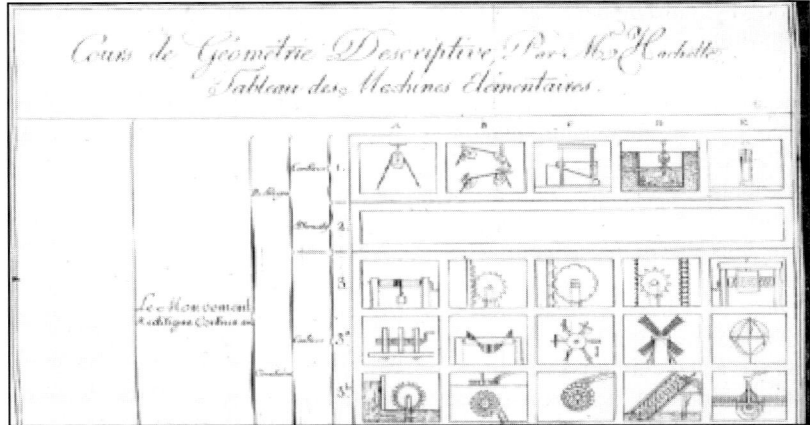

Fig.10: The table for classification of mechanisms in the "Traite elementaire des Machines" by Jean Nicolas Pierre Hachette published in 1811.

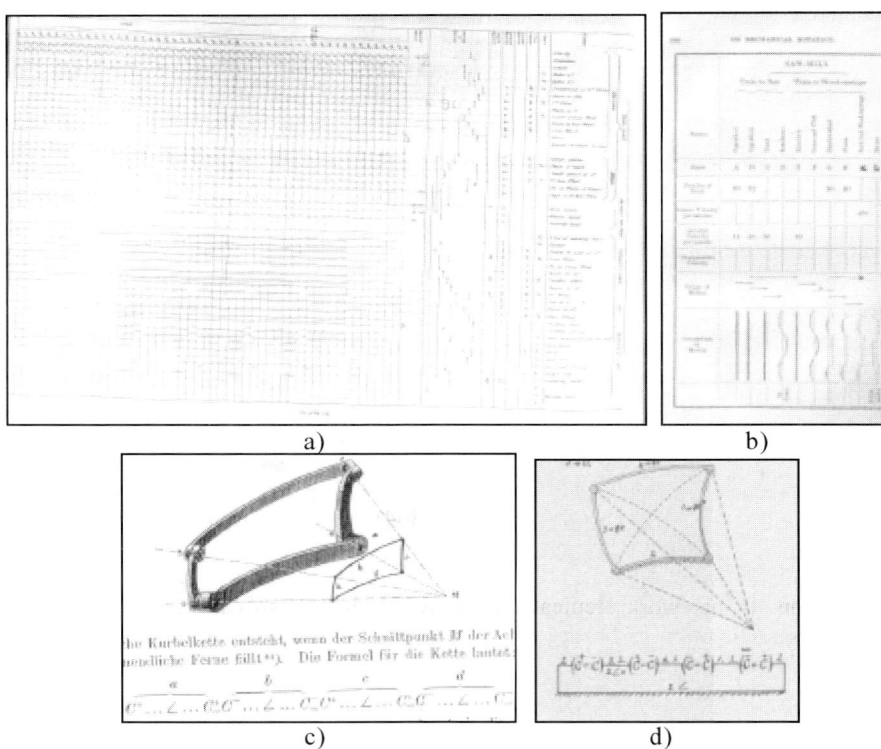

Fig.11: Examples of notations for mechanism classification by: a) Charles Babbage in (1826); b) Robert Willis in (1841); c) Franz Reuleaux in (1875); d) Francesco Masi in (1883).

In Fig.11b), (Willis 1841), the example for a saw-mill mechanism is reported as proposed by Willis to explain his notation and its practical application. Willis notation is based on lines indicating direction of motion between connected mechanism links and makes use of word and numbers to identified the types of mechanism element and its size or characteristic, respectively.

Reuleaux conceived a notation by looking at the chemical formulas. Thus, he assigned a character symbol to each basic element of mechanisms and then the kinematic pairs are characterized with signs to indicate the type of relative motion. A sequence of the characters and symbols for the architecture of mechanism assembly with further signs for the kinematics of the pairs give a complete synthetic description of a mechanism. An example as proposed by Reuleaux is reported in Fig.11 c), (Reuleaux 1875). A further improvement of this notation has been elaborated by Francesco Masi and an example for a spherical four-bar linkage is illustrated in Fig. 11 d), (Masi 1883), in which several more details are given but with the structure of Reuleaux notation of Fig.11 c).

Improvements and Enlargements of Classification Criteria

Monge's approach was considered for long time, even as a first view form which new classifications have been derived. In (1841) Robert Willis proposed to classify mechanisms in three classes depending on the value of transmission ratio and in five divisions depending on the type of contacts among basic components. He considered the classes as functions of the velocity-ratio as constant as constant or varying with constant or changing direction of the motion, and he classified the divisions in term of "rolling contact, sliding contact, wrapping connectors, link-work, reduplication". Thus, he analyzed mechanisms within the classification frame with a common view and notation that can be recognized as a first modern kinematic schematization.

Willis combined the description of mechanism with a mathematical analysis by deducing and using formulas, although still with a specific formulation for specific mechanism architecture. Willis's classification was successfully accepted in brief time all around the industrialized world in teaching textbooks and professional handbooks, like for example in (Giulio 1846), (Rankine 1887) (Robinson 1896), and its notation and concepts were used for long time and still can be found in current approaches.

In (1861) Laboulaye enlarged the Monge's classification by adding a classification of basic elements in the form of rod links, screws, and plane systems. Several further studies were proposed on the basis of Monge's and Willis's classifications in order to have a more exhaustive unifying principle for the analysis of mechanisms. Relevant examples are the works (De La Goullepierre 1864), (Belanger (1864), and (Morin 1872) developed mainly along the French tradition.

Finally, in (1875) Reuleaux conceived the modern TMM view for mechanism classification. In his work Reuleaux approached the classification and catalogue of mechanisms as function of basic components of machinery by introducing an analysis of kinematic architectures and relative motion among mechanism links. Indeed, he introduced the concept and corresponding notation for kinematic chains, kinematic pairs, and mechanism links, so that he could reduce the wide variety of mechanisms to few basic combinations of pairs and few classes of kinematic chains. For these last results he introduced and used successfully the concepts of kinematic inversion that gives the possibility to consider several mechanisms as belonging to the same kinematic architecture chain. The Reuleaux's approach was immediately accepted as a milestone work and it was translated in many languages. Some refinements were also proposed to obtain more practical application even for design purposes, like in the brilliant work by Francesco Masi in (1883) in which he could classify up to 10,362,600 composed quaternary mechanisms. From practical viewpoint Reuleaux catalogued mechanisms in six classes: screw mechanisms, crank mechanisms, gear mechanisms, pulley mechanisms, cam mechanisms, and locking mechanisms. This classification is till used today both for teaching and practice purposes.

However, collections of mechanisms were still of great interest for designers' practice, as for example the success of (Hiscox 1899) illustrates. In this work mechanism drawings are still illustrated with a short explanatory test, but kinematic details are included and emphasis is addressed to mechanisms and not to whole machinery.

Kinematics was deepened and results were used also for classification purposes of mechanisms, whose variety increased day by day. Very significant is the table in Fig.12 from the work by Lorenzo Allievi in (1895), who applied the theory that is outlined by Ludwig Burmester in (1888), in order to classify mechanisms with respect to geometric and kinematic properties of coupler curves.

The concept of kinematic chain was further exploited by classifying mechanisms with a unifying principle that in (1913) Assur recognized in the identification of basic chains with basic kinematic behaviors. By using those basic chains, today named as Assur Groups, it is possible to assemble any mechanism but particularly to deduce the corresponding Kinematics formulation by using an assembling of expressions for used Assur Groups, as shown in the example of Fig.13. In this classification procedure one can recognize a very early use of the concept of Graph Theory for mechanism analysis.

In (1943) Rudolf Franke attempted a further extension of the concept of a mechanism by including electrical systems, fluidic systems, and mechanical systems whose relative motions of the parts do not depend of the connections only. Franke classified mechanisms as: constrained mechanisms, which are 1 d.o.f. because of the relative motion ensured by kinematic pairs; partially constrained mechanisms whose relative motions depend of external actions like spring and inertia forces; mechanical systems in which the d.o.f.s change during the operation.

Another basic means for general mechanism classification can be recognized in the mobility criterion by which the motion capability of a mechanism is synthetically evaluated by the number of degrees of freedom (d.o.f.) for the mechanism motion. The mobility criterion is based on Grubler's formula, (Grubler 1917) and its extension and modifications, which are known as Kutzbach criterion and modified Kutzbach formula. Nowadays the mobility degree of freedom of a mechanism is used as a first classification means, even for design purposes.

Modern Handbooks with Mechanism Classification

The use and usefulness of machine and mechanism catalogues have been persistent over the time because of their practical help in Mechanism Design. Of course, the approach and presentation have evolved as depending of the conceived variety of mechanisms and acquired knowledge on Theory and Practice in Mechanism Design. Thus, modern handbooks have been elaborated not for full machinery but for main mechanisms for specific applications either for a defined task or as a machine component.

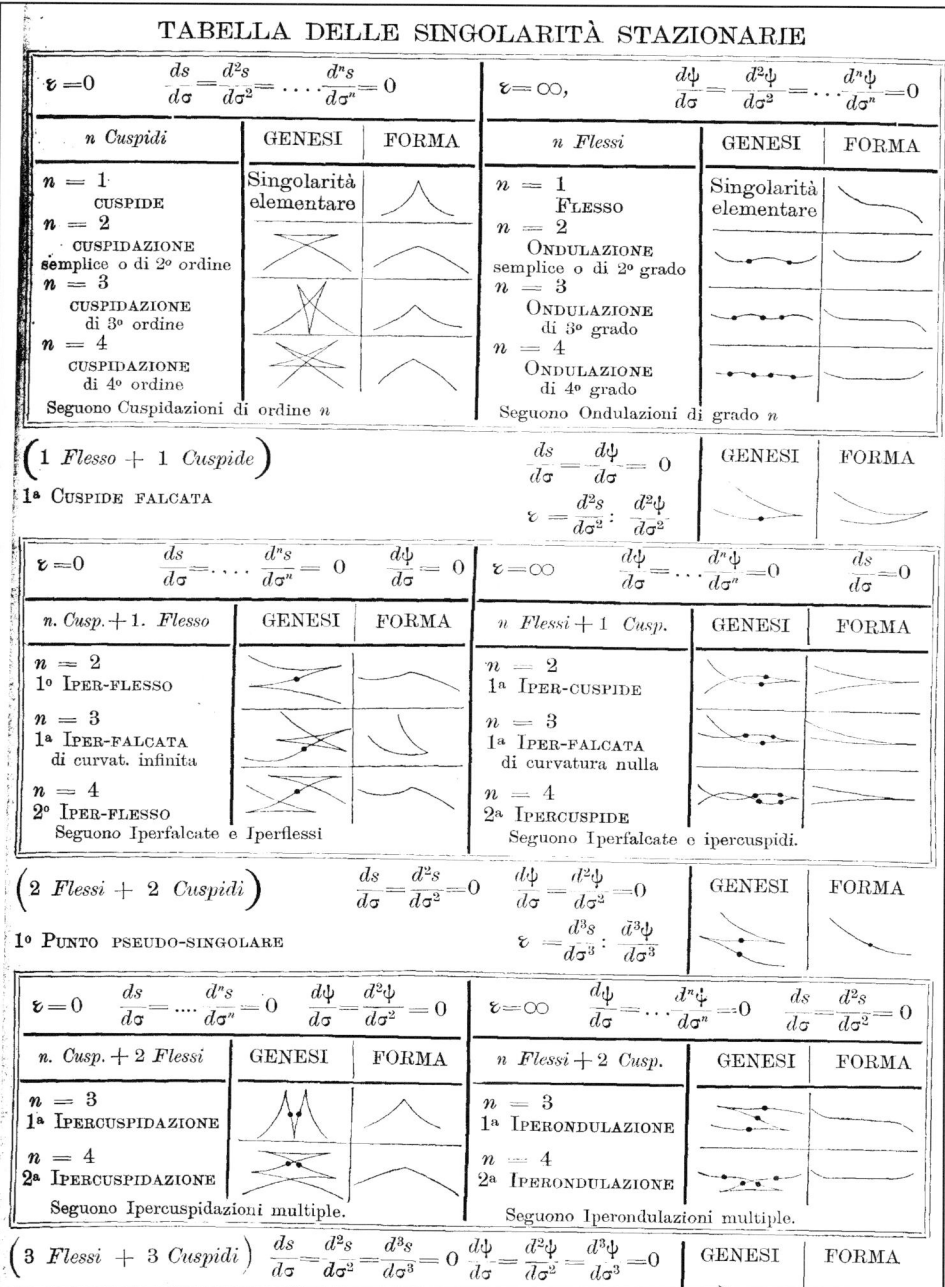

Fig.12: Part of the table for a classification of coupler curves by Lorenzo Allievi published in (1895).

298

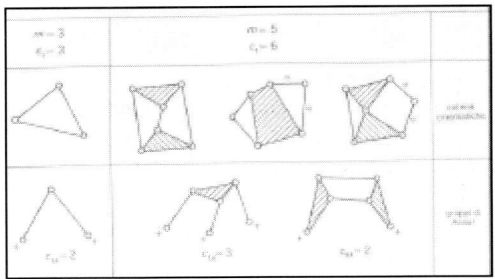

Fig.13: An example of Assur groups in (Ghigliazza and Galletti 1986).

In the example of Fig.14, the handbook by Chironis in (1961) is illustrated with examples of mechanical systems. Each mechanism is shown with basic features on mechanical design and operation, with hints on kinematical schemes and characteristics. A short test helps for a full understanding of the figure yet (like in Theatrum Machinarum !).
However today those handbooks have not a wide diffusion or at least there is not a relevant new production of them since there are already fundamental works with exhaustive collections of existing mechanisms. Very significant is the Encyclopedic work by Ivan I. Artobolevski in (1975-80), in which one can recognize a synthesis of theoretical classifications with practical catalogues of mechanisms.
The classification purpose of Artobolevski's work is synthesized in Table 1of his work, whose a part is shown in Fig.15, in which basic architecture features of mechanisms are used together with mechanism types.
The catalogue purpose is reported in Table 2, in which the mechanisms are listed depending of characteristic applications of mechanisms, but still considering the classification view of Table 1.

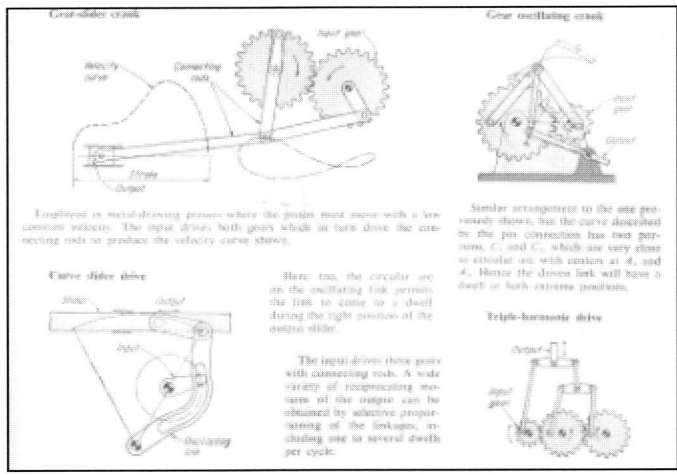

Fig.14: Examples from the handbook by Nicholas F. Chironis in (1961).

Even the mechanism presentation can be seen as fusion of past and modern approaches for mechanism analysis. Kinematic characteristics of mechanisms are outlined by identification of significant points and trajectories. In the enclosed test for each mechanism a description of operation and design characteristics clarifies the interest for the mechanism. The mechanism sheet is completed with an alphabetic code for classification purposes and a numerical code for catalogue aims. The name of the mechanism recognizes also a designer's paternity when possible.

Classifications for Computer Uses

The new technology of computer calculation has influenced all the Engineering disciplines, and analysis and design approaches have been re-formulated for computer calculations. Thus, even mechanism classification has been reconsidered with the aim to achieve algorithms for numerical and/or automatic type synthesis of mechanisms.

Table 1

CLASSIFICATION OF MECHANISMS
BASED ON STRUCTURAL FEATURES

Group No.	I			
Group name	Elements of Mechanisms			
Group index	EM			
	No.	Name	Subgroup Index	Mechanism No.
	1.	Kinematic pairs	KP	1 through 54
	2.	Movable joints	MJ	55 through 119

Group No.	II			
Group name	Simple Lever Mechanisms			
Group index	SL			
	No.	Name	Subgroup Index	Mechanism No.
	1.	Lever mechanisms	L	120 through 162
	2.	Gripping, clamping and expanding mechanisms	GC	163 through 245
	3.	Balance mechanisms	B	246 through 251
	4.	Brake mechanisms	Br	252 through 257
	5.	Stop, detent and locking mechanisms	SD	258 through 334

Fig.15: Classification of mechanisms by Ivan I. Artobolevski in (1975-80).

300

A fundamental new approach has been developed in the 1950's by using modeling and mathematics of graphs in order to obtain formulation that is suitable for computer implementation. A synthetic view of the method is reported in Fig.16, (Tsai 2001) from a very recent text on Mechanism Design. But even in this modern abstract mathematical representation of mechanisms, schemes for kinematic chains and their classifications are needed to complete a modern design procedure but mainly the engineering interpretation, as shown in Fig.16.

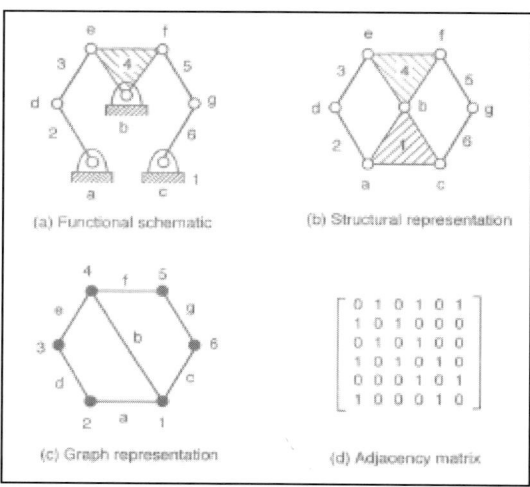

Fig.16: A mechanism model and mathematization by graphs by Tsai L.W. in (2001).

Conclusions

The classification of mechanisms has evolved through two main views, namely the study of basic elements and catalogue of the existing devices. The classifications and catalogues of mechanisms have been evolved from pure practical needs to pure theoretical approaches through step by step modifications and enlargements, although concepts have made relevant changes also in mechanism design. Many of the past approaches can be still of current interest and application for teaching, research, and practice, even for the Computer Oriented Technology in the new Millennium. The completeness of the current mechanism classifications that have been proposed and evolved over the time may rise a question: is it possible to conceive new mechanisms for practical applications and to enlarge the mechanism variety?

References

Allievi L, 1895, "Cinematica della biella piana", Regia Tipografia Francesco Giannini & Figli, Napoli.

Angeles J., 1997, "A Fin-de-Siecle View of TMM", Proc. of Int. Conference on Mechanical Transmissions and Mechanisms, Tianjin.

Artobolevsky I.I., 1975-80, "Mechanisms in Modern Engineering", Mir Publ. Moscow, 5 Vols.

Assur L.V., 1913, "Reserches sur la structure et la classification des mecanismes planes articules à paires cinematique inferieures", Annales de l'Institution Polytechnique de St. Petersburg, Vol.21, pp.187-283.

Babbage C., 1826, "On a method of expressing by signs the action of machinery", Philosophical transactions of the Royal Society, London, vol. 116, pp. 250-265.

Bechmann R., 1991, "Villard de Honnecourt, Le pensée technique au XIII siècle et sa communications", Paris.

Beck J. H. (Ed.), 1969, Mariano di Jacopo (il Taccola), "Liber tertius de ingeniis ac aedifitiis non usitatis", Milan.

Belanger, 1864, "Traitè de Cinèmatique", Paris.

Besson J., 1578, "Théatre des instruments Mathématiques et Mécaniques...", Lyon.

Boeckler G.A., "Theatrum Machinarum Novum", Nuremberg, 1661.

Borgnis G.A. , 1818-21, "Traitè complet de mecanique appliquée aux arts", Bachelier, Paris, 9 Vols.

Branca G. , 1629, "Le machine", Roma.

Brunelleschi F., 1420, "Zibaldone", Firenze.

Burmester L., 1888, "Lehrbuch der Kinematik", Leipzig.

Ceccarelli M., 1998, "Mechanism Schemes in Teaching: A Historical Overview", ASME Journal of Mechanical Design, Vol.120, pp.533-541.

Ceccarelli M., 2000, "Italian Kinematic Studies in XIXth Century", International Symposium on History of Machines and Mechanisms - Proceedings of HMM2000, Kluwer, Dordrecht, pp.197-206.

Ceccarelli M., 2001, "The Challenges for Machine and Mechanism Design at the Beginning of the Third Millenium as Viewed from the Past", Proceedings of Brazilian Conf. on Mechanical Engineering, Uberlandia, Vol.20, pp.132-151.

Chasles M., 1837, "Apercu historique sur l'origin et le développement des méthodes en géométrie ...", Mémoires couronnés par l'Académie de Bruxelles, Vol.11. (2nd Ed., Paris, 1875).

Chironis F., 1961, "Mechanisms, Linkages, and Mechanical Controls, New York, McGraw-Hill.

Cianchi M. , 1984, "Le Macchine di Leonardo", Becocci Publ., Firenze.

Coulomb C.A., 1821, "Theoriè des machines simples", Bachelier Libraire, Paris.

D'Alembert J.B., Diderot D., 1774, "Recueil de Planches, sur les Sciences, les Arts Liberaux, et les Arts Mechaniques", 3° Edition, Livourne.

De la Goullepierre H. , 1864, "Traité des mecanìsmes", Paris.

Del Monte G. , 1577, "Mechanicorum liber", Pesaro.

Dimarogonas A.D., 1993, "The Origins of the Theory of Machines and Mechanisms", in Modern Kinematics – Developments in the Last Forty Years Ed. by A.G. Erdman, Wiley, New York, pp.3-18.

Ferguson E.S., 1962, "Kinematics of Mechanisms from the Time of Watt", Contributions from the Museum of History and Technology, Washington, paper 27, pp. 186-230.

Franke R.,1943, "Vom Aufbau der Getriebe", VDI, Vol.1, 1943; (Vol.2, 1951).

Frontini S.J., 1930, "De Aquaeductu Urbis Romae", codex Casinis 361, reprinted in Montecassino Abbey.

Galilei G., 1593, " Le Meccaniche", reprinted in Opere di Galileo Galilei Edited by F.

302

Brunetti, Torino, 1964.

Galluzzi p. (Editor) 1991, "Prima di Leonardo - Cultura delle macchine a Siena nel Rinascimento", Electa, Milan.

Ghigliazza R. and Galletti C.U, 1986, "Meccanica applicata alle macchine", UTET, Torino.

Giulio C.I., 1846, "Sunti delle Lezioni di Meccanica applicata alle arti", Tipografia Pomba, Torino.

Grubler M., 1917, "Getriebelehre", Springer, Berlin.

Koetsier T., 1999, "The Story of Archimedes and the Screw", Vrije Universiteit, Amsterdam, Rapport nr.WS-523.

Hachette J.N.P., 1811, "Traitè elementaire des machines", Paris.

Hain K., , 1967, "Applied Kinematics", McGraw-Hill, New York.

Hartenberg R.S. and Denavit J., 1956, "Men and Machines ... an informal history", Machine Design, May 3, 1956, pp.75-82; June 14, 1956, pp.101-109; July12, 1956, pp.84-93.

Hiscox G.D., 1899, "Mechanical Movements", NewYork.

Laboulaye C., 1861, "Traitè de Cinèmatique ou theoriè des mecanismes", Paris.

Lanz J.M. and Betancourt A., 1808, "Essai sur la composition des machines", Paris.

Leupold J., 1724, "Theatrum Machinarum", Leipzig.

Masi F., 1883, "Manuale di Cinematica Applicata", Zanichelli, Bologna.

Morin A., 1872, "Notions geometriques sur les mouvements et leurs transfromations", Libraiere Hachette, Paris.

Nolle H., 1974, "Linkage Coupler Curve Synthesis: A Historical Review – I and II", IFToMM Journal Mechanism and Machine Theory, Vol.9, n.2, pp.147-168 and pp.325-348.

Poppe J.H.M., 1803, "Encyclopaedie des gesammten Machinenmesens", Leipzig

Ramelli A., 1588, "Le diverse et artificiose machine", Paris.

Rankine M.W.J., 1887, "Manual of Machinery and Millwork", London.

Reuleaux F., 1875, "Theoretische Kinematic", Braunschweig.

Robinson S.W., 1896, " Principles of Mechanisms, NewYork.

Stigliola C.A., 1597, "Degli Elementi Mechanici", Stamperia PortaReale, Naples.

Tasi L.W., 2001, " Mechanism Design: Enumeration of Kinematic Structures according ot Function", CRC Press, BocaRaton

Vitruvius P. M. , 1511, "De architectura" edited by Fra Giocondo, Verona.

Willis R., 1841, "Principle of Mechanism", London.

ON THE HISTORY OF FOCAL MECHANISMS
AND THEIR DERIVATIVES

Evert Dijksman
Mechanism Design
Luikersteenweg 538, 3920 Lommel, Belgium
evert.dijksman @ belgacom.net

ABSTRACT- It all started with the invention of Peaucellier's *Compound Compass*, and a century later with Kempe's *complete compound* of which the focal linkage, among other cases, just appeared as a particular case; two conditions for degeneration then being necessary. Mathematical exact straight-line mechanisms came out of it, whereas a special case of the complete compound gave rise to a straight-guided body-motion in any required direction. Kempe's complete compound lead also to the most generalized inversor of which Peaucellier's inversor was only a very special case. Application of the focal linkage finally resulted into a practical design for (un)symmetrical steering mechanisms for cars or vehicles with symmetrically or eccentrically installed steering-wheels.

KEYWORDS: Complete Compound, focal linkage, inversion mechanism, rectilinear bar-motion, steering mechanism

1. COMPLETE COMPOUND

In May **1878** A.B.Kempe (ref.[**6**]) found this very fascinating, though overconstrained eight-piece linkage mechanism, consisting of two interconnected *"conjugated four-piece"* linkages with a common *"connecting diagram"* (See the *"complete compound"* of figure **1** containing *two* redundant turning-joints) The mechanism was interconnected by *four* turning-joints, each belonging to corresponding links of the two conjugated four-piece linkages. Kempe further investigated particular cases in which the connecting diagram, common to the two conjugated four-piece linkages, turned into parallelograms or anti-parallelograms.

But, the most interesting case occurred, when one of the conjugated four-piece linkages contracted into a singular point, and so turned this point into a *triple* joint, being inter-connected by binary bars to all four sides of the remaining four-bar linkage. (See figure **2** being still overconstrained as it suffices to change the triple joint into a double joint.)

Note, that if the adjacent angles at the *connected* joints of the two conjugated four-piece linkages are equal, or each other's supplement as in the complete compound of figure **1**, whereas the connected joints are to be found at the very sides of *one* quadrilateral, the same occurs for the remaining conjugated quadrilateral. Hence, the complete compound

304

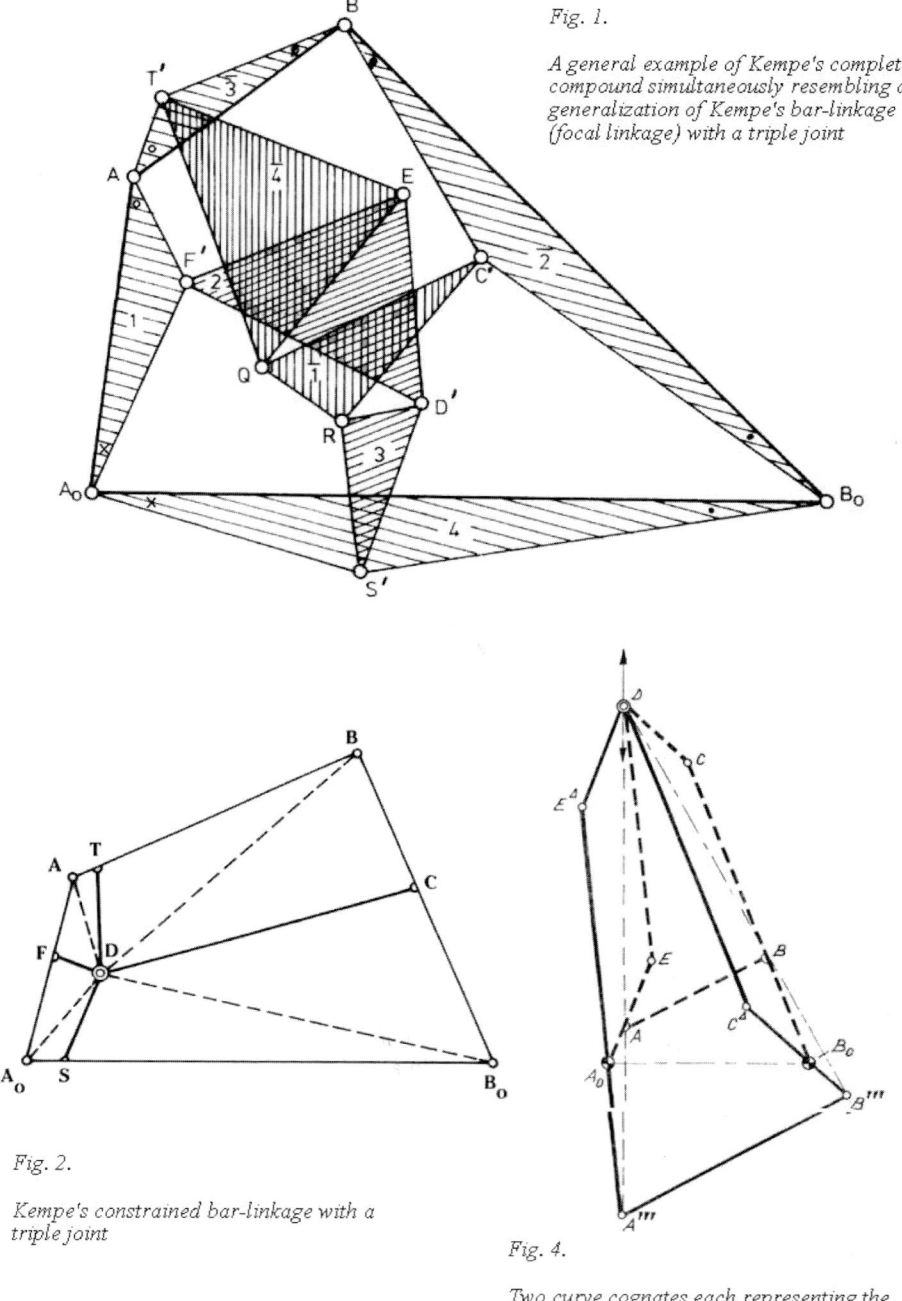

Fig. 1.

A general example of Kempe's complete compound simultaneously resembling a generalization of Kempe's bar-linkage (focal linkage) with a triple joint

Fig. 2.

Kempe's constrained bar-linkage with a triple joint

Fig. 4.

Two curve cognates each representing the 2nd straight-line mechanism of Hart (1877)

degenerates into a *bar*-linkage, generally not containing a triple joint. (See also § 5 of ref.[10] and figure **16** demonstrating part of the complete **bar**-linkage.) A further degeneration may finally turn the then conjugated quadrilateral into the already mentioned triple-joint.

2. FOCAL LINKAGE

In a separate, voluminous paper (Ref.[7]) Ludwig Burmester showed in **1893** that the simplified form of figure **2** appeared to have *focal* properties which gave him the opportunity to name this form the *focal linkage*. In this focal mechanism, originally invented by Mr.Kempe, the triple joint was interconnected to all four sides of a *random* four-bar. Burmester showed that all conic sections that are to be inscribed in the four-bar, lead to a *locus* of focal points all to be utilized as a triple joint of a different focal linkage. So, for each four-bar there is still an infinite number of focal linkages. Clearly, the solutions appear in pairs, as generally, the conic sections, such as the ellipses or the hyperbolas, for instance, have two focal points.

Burmester didn't treated the complete general compound of Mr. Kempe in which the common turning-joints of the two conjugated four-piece linkages were in fact the vertices of triangular links.

However, by using Cognate Theory it is quite easy to find Mr. Kempe's general compound, starting from the focal linkage. This may be done by successively *stretch-rotating* each of the inner four-bars, contained in the focal linkage, about the turning-joints of the overall, initial four-bar (ref.[10].) (The complex multiplication-factor of the first stretch-rotation represents a free choice for its amplitude and its argument, whereas the three remaining stretch-rotations are all dependent on the first. Hence, this gives *two* additional design-degrees of freedom in comparison to the focal linkage. Thus, for a given four-bar, we then totally have *three* additional design-degrees of freedom and so, an infinite to the third number of possibilities to arrive at Mr. Kempe's generalized compound, demonstrated in figure **1**.)

3. STRAIGHT-LINE MECHANISMS

In the 18th and 19th century, many kinematicians such as James Watt (**1784**), Peaucellier (**1864**), Tschebychev, Hart (**1874**), Kempe (refs.[3] & [5]) and Sylvester have searched for the possibilities to mechanically produce a straight-line, the main application being to transform a straight motion into a circular one, at the time being important, for instance, for steam engines in locomotives.

James Watt (with his "*Parallel Motion*") & Tschebychev (with his symmetrical, but crossed four-bar) produced *approximate* solutions, whereas the others were after the mathematical *exact* ones. Though the approximate ones produce coupler curves which are a bit deviating from the straight-line, they still have the practical advantage of utilizing a lower number of links and turning-joints.(Instead of 6 bars as, for instance, in the case of *Hart's 1st* – or *2nd straight-line mechanism* (ref.[4] and figures **3** & **4**), we then have merely 4 bars in use to produce the straight-line.)

4. INVERSION MECHANISMS

Peaucellier was the first to find his exact straight-line mechanism, based on the principle of *geometric inversion*. It contained a kite and a rhomb as a basic form for application. Totally, the mechanism was an *eight-bar,* six years later independently derived by L.Lipkin of St.Petersburg at the 22nd of dec. in **1870** (figure **3**).

Mr. Peaucellier's *Compound Compass,* as he named his discovery, further gave him the

reward of promotion to the rank of lieutenant-colonel in the french army.
A century later, Mr. Harry Hart even found *two* different, mathematical exact, straight-line mechanisms, both being only *six-bar* ones.

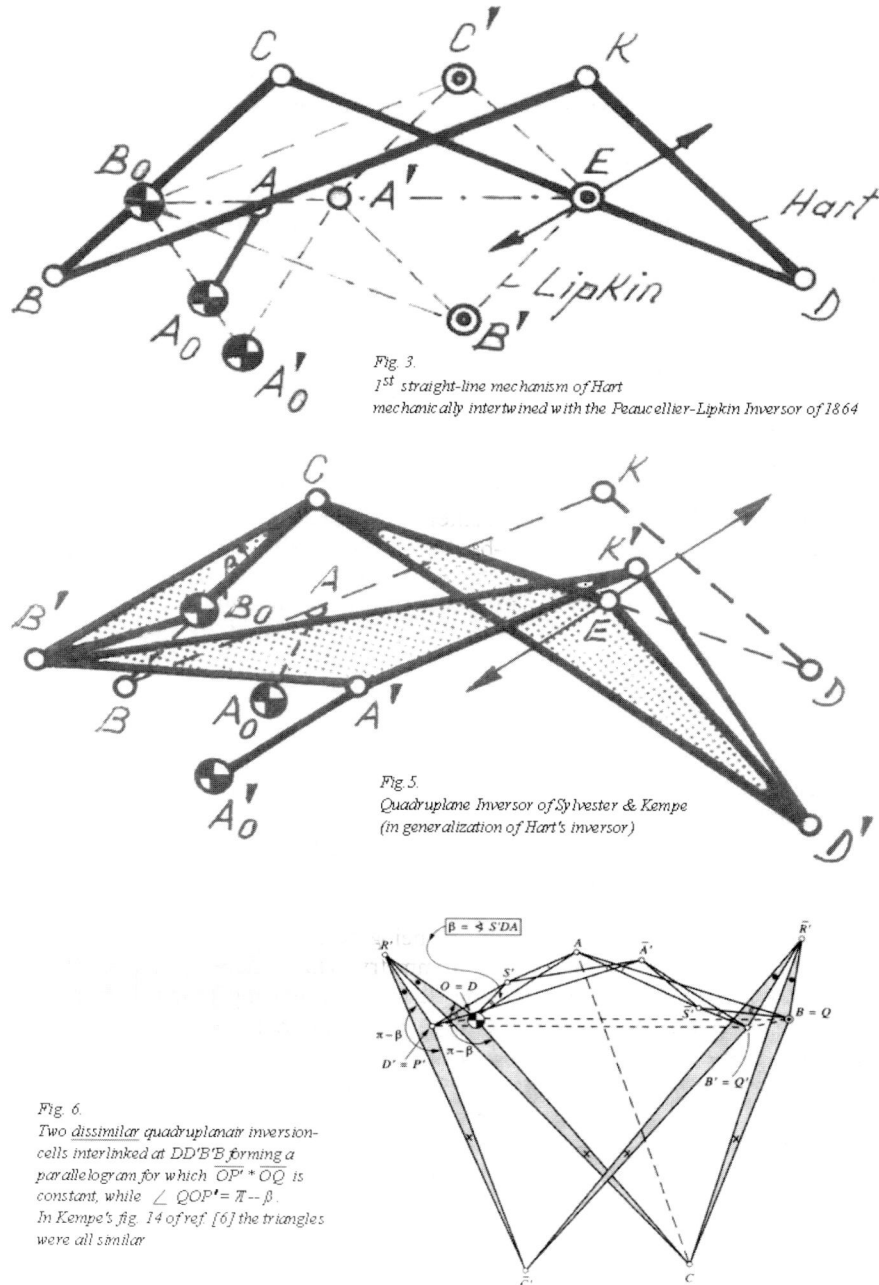

Fig. 3.
1st straight-line mechanism of Hart
mechanically intertwined with the Peaucellier-Lipkin Inversor of 1864

Fig. 5.
Quadruplane Inversor of Sylvester & Kempe
(in generalization of Hart's inversor)

Fig. 6.
Two dissimilar quadruplanair inversion-cells interlinked at DD'B'B forming a parallelogram for which $\overline{OP'} * \overline{OQ}$ is constant, while $\angle\ QOP' = \pi - \beta$.
In Kempe's fig. 14 of ref [6] the triangles were all similar

The first was an *inversion mechanism* with an anti-parallelogram as a basic inversion-cell (figure **3**), whereas the second possessed an *arbitrary* four-bar, to which a dyad linkage was adjoined (figure **4**).

Mr. Kempe (ref.[**6**].) showed that the latter presented a special case of his compound with a triple turning-joint, the joint then imaginarily connected to the infinity-point of the fixed bar.

In **1875**, Kempe (ref.[**3**].) further proved that both, the inversion linkages of Peaucellier and Hart (the latter being the anti-parallelogram one) were to be derived from his so-called *"self-conjugate sextilateral"*.

Much later, we found that the inversion mechanisms of Peaucellier & Hart are to be intertwined in a mechanical way, and so appear to be directly related through *mechanical cognation* (ref.[**9**] and figure **3**). The cognation also lead to entirely new inversion linkages still mechanically related to the famous ones of Peaucellier and Hart.

Both, Kempe and Sylvester, simultaneously generalized Mr. Hart's first inversor to the so-called *quadruplanar inversor of Sylvester and Kempe* (ref.[**5**]). (The sides of the anti-parallelogram thereby formed the bottom sides of (reflected) similar triangular links with apices forming a parallelogram of varying lengths (See figure **5**).)

Two quadruplanar inversion cells meeting the same parallelogram formed by the then common apices (figure **6**) lead to the *most generalized inversion cell* to be built with an *arbitrary* four-bar and ditto vertex of a triangular link (See figure **7** and the refs. [**12** & **15**]). The new cell is comparable to Peaucellier's cell as it contains the same number of links and turning-joints. Besides, Peaucellier's cell appears to be a very particular case to be retrieved when the arbitrary four-bar turns into a kite and the vertex of a triangular link goes to a double-joint of kite and rhomb. Thus, the most generalized inversion cell appears to be a fourfold generalization of Peaucellier's inversion cell (See figure **8** of ref.[**15**]).

5. RECTILINEAR BAR-MOTION

Kempe (Part III of ref.[**5**]) also found 8- and 10-bar linkage mechanisms generating the mathematical exact motion of a bar.

See his device of figure **8** utilizing the quadruplanar inversor of Sylvester and Kempe. (figure **9**). Note, however, that his solution of figure **26** of ref.[**5**] with merely six links represented only an approximate solution. Other approximate 6-bar solutions employing Kurt Hain's method are demonstrated in ref.[**17**].

A four times overconstrained, mathematical exact 8-bar linkage comes into view with two random quadrilaterals simultaneously being reflected equal as well as being perspective from a point T at the axis of symmetry. The center T of perspectivity then has to join the focal curve of each quadrilateral, the curve being a locus of points seeing opposite sides under equal angles or under angles that are each other's supplement.

The complete compound, so composed, then contains four common turning-joints, of which there are two redundant. (See figure **10** or figure **1** of ref.[**14**])

In case *one* of the common joints goes to infinity, an 8-bar linkage emerges having a bar moving perpendicular to the fixed link. (See figure **11** or figure **9** of ref.[**14**])

Instead of anti- or contra-parallelogram based 8-bar linkages with mathematical exact rectilinear bar-motion, it appears that more general 8-bar solutions do exist. Though

308

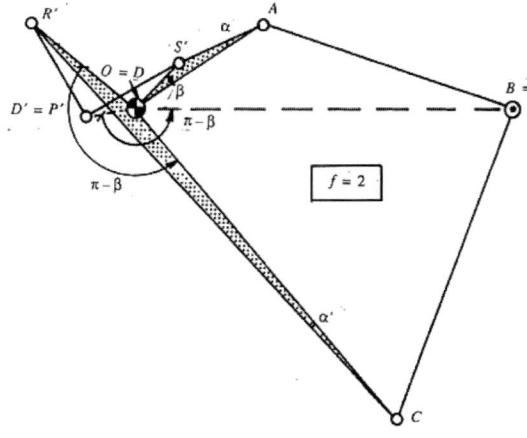

Fig. 7.

New Inversion Cell with a random

4-bar ABCD and a random joint S'.

$\overline{OP'} * \overline{OQ}$ = constant

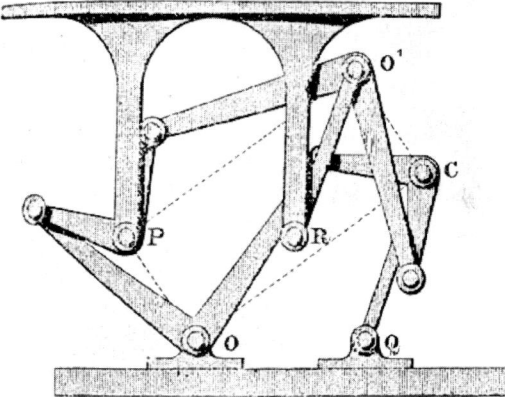

Fig. 8.

Kempe's traversing table constituting

the Sylvester-Kempe parallel motion

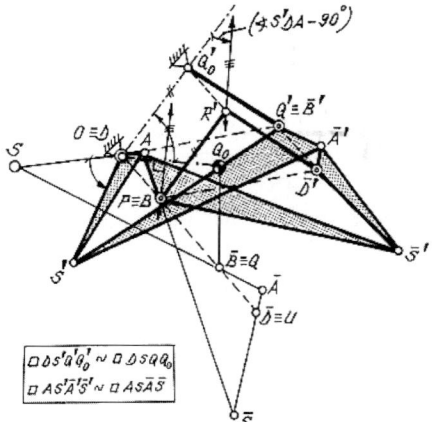

Fig. 9.

8-bar linkage mechanism containing a

bar PR' moving in a invariable but

oblique direction with the frame (see

fig. 27 from ref [5] for application)

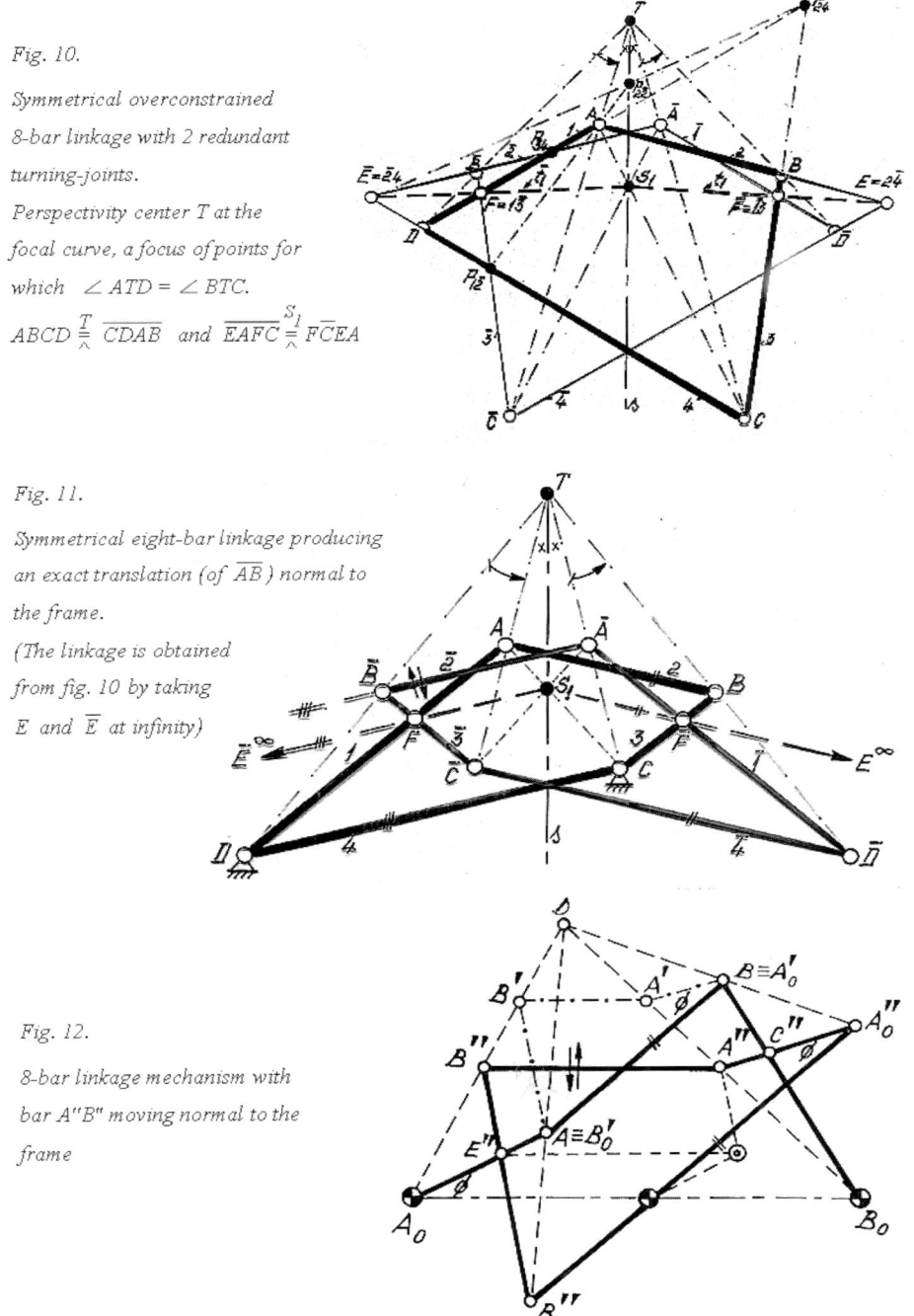

Fig. 10.

Symmetrical overconstrained 8-bar linkage with 2 redundant turning-joints.

Perspectivity center T at the focal curve, a focus of points for which $\angle ATD = \angle BTC$.

$ABCD \overset{T}{\underset{\wedge}{=}} \overline{CDAB}$ *and* $\overline{EAFC} \overset{S_1}{\underset{\wedge}{=}} \overline{FCEA}$

Fig. 11.

Symmetrical eight-bar linkage producing an exact translation (of $\overline{AB}$) normal to the frame.

(The linkage is obtained from fig. 10 by taking E and $\overline{E}$ at infinity)

Fig. 12.

8-bar linkage mechanism with bar $A''B''$ moving normal to the frame

L.D.Ruzinov (ref.[18]) didn't thought this possible in his time (**1968**), one may nonetheless design such a solution with completely *arbitrary* four-bars. (See the figures

310

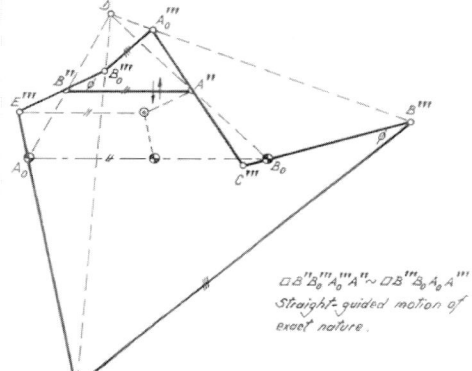

Fig. 13.

8-bar coupler cognate having bar A"B"
moving normal to the frame.

$\square\ B''B_0\ '''A_0'''\ A'' \sim \square\ B'''B_0\ A_0\ A'''$

$\square B''B_0\ A_0'''\ A'' \sim \square B'''B_0\ A_0\ A'''$
Straight-guided motion of
exact nature.

Fig. 14.

Random choice for E''' (or for C''')
Generalized 8-bar with link A'''B'''
moving normal to B_0F'''
Special case of Kempe's complete
compound.

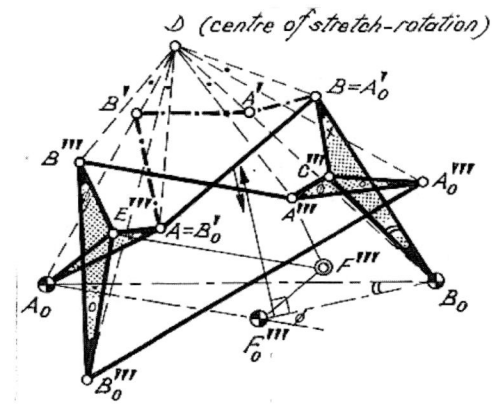

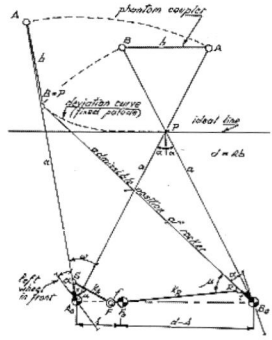

Fig. 15.

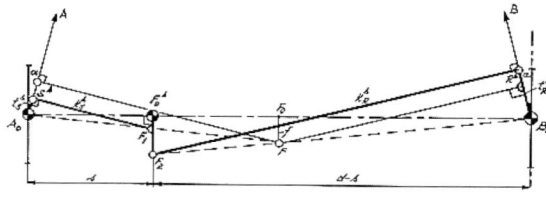

Fig. 16.

Non-symmetrical Watt-II steering mechanism, derived from
*a symmetrical **focal** linkage after multiplication of two four-*
*bars. (The resulting **bar**-linkage resembles part of Kempe's*
complete compound with flattened triangles.)

"b over a" corresponds with "f over k_R" w being an angle of wheel-turning. Unsymmetrical Watt-II
steering mechanism as part of a focal mechanism producing a symmetrical wheel motion

12 & **13** and ref.[**19**]). Above that, one still has an infinite row of possibilities, whereas the mechanism-solutions appear in pairs, as always *two* eight-bar *coupler cognates* exist, the initial one included. Besides, generalization is further feasible by allowing triangular links in the design. (See figure **14** and ref.[**19**].) This has the practical advantage of possibly choosing the fixed direction of the rectilinear motion in which the bar has to move. Thus, apart from the free choice of the initial four-bar, one can additionally choose the magnitude as well as the direction of the so straight-guided motion.

One finally remarks that the configuration of figure **14** of ref.[**19**] is in fact a special case of Mr. Kempe's complete compound eight-bar linkage, as it meets the condition that *one* common joint of the two conjugated four-piece linkages, comprised in the compound, has to stay at infinity.

6. APPLICATION

In order to find six-bar (Watt-II) steering mechanisms for cars, the utilization of the *focal* linkage comes in very handy. The reason is, that the crossed double rocker giving the best possible solution to meet Ackermann's Principle for cars, doesn't have applicable dimensions for direct use.

The focal mechanism, however, having an optimized double rocker as initial four-bar, represents an overconstrained eight-bar, from which the much lesser space-taking Watt-II steering mechanism may be extracted. The result is highly applicable. (See ref.[**20**] and figure **15**).

Though the crossed double rocker has to be *symmetrical* on the grounds of having equal right- and left turnings for the car, the *total* focal mechanism, and so the Watt-II steering mechanism, doesn't have to be symmetrical. This freedom leads to the **un**symmetrical (Watt-II) steering linkages as demonstrated in the figures **15** & **16**. (In the last figure stretch-rotation has been used in order to obtain the unsymmetry for cars having a steering-wheel not on their symmetry-axis.)

REFERENCES

[1] Peaucellier,A., *Note sur une question de geometrie de compas*, Nouvelle Annales de Mathématique, Sér.II,Tome III, p.344 (**1864**); and Sér.II, Tome XII, p.71-78, **1873**.

[2] Hart,H., *On certain Conversions of Motion*, Messenger of Mathematics, Vol.IV,p.82, **1874**.

[3] Kempe,A.B., *On a general method of producing exact rectilinear motion by linkwork,* Proceedings of the Royal Society of London 23 (**1875**) p.565-577.

[4] Hart,H.,*On some cases of Parallel Motion*, Proceedings of the London Mathematical Society 8 (**1877**), p. 286-289.

[5] Kempe,A.B., *How to draw a Straight-Line*, Nature (**1877**) Vol.XVI, Part II,p.86-89 & Part III, p.125-127.

[6] Kempe,A.B. *On conjugate four-piece linkages*, Proc. London Mathematical Society 9, p.133-147 (**1878**)

[7] Burmester,L., *Die Brennpunktmechanismen*, Z.Math.Phys. **38** (4), S.193-223 (**1893**).

[8] Wunderlich,W.,*On Burmester's focal mechanism and Hart's straight-line motion*, J. of Mechanisms, **3**, p.79-86 (**1968**).

[9] Dijksman,E.A., *A strong relationship between new and old inversion mechanisms,*

312

Trans. ASME, Ser.B.,J.Eng.Industry, **93**, (**1971**), p.334-9.

[10]Dijksman,E.A., *Kempe's (focal) Linkage Generalized, particularly in connection with Hart's 2nd straight-line mechanism,* Mechanism and Machine Theory, **1975**, Vol.10, pp.445-460.

[11] Dijksman,E.A. **Motion Geometry of Mechanisms,** Cambridge University Press, Cambridge, London, New York, Melbourne (**1976**), Chapters 8 & 9.

[12] Dijksman,E.A., *A way to generalize Peaucellier's Inversor,* Proceedings 6th Int. Symposium on Linkages and CAD-Methods, (Theory and Practise of Mechanisms), Vol.1, Paper 10, pp. 73-82, SYROM'93, Bucure di, România, June **1993**.

[13]Artobolevskii,I.I., *An Inversor Mechanism,* DAN,USSR, 104, Nr. 6, **1955**.

[14] Dijksman,E.A., *Overconstrained Linkages to be derived from Perspectivity and Reflection,* Proc. 7th World Congress on TMM, Sevilla (Spain), Vol.1 (**1987**),pp.69-73

[15] Dijksman,E.A., *A general Inversion cell, obtained from two random contra-Parallelogram Linkages, interconnected at the vertices of a Parallelogram.*I Congreso Iberamericano de Ingeniería Mecánica, Madrid, 21-24, Sept.**1993**, Vol.3, pp.401-406 Mechanism and Machine Theory, Vol.29, No.6,pp.793-801, **1994**.

[16] Dijksman,E.A. *A fourfold Generalization of Peaucellier's Inversion Cell,* Meccanica **31**, p.407-420 (**1996**) Kluwer Academic Publishers, printed in The Netherlands.

[17] Dijksman,E.A. *Lifting linkage mechanisms of the Watt-I type with an almost rectilinear translating bar (When Linearizers are to be designed with Watt-I six-bars),* 9th World Congress on TMM, Proceedings Politecnico di Milano,Italy, August 29/Sept.2, **1995**, Vol.1, pp.3-7.

[18] Ruzinov,L.D., **Design of Mechanisms by Geometric Transformations** (translated from the Russian into English), Iliffe Books Ltd., London (**1968**).

[19] Dijksman,E.A., ***True*** *straight-line linkages having a rectilinear translating bar,* Advances in Robot Kinematics and Computational Geometry, 4 ARK, Ljubljana, July 6, 1994. Kluwer Academic Publishers (**1994**), p.411-420.

[20] Dijksman,E.A., *Do the front-wheels of your car really meet Ackermann's Principle when driving in a bend of the road ?* PRASIC 2002, Vol.I, Mecanisms & Tribology, pp.103-108, 7-8 Nov. **2002**, Bra ìov, România. (ISBN 973-635-064-9)

FIGURE-CAPTIONS

Fig.1. from fig.8 of ref.[10] *A general example of Kempe's complete compound, simultaneously resembling a generalization of Kempe's **bar**-linkage (focal linkage) with a triple joint*

Fig.2 from fig.1 of ref.[10] *Kempe's constrained bar-linkage with a triple joint*

Fig.3 from fig.1 of ref.[19] *1st straight-line mechanism of Hart mechanically intertwined with the **Peaucellier**-Lipkin Inversor of **1864***

Fig.4 from fig. 2 of ref.[19] *Two curve cognates each representing the 2nd straight-line mechanism of Hart (**1877**)*

Fig.5 from fig.3 of ref.[19] *Quadruplane Inversor of Sylvester & Kempe (in generalization of Hart's inversor)*

Fig. 6 from fig.3 of ref.[15] *Two <u>dissimilar</u> quadruplanar inversion-cells interlinked at DD'B'B forming a parallelogram for which*

$$\overline{OP'} \square \overline{OQ} = \text{constant, while } \square\, QOP^{\square} = \square\square\, \square$$

In Kempe's fig.14 of ref.[6] the triangles were all similar.

Fig.7 from fig.4 of ref.[15] *New Inversion Cell with a random 4-bar ABCD and a random joint $S^{t\,z}$.*

$$\overline{OP^{\square}} \square \overline{OQ} = \text{constant}$$

Fig.8 from fig.27 of ref.[5] *Kempe's traversing table constituting the Sylvester-Kempe parallel motion*

Fig.9 from fig.6 of ref.[19] *8-bar linkage mechanism containing a bar $PR^{t\,z}$ moving in an invariable but oblique direction with the frame. (See fig. 27 from ref.[5] for application.)*

Fig.10 from fig.1 of ref[14] *Symmetrical overconstrained 8-bar linkage with 2 redundant turning-joints*
Perspectivity-center T at the focal curve, a locus of points for which

$$\square\, ATD = \square\, BTC.$$

$$ABCD \underset{\square}{\overset{T}{=}} \overline{CDAB} \ \& \ \overline{EAFC} \underset{\square}{\overset{S_1}{=}} \overline{FCEA}$$

Fig. 11 from fig.9 of ref.[14] *Symmetrical eight bar-linkage, producing an exact translation (of $\overline{AB}$) normal to the frame.*

(The linkage is obtained from figure 10 by taking $\overline{E}$ and E at infinity)

314

Fig.12 from fig.4 of ref.[19] *8-bar linkage mechanism with bar A" B" moving normal to the frame*

Fig.13 from fig.10 of ref.[19] *8-bar coupler cognate having bar A"B" moving normal to the frame.*
$$1 \ B"B_0{}'''A_0{}'''A" \sim 1 \ B'''B_0A_0A'''$$
straight-guided bar-motion of exact nature

Fig.14 from fig. 12 of ref.[19] *Random choice for E''' (or for C''')*
Generalized 8-bar with link A'''B''' moving normal to B_0F'''
Special case of Kempe's complete compound

Fig.15 from fig.4 of ref.[20] *"b over a" corresponds with "f over k_R",*
w being an angle of wheel-turning
Unsymmetrical Watt-II steering mechanism as part of a ***focal*** *mechanism producing a symmetrical wheel-motion*

Fig.16 from fig.5 of ref.[20] *Non-symmetrical Watt-II steering mechanism, derived from a symmetrical **focal** linkage after multiplication of two four-bars. (The resulting **bar**-linkage resembles part of Kempe's complete compound with flattened triangles.)*

HISTORICAL EVOLUTION OF THE POLE-THEORY

Kurt Luck,
IFToMM TC Linkages & Cams, Honorary Member
e-mail: luck@mfk.mw.tu-dresden.de

Willi Rehwald,
IFToMM TC Linkages & Cams, Honorary Member
e-mail: axel.rehwald@fgk-ant.de

KEYWORDS : planar motion, polodes, *Bresse'* circles, geometrical and kinematic poles, geometrical and kinematic transfer functions.

INTRODUCTION

The paper includes an investigation with respect to the historical evolution of the pole-theory. *De-La Hire* and *Bresse* [1] did at first scientific research in this field; the inflexion circle was found by *De-La Hire about 1706*. *Burmester* investigated in his famous "Lehrbuch der Kinematik" [2] this subject geometrically and defined the corresponding theoremes exactly. The planar motion was investigated numerically by *Krause* [3] with respect to pole theory, curvature theory, circle-point curves, *Ball's* point etc. But all these investigations are based on geometry, not upon time t. Therefore we have to understand motion as position-change.

The graphical simulation by circle and ruler was applied till about 1960. Then the computer became more and more an important tool, to solve problems numerically and very quickly. *Rehwald* [6, 7] investigated the analysis of planar motion (position-change) in general and created a modern pole-theory, which will be presented in this paper.

EARLY INVESTIGATIONS AND TECHNIQUES
POLES AND POLODES

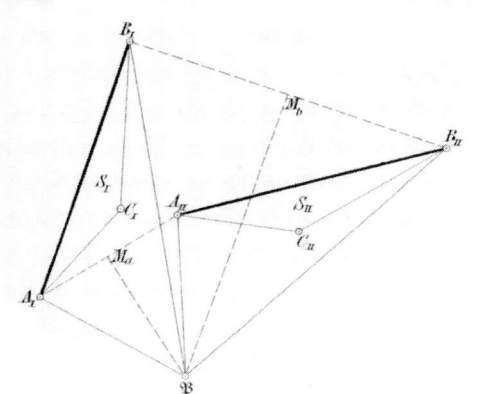

Fig.1: Two discrete positions of a rigid body

Burmester did his research very thoroughly and carefully. Fig. 1 demonstrates the planar position-change of a rigid body, which is represented by two points A and B in two positions with respect to the basic system. Point A has moved from position A_I to A_{II} and point B has moved from position B_I to B_{II}. The perpendicular bisectors of the line segments $\overline{A_I A_{II}}$ and $\overline{B_I B_{II}}$ intersect at the basic system in the centre of rotation P, also called the *virtual pole*. $A_{II}B_{II}$ can be obtained from $A_I B_I$ by a rotation round about P. As soon as position II approaches to position I, the

315

connecting line A_IA_{II} approaches to the tangent of path α at A and likewise B_IB_{II} to the tangent of path β at B (fig. 2). The intersection of both path-normals at points A and B

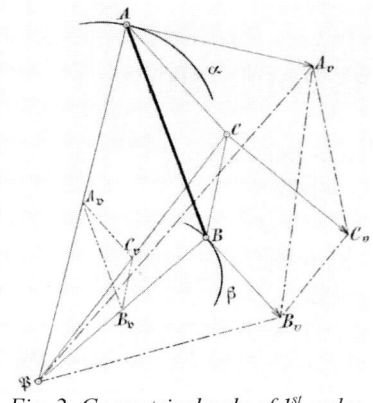

Fig. 2: Geometrical pole of 1ˢᵗ order

is called the instantaneous pole P or geometrical pole of 1st order. Both infinitely separated positions can be transfered by an infinite rotation at the pole P. If there are other points C; D etc. besides the points A and B at the rigid body, the path-normals PA; PB; PC; PD etc.; the so-called pole-rays all intersect in the corresponding instantaneous pole P, fig.2.

Figure 3 represents the planar motion by separated postions A_1B_1; A_2B_2; A_3B_3; A_4B_4 etc. including the corresponding instantaneous poles P_1; P_2; P_3; P_4 at the fixed polode π in the basic system. The reverse motion delivers at the moving plane the corresponding poles P_1'; P_2'; P_3'; P_4' etc. which can be obtained graphically by respecting the following equations:

$$\begin{aligned}
1\ \mathbf{A}_1B_1P_1' &\cong 1\ \mathbf{A}_1B_1P_1 \\
1\ \mathbf{A}_1B_1P_2' &\cong 1\ \mathbf{A}_2B_2P_2 \\
1\ \mathbf{A}_1B_1P_3' &\cong 1\ \mathbf{A}_3B_3P_3
\end{aligned} \qquad (1)$$

. .   . .

The planar motion can be realized by rolling of these two polodes one upon the other. If the moving polode p rolls upon the fixed polode π without sliding, we get the motion with respect to the frame (basic system). The reverse motion is characterized by the frame-fixed polode π, which rolls upon the resting polode p.

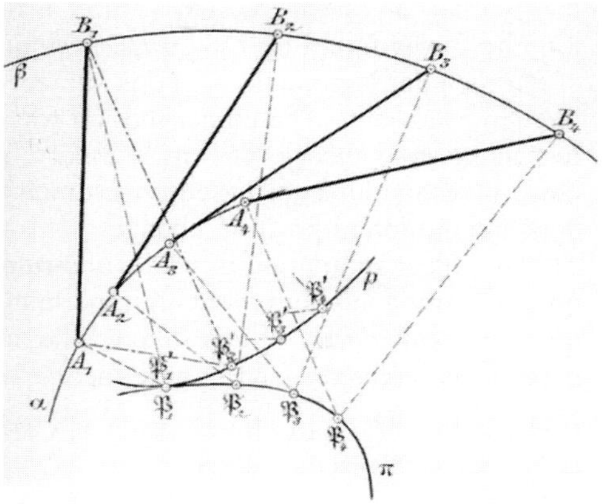

Fig. 3: Planar motion of rigid body AB, fixed polode π ; moving polode p

CURVATURE AND INFLEXION CIRCLE

Three discrete positions S_I; S_{II}; S_{III}; of the moving System S, with respect to the basic system Σ are given, fig. 4, including the three positions f_I; f_{II}; f_{III} of the circle f (with centre point F) at S. The oscillation points between the three congruent circles f_I; f_{II}; f_{III} and a circle φ at the basic system Σ may be called A_I; B_{II}; C_{III} corresponding to the same side. The centre points F_I; F_{II}; F_{III} of these circles itself are situated on a circle φ', concentrically to the circle φ round about the centre point Φ, which is the intersection of the lines F_IA_I; $F_{II}B_{II}$; $F_{III}C_{III}$. Further we choose a line t ,which includes the unknown poles $P_{I,II}$; $P_{II,III}$ of the discrete positions S_I, S_{II} and S_{II}, S_{III}. The perpendicular bisectors of the line segments

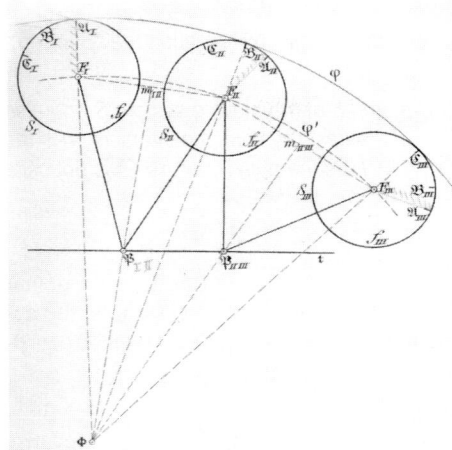

Fig. 4: Three positions; centre point ı -

$\overline{F_I F_{II}}$ and $\overline{F_{II} F_{III}}$ intersect in the centre point Φ. Now the positions of the moving system S are defined by rotating the segment $P_{I,II}F_I$ to $P_{I,II}F_{II}$ and then the segment $P_{II,III}F_{II}$ to $P_{II,III}F_{III}$. In this way the circle f moves from f_I to f_{II} and then to f_{III}; including the corresponding oscillation points A_I; B_{II}; C_{III} and the homologous points A_{II}; A_{III}; B_I, B_{III}; C_I;C_{II}. The realization of these moving positions S_I; S_{II}; S_{III} demands only the centre points F_I; F_{II}; F_{III} and the straight line t. These relations are also valid, if the three finitely separated positions S_I; S_{II}; S_{III} become infinitely separated, i.e. If A_I; B_{II}; C_{III} and also F_I; F_{II}; F_{III} become infinitely separated, then the circle ϕ' transfers into the circle of curvature with respect to the coupler curve of point F, and the straight line t transfers into the pole tangent. Looking to the circle f as a circle of curvature of a curve at the moving system S, then the circle ϕ transfers into the circle of curvature of the corresponding envelope-curve.

Taking another point L as centre point of a curve at the system S into consideration, with L_I; L_{II}; L_{III} as the corresponding inifinitely separated positions of L, then the centre point Λ of the circle through these three points is simultaneously the centre of curvature with respect to the coupler curve of point L and the corresponding envelope-curve λ. *Burmester* defined the following

Theorem: *Two corresponding centres of curvature and the attached pole tangent form an equivalent with respect to three infinitely separated positions of a moving system and define also the corresponding positions of any attached centre of curvature with respect to all sliding curves.*

The centre point F of a curve f at the moving system S, further the corresponding centre of curvature Φ and the attached envelope-curve ϕ at the basic system Σ, also the pole tangent t are given, see fig. 5. Considering any centre of curvature of one sliding curve,

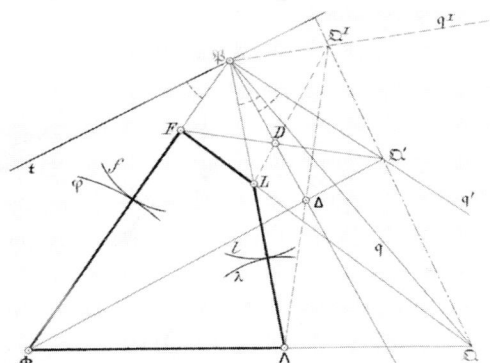

it is possible to find the corresponding centre point of the other curve. L is a given centre of curvature of a curve l at system S and searched is the corresponding centre of curvature Λ of the attached system curve λ. The graphical solution needs the following steps, see fig. 5:

❖   *Draw the straight line PL,*
❖ $\angle FPt = - \angle LPq$
❖   *Intersection of lines Pq and FL* $\Rightarrow Q$ *(collineation point)*

Fig. 5: Bobillier's theorem

❖ *Intersection of lines* ΦQ *and* $PL \Rightarrow \Lambda$ *(centre of curvature).*

Λ is the centre-point of the envelope-curve λ. On the basis of this investigation the *"Bobilier's theorem"* is defined as follows:

Connect according to fig.5 the two centres of curvature F and L of the moving system S and also the two corresponding centres of curvature Φ and Λ of the basic system; the intersection is the collineation point Q. Intersection of the lines ΦF and ΛL is the pole P. $\angle \Phi Pt$ is opposite equal to the angle $\angle \Lambda PQ$.

BRESSE′CIRCLES AND ACCELERATION POLE

Burmester investigated the relationships between the accelerations of points at a planar moving system S, see fig. 6. The instantaneous pole P, of the moving system S, the acceleration-pole J, velocity AA_v and acceleration AA_j of the point A are given. According to the equation

$$1 \; \mathfrak{R}AA_v \approx 1 \; \mathfrak{R}JJ_v \qquad (2)$$

we get the velocity JJ_v of J. Because the acceleration-pole J has instantaneously no acceleration, its velocity JJ_v is constant in the same direction. The acceleration PP_j of the timely resting pole P, which starts moving at the next moment, can be obtained by using the following equation:

$$\Delta JPP_j \approx \Delta JAA_j \qquad (3)$$

With respect to the following facts:

$$\angle JAA_j = \angle \Theta \; ; \quad \angle A_v AA_j = \angle \lambda \; ;$$

$$\angle PAJ = \angle \sigma ; \quad \angle PAA_v = 90°$$

we obtain: $\lambda = 90° - \Theta - \sigma \qquad (4)$

Conclusion: *The angle λ is constant for such system-points A, lying on the circle k^λ, which is defined by the peripherical angle $\sigma = 90° - \Theta - \lambda$ above the chord PJ.*

Considering the cases $\lambda = 0$, $\lambda = 180°$

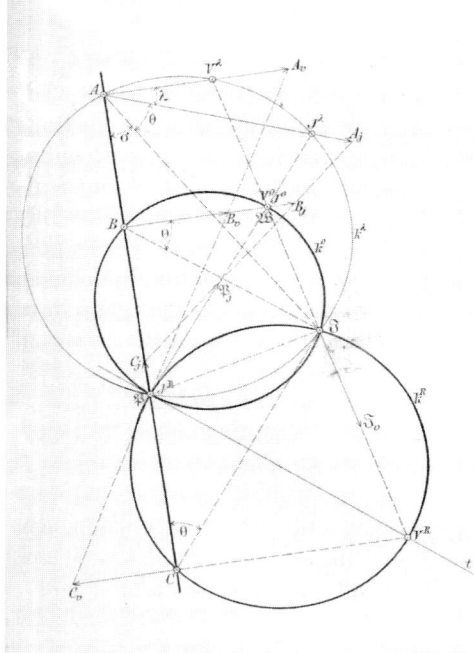

Fig 6: Bresse ′circles of 1st and 2nd kind

velocity and acceleration of each point of the circle k^0 are lying in one line. Therefore the normal acceleration of all system-points, lying on these circle k^0 is zero. Such inflexion-points have three points coincident with the coupler curve, lying infinitesimally close together in one line, which is the tangent to the curve. The circle k^0 is called *"Inflexion circle"*, according to the

Theorem: *The "Inflexion circle" is the geometrical locus of all points at the moving system, which have instantaneously no normal acceleration; it is also called "Bresse' circle of 1st kind".*

The directions of velocities and accelerations of all points, lying on the inflexion circle, intersect in the inflexion-pole W, which is diametrical opposite to the instantaneous pole P. Figure 6 includes velocity and acceleration of the inflexion point B. Considering the case $\lambda = 90°$, velocity and acceleration of each point of the corresponding circle $k^{R_1} k^{R_0}$ are perpendicular to each other. Therefore the tangential acceleration of all system-points, lying on these circle k^R, is zero. During two, one upon another infinite time-elements such points pass through the same two way-elements; therefore this circle is called *"Gleichenkreis or Normal circle"*. Both, inflexion circle and normal circle intersect perpendiculary.

Theorem: *The "Normal circle" is the geometrical locus of all points at the moving system, which instantaneously have no tangential acceleration; it is also called "Bresse' circle of 2nd kind".*

Both, the inflexion circle and the Bresse'circle of 2nd kind intersect in the instantaneous pole P (velocity is zero) and the accelaration pole J (acceleration is zero). The directions of acceleration of all that system-points at k^R intersect in the instantaneous pole P and its directions of velocity intersect in the point V^R, which is diametral opposite to the pole P. The diameter PV^R of the Bresse' circle of 2nd kind is perpendicular to the diameter PW of the inflexion circle; therefore PV^R coincides with the pole tangent Pt. The velocity of point C of the normal circle is demonstrated in fig.6.

BALL'S POINT

Several investigations with respect to *Ball's point* are carried out in [4, 5, 9, 10]. A point with interesting properties occurs at the intersection of the cubic of stationary curvature with the inflexion circle; it is called *Ball's point*. A point of a coupler coincident with *Ball's point* describes a path which is approximately a straight line, because it has stationary curvature and is located at an inflexion point of its path.

MODERN ANALYSIS OF PLANAR MOTION (POSITION-CHANGE)

This chapter introduces a method to find geometrical and kinematic poles using the analytical way. The mathematical algorithms are the basis to use the modern computer-technique also in the design process. Practical applications, e.g. cutting of prismatic workpieces, are demonstrated. The mathematical connection between the geometrical poles and the wellknown *Bresse circles* is formulated by geometry and differential geometry. The solution is carried out by the computer-programme KOSIM [8].

GEOMETRICAL AND KINEMATIC TRANSFER FUNCTIONS

By constrained linkages only the value of one coordinate can be chosen free, all other coordinates e.g. the relative rotating angles between two links and also the relative local-vectors to a coupler curve are nonlinear functions of the independent variable φ_1 which itself is a nonlinear function of the time. This nonlinear function between φ_1 and the time t is determined by the parameters of the linkage, by the impressed forces and impressed moments, by the inertia forces and moments of inertia of the links, the relative coordinates of the center of gravity etc.

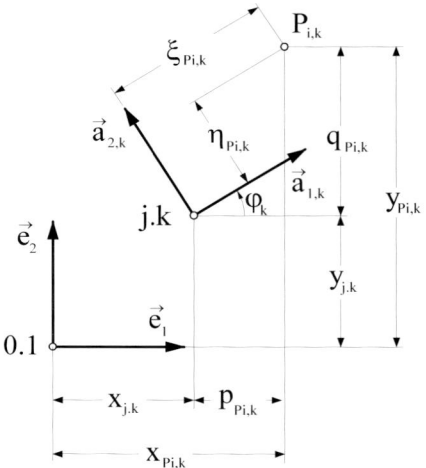

Fig. 7: Coordinates of a link-point at the frame

The turning angle of any link (k) relative to the frame is

$$\varphi_k = f(\varphi_1(t)) \tag{1}$$

and the local-vector to any point $P_{i,k}$ of any link (k) relative to the frame is:

$$\vec{x}_{Pi,k} = f(\varphi_1(t)) \tag{2}$$

This functions derivated with respect to the independent variable φ_1 deliver the "Geometrical Transfer-Function -GTF", the derivation with respect to the time t results in the "Kinematic Transfer-Function - KTF":

$$\frac{d\varphi_k}{d\varphi_1} = \varphi_k^{(1)} \Rightarrow GTF1; \quad \frac{d\varphi_k}{dt} = \frac{d\varphi_k}{d\varphi_1}\frac{d\varphi_1}{dt} \tag{3}$$

$$= \varphi_k^{(1)}\varphi_1^{[1]} \Rightarrow KTF1 \quad \text{of angle } \varphi_k$$

$$\frac{d\vec{x}_{Pi,k}}{d\varphi_1} = \vec{x}_{Pi,k}^{(1)} \Rightarrow GTF1; \quad \frac{d\vec{x}_{Pi,k}}{dt} = \frac{d\vec{x}_{Pi,k}}{d\varphi_1}\frac{d\varphi_1}{dt} = \vec{x}_{Pi,k}^{(1)}\varphi_1^{[1]} = \vec{x}_{Pi,k}^{[1]} \Rightarrow KTF1 \quad \text{of point } P_{i,k} \tag{4}$$

This nonlinear functions have differential-quotients of any order. The repeated differentiation by the time t produces the following equations:

$$\vec{x}_{Pi,k}^{[2]} = \vec{x}_{Pi,k}^{(1)}\varphi_1^{[2]} + \vec{x}_{Pi,k}^{(2)}\varphi_1^{[1]2} \Rightarrow KTF2 \tag{5}$$

as the KTF2, kinematic transfer-function of the 2nd order corresponding to the acceleration,

$$\vec{x}_{Pi,k}^{[3]} = \vec{x}_{Pi,k}^{(1)}\varphi_1^{[3]} + 3\vec{x}_{Pi,k}^{(2)}\varphi_1^{[1]}\varphi_1^{[2]} + \vec{x}_{Pi,k}^{(3)}\varphi_1^{[1]3} \tag{6}$$

as the KTF3, kinematic transfer-function of the 3rd order, etc.

GEOMETRICAL AND KINEMATIC POLES / BRESSE' CIRCLES

In a discrete position of a constrained linkage the "geometrical pole of the $^{v\text{th}}$ order" is that point at the coupler plane, which correspondent local-vector GTF$^{(v)}$ vanishes. The "kinematic pole of the $^{v\text{th}}$ order" is that point at the coupler plane, which correspondent local-vector KTF$^{(v)}$ vanishes. The geometrical pole of any order of the same coupler, of the same linkage in the same position is always the same, but the kinematic pole of any order in the same position is not the same, if the motion-status of the linkage varies.

The geometrical pole of 1st order $G_{1,k}$ is defined by: $\quad \vec{x}_{G1,k}^{(1)} = 0 \tag{7}$

GTF1 = 0 informs that GTF0 is stationary in this position, that means however, that the coupler curve of the Pole of 1st order has a cusp in this position.

The local-vector of any coupler point (fig. 7) is the sum of the vector to point j.k at the basic sytem and the vector to point $P_{i,k}$ at the moving system.

$$\vec{x}_{Pi,k} = \vec{x}_{j,k} + \vec{a}_{1,k}\xi_{Pi,k} + \vec{a}_{2,k}\eta_{Pi,k} \tag{8}.$$

The unit vectors of the moving system are defined by the following equations:

$$\left.\begin{array}{l}\vec{a}_{1,k} = \vec{e}_1 \cos\varphi_k + \vec{e}_2 \sin\varphi_k; \\ \vec{a}_{2,k} = -\vec{e}_1 \sin\varphi_k + \vec{e}_2 \cos\varphi_k\end{array}\right\} \tag{9}$$

We create the auxiliary variables

$$p_{Pi,k} = \xi_{Pi,k} \cos\varphi_k - \eta_{Pi,k} \sin\varphi_k; \quad q_{Pi,k} = \xi_{Pi,k} \sin\varphi_k + \eta_{Pi,k} \cos\varphi_k \tag{10}$$

and get the GTF0

$$\vec{x}_{Pi,k} = \vec{x}_{j,k} + \vec{e}_1 p_{Pi,k} + \vec{e}_2 q_{Pi,k} \tag{11}$$

Derivation of GTF0 by φ_1 results in GTF1 as follows:

$$\vec{x}_{Pi,k}^{(1)} = \vec{x}_{j,k}^{(1)} - \varphi_k^{(1)}\left(\vec{e}_1 q_{Pi,k} - \vec{e}_2 p_{Pi,k}\right) \tag{12}$$

The GTF1-vector vanishes, if its components are zero, i.e:

$$0 = x_{j,k}^{(1)} - \varphi_k^{(1)} q_{G1,k}; \quad 0 = y_{j,k}^{(1)} + \varphi_k^{(1)} p_{G1,k} \tag{13}$$

Equation (13) produces

$$\left.\begin{array}{l}q_{G1,k} = \dfrac{x_{j,k}^{(1)}}{\varphi_k^{(1)}} = \xi_{G1,k} \sin\varphi_k + \eta_{G1,k} \cos\varphi_k \\[4mm] p_{G1,k} = -\dfrac{y_{j,k}^{(1)}}{\varphi_k^{(1)}} = \xi_{G1,k} \cos\varphi_k - \eta_{G1,k} \sin\varphi_k\end{array}\right\} \tag{14}$$

The coordinates of $G_{1,k}$ are:

$$\left.\begin{array}{l}x_{G1,k} = x_{j,k} + p_{G1,k}; \quad y_{G1,k} = y_{G1,k} + q_{G1,k} \qquad \text{at the frame and} \\[4mm] \xi_{G1,k} = \dfrac{x_{j,k}^{(1)} \sin\varphi_k - y_{j,k}^{(1)} \cos\varphi_k}{\varphi_k^{(1)}}; \quad \eta_{G1,k} = \dfrac{x_{j,k}^{(1)} \cos\varphi_k + y_{j,k}^{(1)} s\varphi_k}{\varphi_k^{(1)}} \quad \text{at link (k)}\end{array}\right\} \tag{15}$$

The geometrical pole of 1st order for the relative positions of two links exists always. Its coordinates become infinite, if the GTF1 of the turning angle vanishes, i.e. crank and rocker are parallel.

The kinematic pole of 1st order $K_{1,k}$ is defined by:

$$0 = \vec{x}_{K1,k}^{(1)} \varphi_1^{[1]} \tag{16}$$

A kinematic pole of 1st order exists only in that case, if link (1) rotates, i.e. the KTF1 of the independent variable is not zero (KTF1 $\neq 0$). If a linkage is resting ($\varphi_1^{[1]} = 0$), then all linkage-points are kinematic poles of 1st order. The geometrical pole of 1st order and the kinematic pole of 1st order are identical points, look at equation (16). If KTF1 is not zero, this equation can only be satisfied, if GTF1=0; but this is the condition for a geometrical pole of 1st order.

In a discrete position of a linkage, the coupler point with vanishing GTF2-vector is geometrical pole of 2nd order and the point with vanishing KTF2-vector is kinematic pole of 2nd order. The geometrical pole of 2nd order G2,k is defined by:

$$0 = \vec{x}_{G2,k}^{(2)} = \vec{x}_{j,k}^{(2)} - \vec{e}_1\left(q_{G2,k}\varphi_k^{(2)} + p_{G2,k}\varphi_k^{(1)2}\right) + \vec{e}_2\left(p_{G2,k}\varphi_k^{(2)} - q_{G2,k}\varphi_k^{(1)2}\right) \tag{17}$$

with the components:

322

$$0 = x_{j,k}^{(2)} - q_{G2,k}\varphi_k^{(2)} - p_{G2,k}\varphi_k^{(1)2}; \quad 0 = y_{j,k}^{(2)} + p_{G2,k}\varphi_k^{(2)} - q_{G2,k}\varphi_k^{(1)2} \tag{18}$$

From both equations we get the following result:

$$p_{G2,k} = \frac{x_{j,k}^{(2)}\varphi_k^{(1)2} - y_{j,k}^{(2)}\varphi_k^{(2)}}{\varphi_k^{(2)2} + \varphi_k^{(1)4}}; \quad q_{G2,k} = \frac{x_{j,k}^{(2)}\varphi_k^{(2)} + y_{j,k}^{(2)}\varphi_k^{(1)2}}{\varphi_k^{(2)2} + \varphi_k^{(1)4}} \tag{19}$$

Respecting the additional variables

$$\left.\begin{array}{l} p_{G2,k} = \xi_{G2,k}\cos\varphi_k - \eta_{G2,k}\sin\varphi_k \\ q_{G2,k} = \xi_{G2,k}\sin\varphi_k + \eta_{G2,k}\cos\varphi_k \end{array}\right\} \tag{20}$$

we get the following coordinates of the geometrical pole of 2nd order of link (k).

$$\xi_{G2,k} = p_{G2,k}\cos\varphi_k + q_{G2,k}\sin\varphi_k; \quad \eta_{G2,k} = q_{G2,k}\cos\varphi_k - p_{G2,k}\sin\varphi_k \tag{21}$$

The derivation of equation (16) produces

$$0 = \bar{x}_{K2,k}^{(1)}\varphi_1^{[2]} + \bar{x}_{K2,k}^{(2)}\varphi_1^{[1]2} \tag{22}$$

which determines the kinematic pole of 2nd order K2,k. From this equation we see, that this kinematic pole of a resting linkage is indefinite, i.e. all points of such linkages are kinematic poles of 2nd order.

We investigate the following cases:

1. $0 = \varphi_1^{[1]} = \varphi_1^{[2]} \Rightarrow$

   K2,k: each link-point

2. $0 \neq \varphi_1^{[1]} \quad 0 = \varphi_1^{[2]} \Rightarrow$

   G2,k = K2.k

3. $0 \neq \varphi_1^{[2]}; \quad 0 = \varphi_1^{[1]}$

   $\Rightarrow$ G1,k = K2,k

4. According to equation (22) the vector-sum can only vanish, if the vectors GTF1 and GTF2 are collinear [6]. There-

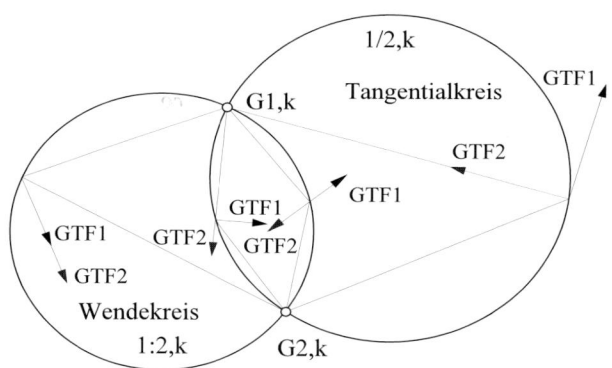

Figt. 8: Bresse' Circles 1:2,k and 1/2,k;

fore the kinematic pole is fixed on the *Bresse' Circle of 1st kind;1:2,k,* (fig. 8). The sign of KTF2 ($\varphi_1^{[2]}$) decides the arc-piece of the *Bresse Circle* according to the directions of the vectors GTF1 and GTF2, either negative (same direction) or positive (opposite direction), see fig. 8.

The geometrical pole of 3rd order of any link (k) with respect to the frame is that point of the link, which GTF3-vector vanishes; the kinematic pole is defined through KTF3=0; further research work has been done by *Rehwald*, [6, 7].

In each linkage-position (discrete value of independent variable) the local-vectors to (∞^2) coupler points of a constrained mechanism have collinear GTF-vectors of any two orders. Those local-vectors which have collinear GTF-vectors of the same two orders, are fixed at the periphery of a *Bresse Circle of 1st kind;* the centre point and the radius of this circle changes from one to another position of the mechanism. If the GTF-vectors are of vth and μ^{th} order, then we have the *Bresse Circle of 1st kind; $v : \mu, k$*.

Both vectors are collinear, therefore their vectorproduct vanishes:

$$0 = \mathbf{x}_{v:\mu,k}^{(v)} \times \mathbf{x}_{v:\mu,k}^{(\mu)}$$

$$0 = \left(x_{j.k}^{(v)} - B_{1,v,k}q_{v:\mu,k} - B_{2,v,k}p_{v:\mu,k}\right)\left(y_{j.k}^{(\mu)} + B_{1,\mu,k}p_{v:\mu,k} - B_{2,\mu,k}q_{v:\mu,k}\right)$$

$$- \left(y_{j.k}^{(v)} + B_{1,v,k}p_{v:\mu,k} - B_{2,v,k}q_{v:\mu,k}\right)\left(x_{j.k}^{(\mu)} - B_{1,\mu,k}q_{v:\mu,k} - B_{2,\mu,k}p_{v:\mu,k}\right) \qquad (23)$$

$$0 = x_{j.k}^{(v)}y_{j.k}^{(\mu)} - y_{j.k}^{(v)}x_{j.k}^{(\mu)} + \left(p_{v:\mu,k}^2 + q_{v:\mu,k}^2\right)\left(B_{1,v,k}B_{2,\mu,k} - B_{2,v,k}B_{1,\mu,k}\right)$$

$$- p_{v:\mu,k}\left(B_{2,v,k}y_{j.k}^{(\mu)} - B_{1,\mu,k}x_{j.k}^{(v)} + B_{1,v,k}x_{j.k}^{(\mu)} - B_{2,\mu,k}y_{j.k}^{(v)}\right)$$

$$- q_{v:\mu,k}\left(B_{1,v,k}y_{j.k}^{(\mu)} + B_{2,\mu,k}x_{j.k}^{(v)} - B_{2,v,k}x_{j.k}^{(\mu)} - B_{1,\mu,k}y_{j.k}^{(v)}\right)$$

Several transformations result in the coordinates of centre-point and radius of the Bresse´ Circle of 1st kind:

$$p_{Mv:\mu,k} = \frac{B_{2,v,k}y_{j.k}^{(\mu)} - B_{1,\mu,k}x_{j.k}^{(v)} + B_{1,v,k}x_{j.k}^{(\mu)} - B_{2,\mu,k}y_{j.k}^{(v)}}{2\left(B_{1,v,k}B_{2,\mu,k} - B_{2,v,k}B_{1,\mu,k}\right)} \qquad (24)$$

$$q_{Mv:\mu,k} = \frac{B_{1,v,k}y_{j.k}^{(\mu)} + B_{2,\mu,k}x_{j.k}^{(v)} - B_{2,v,k}x_{j.k}^{(\mu)} - B_{1,\mu,k}y_{j.k}^{(v)}}{2\left(B_{1,v,k}B_{2,\mu,k} - B_{2,v,k}B_{1,\mu,k}\right)} \qquad (25)$$

$$r_{v:\mu,k}^2 = \frac{y_{j.k}^{(v)}x_{j.k}^{(\mu)} - x_{j.k}^{(v)}y_{j.k}^{(\mu)}}{B_{1,v,k}B_{2,\mu,k} - B_{2,v,k}B_{1,\mu,k}} + p_{Mv:\mu,k}^2 + q_{Mv:\mu,k}^2 \qquad (26)$$

In each linkage-position (discrete value of independent variable) the local-vectors to (∞^2) coupler-points of a constrained mechanism have orthogonal GTF-vectors of any two orders. Those local-vectors which have orthogonal GTF-vectors of the same two orders, are fixed at the periphery of a *Bresse Circle of 2nd kind;* the centre point and the radius of this circle changes from one to another position of the mechanism. If the GTF-vectors are of vth and μ^{th} order, then the *Bresse Circle of 2nd kind; $v/\mu, k$* results in:

$$0 = \mathbf{x}^{(\nu)}_{\nu/\mu,k} \cdot \mathbf{x}^{(\mu)}_{\nu/\mu,k}$$

$$0 = \left(x^{(\nu)}_{j.k} - B_{1,\nu,k}q_{\nu/\mu,k} - B_{2,\nu,k}p_{\nu/\mu,k}\right)\left(x^{(\mu)}_{j.k} - B_{1,\mu,k}q_{\nu/\mu,k} - B_{2,\mu,k}p_{\nu/\mu,k}\right)$$
$$+ \left(y^{(\nu)}_{j.k} + B_{1,\nu,k}p_{\nu/\mu,k} - B_{2,\nu,k}q_{\nu/\mu,k}\right)\left(y^{(\mu)}_{j.k} + B_{1,\mu,k}p_{\nu/\mu,k} - B_{2,\mu,k}q_{\nu/\mu,k}\right) \quad (27)$$

$$0 = x^{(\nu)}_{j.k}x^{(\mu)}_{j.k} + y^{(\nu)}_{j.k}y^{(\mu)}_{j.k} + \left(p^2_{\nu/\mu,k} + q^2_{\nu/\mu,k}\right)\left(B_{1,\nu,k}B_{1,\mu,k} + B_{2,\nu,k}B_{2,\mu,k}\right)$$
$$- p_{\nu/\mu,k}\left(B_{2,\nu,k}x^{(\mu)}_{j.k} - B_{1,\mu,k}y^{(\nu)}_{j.k} - B_{1,\nu,k}y^{(\mu)}_{j.k} + B_{2,\mu,k}x^{(\nu)}_{j.k}\right)$$
$$- q_{\nu/\mu,k}\left(B_{1,\nu,k}x^{(\mu)}_{j.k} + B_{2,\mu,k}y^{(\nu)}_{j.k} + B_{2,\nu,k}y^{(\mu)}_{j.k} + B_{1,\mu,k}x^{(\nu)}_{j.k}\right)$$

Analogous to the Bresse' Circle of 1st kind the coordinates of centre-point and radius of the Bresse' Circle of 2nd kind result in:
:

$$p_{M\nu/\mu,k} = \frac{B_{2,\nu,k}x^{(\mu)}_{j.k} - B_{1,\mu,k}y^{(\nu)}_{j.k} - B_{1,\nu,k}y^{(\mu)}_{j.k} + B_{2,\mu,k}x^{(\nu)}_{j.k}}{2\left(B_{1,\nu,k}B_{1,\mu,k} + B_{2,\nu,k}B_{2,\mu,k}\right)} \quad (28)$$

$$q_{M\nu/\mu,k} = \frac{B_{1,\nu,k}x^{(\mu)}_{j.k} + B_{2,\mu,k}y^{(\nu)}_{j.k} + B_{2,\nu,k}y^{(\mu)}_{j.k} + B_{1,\mu,k}x^{(\nu)}_{j.k}}{2\left(B_{1,\nu,k}B_{1,\mu,k} + B_{2,\nu,k}B_{2,\mu,k}\right)} \quad (29)$$

$$r^2_{\nu/\mu,k} = p^2_{M\nu/\mu,k} + q^2_{M\nu/\mu,k} - \frac{x^{(\nu)}_{j.k}x^{(\mu)}_{j.k} + y^{(\nu)}_{j.k}y^{(\mu)}_{j.k}}{B_{1,\nu,k}B_{1,\mu,k} + B_{2,\nu,k}B_{2,\mu,k}} \quad (30)$$

In the special case $\nu = 1$ and $\mu = 2$ the *Bresse Circle 1:2,k* is called *"Wendekreis or Inflexion Circle"* and the *Bresse Circle 1/2,k* is called *"Gleichenkreis or Normal Circle"* of the coupler (k); first investigations have been done by *De La Hire, Bresse* discovered the normal circle. The analytical research-work respecting differential-geometrical variables etc. has been done by *Rehwald* [6] and in [11].

It is very important to know, that the *Bresse' Circles* are not based on acceleration. Geometry and differential-geometry don't include the time t. In general, a lot of *Bresse' Circles of both kinds* exist at each coupler plane according to $\nu:\mu$ bzw. ν/μ; ν and μ can run through the whole row of positive numbers.

BALL'S POINT
The Bresse' Circles of 1st kind 1:2,k,1 . and 1:3,k,1 .intersect in the geometrical pole $G_{1,k,1}$. and in the Ball's point 1:2:3,k,1 ;.see fig. 9. The equations to calculate the coordinates are deduced.

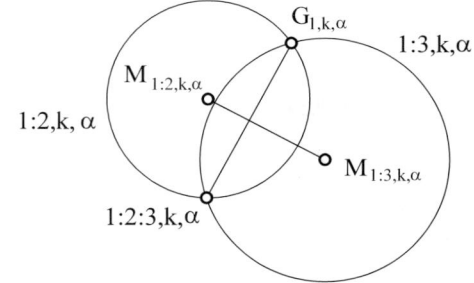

Fig. 9:Bresse' Circles and Ball's Point

According to fig. 10 the coordinates of the centre points of the unpaired joint-elements of the elementary group EG **A2.1** are wellknown as coordinates of centre points of two Bresse´ Circles and also the radii of these circles as length of the two links, see fig. 9 and 10. On the basis of this parameters the angles φ_I and φ_{II} can be calculated by the following equations, see figt.10.

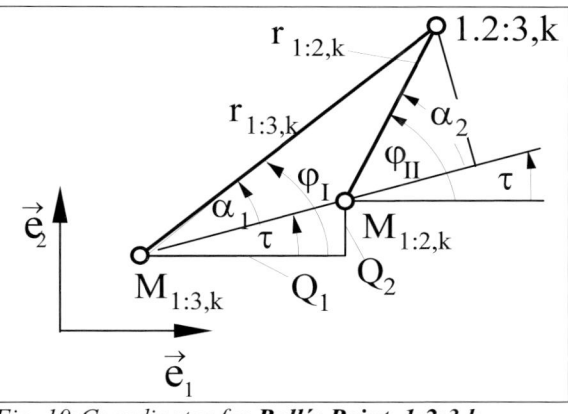

Fig. 10: Coordinates for ***Ball´s Point 1:2:3,k***

$$Q_1 = p_{M1:2,k} - p_{M1:3,k}; \quad Q_2 = q_{M1:2,k} - q_{M1:3,k}; \quad e = \sqrt{Q_1^2 + Q_2^2}; \quad \sin\tau = \frac{Q_2}{e}; \quad \cos\tau = \frac{Q_1}{e};$$

$$r_{1:3,k}^2 = e^2 + r_{1:2,k}^2 + 2er_{1:2,k}\cos\alpha_2 \Rightarrow \cos\alpha_2 = \frac{r_{1:3,k}^2 - r_{1:2,k}^2 - e^2}{2er_{1:2,k}};$$

$$r_{1:2,k}^2 = e^2 + r_{1:3,k}^2 - 2er_{1:3,k}\cos\alpha_1 \Rightarrow \cos\alpha_1 = \frac{r_{1:3,k}^2 - r_{1:2,k}^2 + e^2}{2er_{1:3,k}}$$

$$K_1 = r_{1:3,k}^2 - r_{1:2,k}^2 + e^2; \quad K_2 = r_{1:3,k}^2 - r_{1:2,k}^2 - e^2;$$

$$\cos\alpha_1 = \frac{K_1}{2er_{1:3,k}}; \quad \cos\alpha_2 = \frac{K_2}{2er_{1:2,k}}; \quad \varphi_I = \alpha_1 + \tau; \quad \varphi_{II} = \alpha_2 + \tau;$$

$$\sin\varphi_I = \sin\alpha_1\cos\tau + \cos\alpha_1\sin\tau; \quad \sin\varphi_{II} = \sin\alpha_2\cos\tau + \cos\alpha_2\sin\tau;$$

$$\cos\varphi_I = \cos\alpha_1\cos\tau - \sin\alpha_1\sin\tau; \quad \cos\varphi_{II} = \cos\alpha_2\cos\tau - \sin\alpha_2\sin\tau;$$

$$\sin\alpha_1 = \sqrt{1-\cos^2\alpha_1} = \frac{\sqrt{4e^2r_{1:3,k}^2 - K_1^2}}{2er_{1:3,k}}; \quad \sin\alpha_2 = \frac{\sqrt{4e^2r_{1:2,k}^2 - K_2^2}}{2er_{1:2,k}}; \quad K_3 = \sqrt{4e^2r_{1:3,k}^2 - K_1^2};$$

$$\cos\varphi_I = \frac{K_1Q_1 \mp K_3Q_2}{2e^2r_{1:3,k}}; \quad \sin\varphi_I = \frac{K_1Q_2 \pm K_3Q_1}{2e^2r_{1:3,k}}$$

$$\cos\varphi_{II} = \frac{K_2Q_1 \mp K_3Q_2}{2e^2r_{1:2,k}}; \quad \sin\varphi_{II} = \frac{K_2Q_2 \pm K_3Q_1}{2e^2r_{1:2,k}};$$

$$p_{1:2:3,k} = p_{M1:2,k} + r_{1:2,k}\cos\varphi_{II} = p_{M1:3,k} + r_{1:3,k}\cos\varphi_I;$$

$$q_{1:2:3,k} = q_{M1:2,k} + r_{1:2,k}\sin\varphi_{II} = q_{M1:3,k} + r_{1:3,k}\sin\varphi_I.$$

The Bresse´ Circles 1:2,k and 1:3;k intersect in the geometrical pole of 1st order $G_{1,k,1}$. further 1:2,k and 2:3,k in $G_{2,k,1}$. further 1:3,k and 2:3,k in the geometrical pole of 3rd order and additional in the **Ball´s point 1:2:3,k;** in this point the GTF1-, GTF2- and GTF3-vectors are collinear.

PRACTICAL APPLICATION

The demonstrated theory is the basis to the following facts:

- The coupler curves of geometrical poles of 1st order (G1,k) have one or several cusps.
- The coupler-curves of geometrical poles of 2nd order (G2,k) have one or several straight line-pieces, which are associated with one or several cusps at the GTF1-hodograph. Comment: GTF2 = 0 signifies, that the associated GTF1-curve is stationary and therefore has a cusp in this position. A cusp at the GTF1-hodograph however is associated with a straight line-piece of the coupler curve GTF0.
- The GTF2-hodograph of poles of 3rd order has at least one cusp; GTF3 = 0 means, that the GTF2-curve is stationary and therefore has a cusp in this position. The accompanying GTF1-hodograph has a straight line-piece in the associated position. If the tangent of this straight line-piece crosses the origin of the GTF1-hodograf, then the coupler curve GTF0 of the geometrical pole of 3rd order G3,k describes extremely exact a straight line-piece

1.) Example: The following practical application demonstrates the production of special prismatic work-pieces using the geared linkage **R1.1**-1-0. All numerical calculations are done by the computer-programme KOSIM [8].

The geared linkage **R1.1**-1-0 has the length of 100mm of link (1) and the gear-ratio i = 4, including coupler point ($\xi = -11.111$; $\eta = 0$), which is fixed at the geometrical pole-curve of 2nd order in link (2); solution, see fig. 11.

Solution: Fig. 11: *Principle of regular polygon-turning;* • *Coupler curve of point ($\xi = -11.111$; $\eta = 0$) at (02);* o *GTF1-function of coupler point at (02);* ı *GTF2-function of coupler point at (02).*

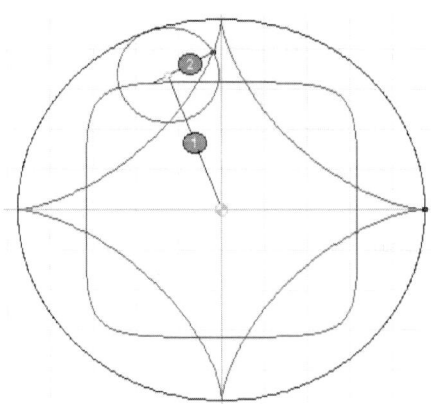
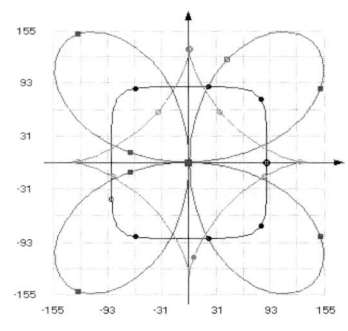

- Coupler curve of the point on (02)   [mm] / [mm]
- GTF1 of the coupler point on (02)   [mm / rad^1] / [mm / rad^1]
- GTF2 of the coupler point on (02)   [mm / rad^2] / [mm / rad^2]

Fig. 11: $\xi_{1.2} = 100$; $r_{2*0} = 33,3333..$; $r_{0*2} = 133,3333..$; $\xi_2 = -11,111...$

2.) Example: The belt-linkage **Z2.1**-1-0 is given by the parameters: $\xi_{1.2} = 100$ mm (unit of length is [mm]); $r_{0*3} / r_{2*3} = 3$; $r_{0*3} = 45$; $r_{2*3} = 15$, see fig.12.

Searched: Geometrical polodes of 1st and 2nd order; Bresse' Circle in position $_1\,_3\!\!= 0^0$; coupler curves with cusps, alternating and one-sense curvature.

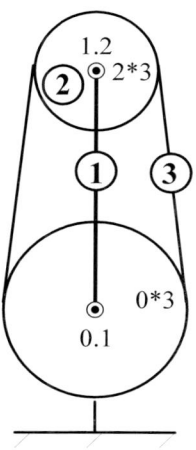

Fig. 12: Linkage-Type **Z2.1**-1-0

Solution: All numerical calculations are done by the computer-programme KOSIM [8]. The polodes of 1st order are circles with radius $r_{G1,0} = 150$ at link (0); $r_{G1,2} = 50$ at link (2). The polodes of 2nd order are circles with radius $r_{G2,0} = 75$ at link (0); $r_{G2,2} = 25$ at link (2). The Bresse' Circle at position $_1\,_3\!\!= 0^0$ in the coordinate system of link (2) results in $\xi_{1.2;2-0} = 12{,}5$; $\check{}_{1\boxed{2}2-0} = 0$ and radius $r_{1:2,2-0} = 37{,}5$. The coordinate system of link (2) demonstates an annulus for such points of link (2), which generate coupler curves of alternating curvature, see fig.13.

geometrische Polkurve erster Ordnung

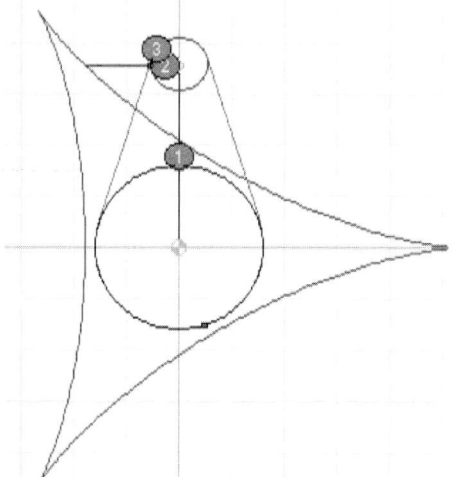

Wendekreis

$G_{2,2}$ $_{1.2}$ $G_{1,2}$

geometrische Polkurve zweiter Ordnung

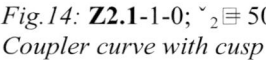

Fig. 13: Annulus in (2)

The following figures demonstrate coupler curves with cusp, straight-line and alternating sense of curvature.

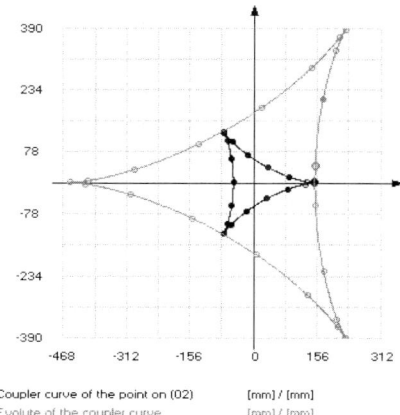

●	Coupler curve of the point on (02)	[mm] / [mm]
○	Evolute of the coupler curve	[mm] / [mm]

Fig.14: **Z2.1**-1-0; $\check{}_2\!\!\equiv 50$
Coupler curve with cusp

Fig. 15: Coupler curve with Cusp and evolute

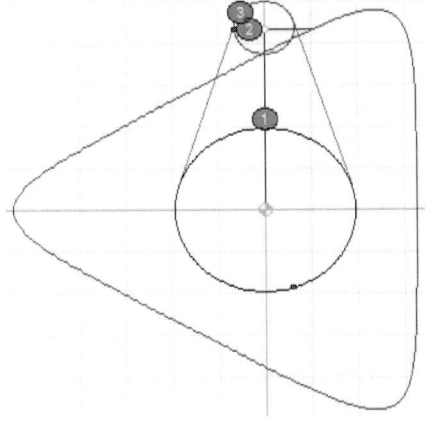

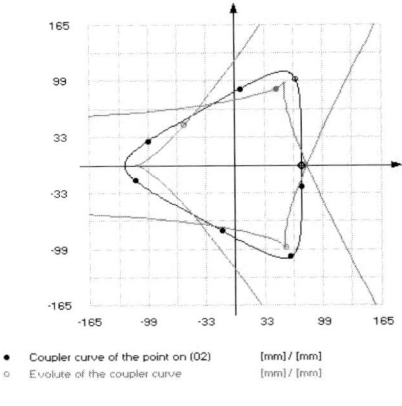

Figt. 16: **Z2.1**-1-0; coupler
curve with straight-line; ˇ$_2$⊟ -25

Fig. 17: Coupler curve and evolute

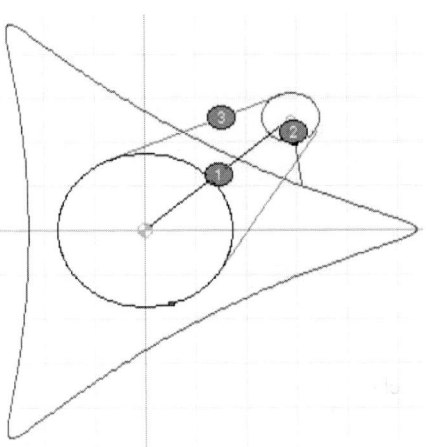

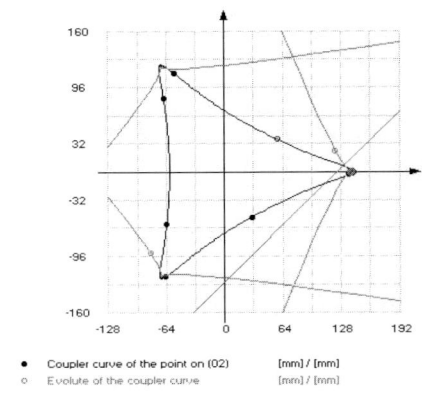

Fig.18: **Z2.1**-1-0; coupler curve
with alternating curvature; ˇ$_2$⊟ 40

Fig. 19: Coupler curve and evolute

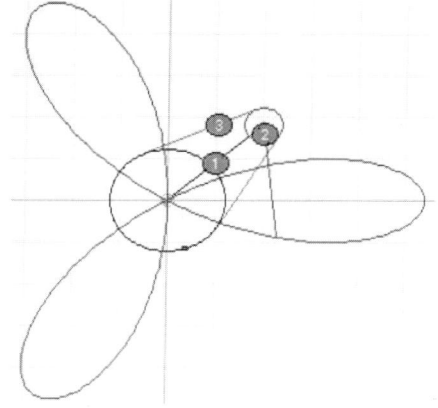

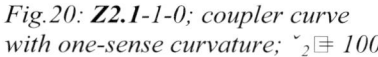

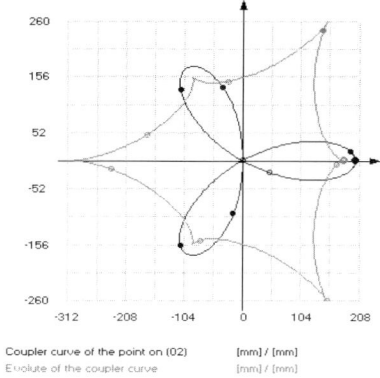

- Coupler curve of the point on (02) [mm] / [mm]
- Evolute of the coupler curve [mm] / [mm]

*Fig.20: **Z2.1**-1-0; coupler curve with one-sense curvature;* $\check{}_2 \boxplus 100$

Fig. 21: Coupler curve and evolute

REFERENCES

[1] Bresse: Memoire sur un theoreme nouveau concernant les mouvements plans, et sur l'application de la cinematique a la determination des rayons de courbure. Journal de l'Ecole polytechnique. Cah.35. p.89, 1853

[2] Burmester, L.: Lehrbuch der Kinematik. Verlag von Arthur Felix, Leipzig, 1888

[3] Krause, M.: Analysis der ebenen Bewegung. Vereinigung Wissenschaftlicher Verleger Walter de Gruyter & Co, Berlin und Leipzig 1920

[4] Dijksman, E.A.: Motion Geometry of Mechanisms. Cambridge University Press, Cambridge, London, New York, Melbourne, 1976

[5] Luck, K.; Modler, K.-H.: Getriebetechnik – Analyse, Synthese, Optimierung. Springer-Verlag Wien, New York, 1990

[6] Rehwald, W.: Analytische Kinematik von Koppelebenen. Otto Krausskopf-Verlag GmbH, Mainz 1972

[7] Rehwald, W.: Kinetische Bewegungsanalyse ebener, ungleichförmig übersetzender Kurbelgetriebe. VDI-Forschungsheft 503, VDI-Verlag GmbH, Düsseldorf 1964

[8] Rehwlad, W.; Luck, K.: KOSIM-Koppelgetriebesimulation. Fortschritt-Berichte VDI, Reihe 1, Nr.:332, ISBN 3-18-333201-9, VDI Verlag GmbH, Düsseldorf 2000

[9] Müller, H.-R.: Kinematik. Sammlung Göschen, Band 584/584a. Verlag Walter de Gruyter & Co, Berlin, 1963

[10] Erdman, A.G.; Sandor, G.N.: Advanced Mechanism Design: Analysis and Synthesis. Volume 1 and 2, Prentice-Hall, Inc., Englewood Cliffs, New Jersey 07632, 1984

[11] Rehwald, W.; Luck, K.: Die Bresse'schen Kreise und weitere Phänomene. Pro ceedings of DRESDEN SYMPOSIUM GEOMETRY; constructive & kinematic, TU Dresden, Institut für Geometrie, 27.02.-01.03.2003, ISBN 3-86005-394-9.

THE REULEAUX MODELS: CREATING AN INTERNATIONAL DIGITAL LIBRARY OF KINEMATICS HISTORY

Francis C. Moon
Sibley School of Mechanical and Aerospace Engineering
Cornell University
Upson Hall, Ithaca, NY, USA 14853
e-mail:fcm3@cornell.edu

ABSTRACT

This paper describes a project to document and preserve a collection of 220 Reuleaux kinematic models and to create a virtual museum of models connecting artifacts, mathematics and historical material relating to kinematics of machines. The digital library will also provide links to other kinematic model collections. This paper will also survey existing model collections related to mechanisms and the theory of machines. One of the goals is bring attention to important kinematic artifacts and help preserve these historic collections.

KEYWORDS: Kinematics, Reuleaux, History, Models.

INTRODUCTION

This paper describes the creation of an Internet-based virtual museum and digital library of kinematic history at Cornell University. This virtual museum is based on Cornell's Collection of Kinematic models including 220 models from the Voigt catalog of Reuleaux models circa 1882, [1], and 30 other historic models. These models reproduce many of the important mechanisms in Franz Reuleaux's 800 model collection in Berlin which was lost during World War II. The Cornell collection has been designated by the American Society of Mechanical Engineers as a *National Landmark Collection.* The virtual museum is named KMODDL or "Kinematic Models for Design Digital Library" (http://kmoddl.library.cornell.edu) and will be part of the United States National Science Digital Library or NSDL. (www.nsdl.edu) The NSDL project will enable users to access mathematical, scientific and technical data and educational material for levels ranging from middle and high school through university and professional applications. The KMODDL is the first NSDL project involving mechanical engineering.

One of the goals of the Cornell project is to incorporate information on international collections of kinematic models especially of a historic nature. The project will also invite specialists in kinematics of mechanisms and the history of machines to contribute information to the website. Towards this end the Author is preparing a catalog of current known collections of kinematic models some of which are summarized in this paper. It is hoped that this paper will become a catalyst to uncover other collections and inspire universities and institutions to document and conserve such collections as part of the history of the evolution of machines and mechanisms.

Figure 1. Four rotary pump/engine mechanisms from the Cornell University Collection of Reuleaux Kinematic Models, [1].

HISTORICAL USE OF MACHINE MODELS

Computer-aided design codes and multi-body dynamics codes provide new tools with which to visualize complex motions of constrained machine components. For centuries however, architects, engineers and mathematicians have used physical models to demonstrate kinematic and dynamic motions in both simple and complex machines and mechanisms.

The use of models in engineering has had, until the last quarter century, a long and useful history. Filippo Brunelleschi (1377-1436), the architect and engineer of the Duomo in Florence is known to have created construction models, including machines. During the 18th century, there appeared collections of 'Physical Cabinets' assembled mainly by royalty and wealthy persons concomitant with the interest in science in the so-called Age of Reason and the Enlightenment. These collections included objects related to mechanics, optics, electricity and magnetism as well as chemistry and related sciences. An example is the Royal collection in Dresden. Another is the Hauck Collection in Denmark. Christopher Polhem (1661-1751) in Sweden created a 'mechanical alphabet' of models for machines, [2]. At the same time new physical artifacts such as clocks, automata, and steam engines inspired the making of models of these devices. James Watt used a model of a Newcomen steam engine to advance the efficiency of steam engines. According to the historian Eugene Ferguson, the emergence of the international fairs of the 19th century led to the creation of many museums of science and technology and the collection of machine models [3].

In the 19th century, Robert Willis (1800-1875) of Cambridge was also known for his kinematic teaching models though few have survived, [4]. Ferdinand Redtenbacher of Karlsruhe,[5], and J. Schubert of Dresden,[6], also created kinematic models for educational use.

Figure 2. Johann A.Schubert model(c.1830), T.U. Dresden.Collection, [6].

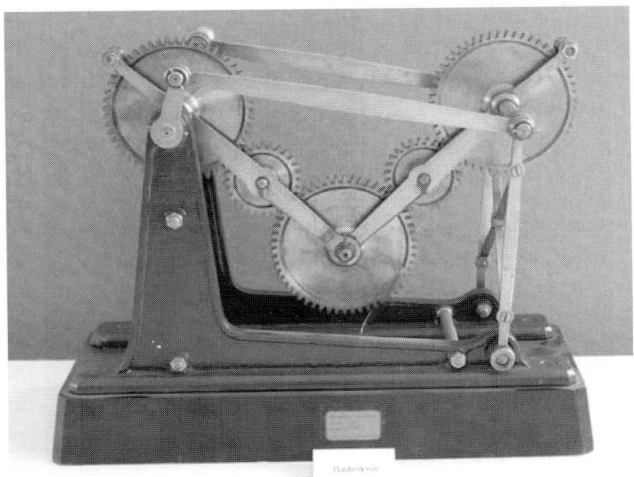

Figure 3. F.Redtenbacher model, (c. 1860) Universität Karlsruhe, [5].

Franz Reuleaux (1829-1905) of Berlin created the world's largest collection of kinematic models at the Technical University of Berlin with over 800 models. His collection was first noted in a report of Robert Thurston on the Vienna Exposition of 1873, [7], though at the time none of the models were reproduced. Reuleaux authorized and supervised the reproduction of approximately 360 mechanisms by the model maker Gustav Voigt in Berlin, [8],[9]. J. Schröder of the Polytechnische Arbeits Institute, Darmstadt also created kinematic models based on the books of Reuleaux, and Redtenbacher,[10] which were later produced by the model works of Peter Koch, [11]. Some of the models of Schröder and Voigt are in collections in Europe, North America and Japan. The Deleuil company of Paris, (c. 1865) produced kinematic models for sale and included two of Robert Willis's designs, [12]. Most of the original Reuleaux in Berlin collection was destroyed in the second world war, though there are 60 originals in the Deutsches Museum in Munich and about a dozen in Hannover.

Model makers often exhibited at international exhibitions and competed for prizes and medals. Deleuil and Schröder exhibited at the London Exhibition of 1851. The Schröder models also won a medal at the 1876 Centennial Exhibition in Philadelphia. Franz Reuleaux was a member of the judging panel which perhaps inspired Reuleaux to produce his own models through Gustav Voigt of Berlin who had worked for Reuleaux, [13]. Reuleaux also exhibited 300 models from his Berlin Collection in the South Kensington Exhibition of Scientific Instruments in London in 1876, [14].(See Figure 4.)

Figure 4. Reuleaux Berlin model exhibited in London 1876. Apparatus for drawing complex curves. (Courtesy Science Museum, London.)

THEORETICAL ROLE OF KINEMATIC MODELS

In an age dominated by mathematical analysis and computer calculations, modern theorists often view physical models as in the realm of technicians that are used to illustrate practical machines and devices for students and a public who lack mathematical sophistication. Reuleaux however was a theoretician and he and other 19th century kinematicians such as Robert Willis used models as a way to encapsulate theoretical and mathematical ideas especially those that involve complex geometry and topology. As modern textbooks use CD's with computer exercises, Reuleaux expected that his models would be taught in conjunction with his famous text on Kinematics of Machinery (1875,1900), [15],[16]. Many of the models have inscribed letters and numbers on links and joints that correspond to the figures in his famous textbook. Among the theoretical ideas that he incorporated in his models were lower and higher kinematic pairs, kinematic inversions, centrode pairs, straight-line and parallel mechanisms, mechanisms based on spherical geometry, topologically equivalent mechanisms based on his symbol representation of kinematic chains, mechanics of belt stability, gear tooth profiles including the involute and epicycloids, roulettes, ratchets and escapements and curves of constant breadth. Many of these ideas and concepts were the subject of deep mathematical study during the 19th century. For example, Reuleaux's extensive treatment of curves of constant width or breadth has been recognized by mathematicians by referring to the curved triangle as the "Reuleaux triangle". Reuleaux had four models illustrating the mathematics of curves of constant width and six others showing their use in positive return mechanisms, [1].

The use of models and visual representation of technical knowledge has been documented by Eugene Ferguson (1978, 1990), [17], [18]. The use of physical models has atrophied in many engineering institutions with the advance of computer simulation. However the widespread use of models in the 19th century is documented in part by the list of model collections described below. To help preserve these valuable collections, Cornell University and the US National Science Foundation has launched a project to document the Reuleaux models in a virtual museum and digital library on the web.

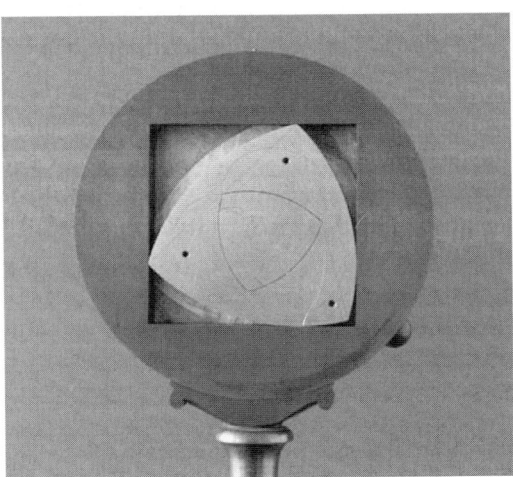

Figure 5. Reuleaux-Voigt model of a curved triangle (curve of constant width) with the moving centrode engraved in the Reuleaux triangle. [Cornell Collection, [1]]

A VIRTUAL MUSEUM OF KINEMATICS: KMODDL

Cornell University was fortunate to have purchased a substantial part of the Voigt-Reuleaux models in 1882. These models were used for teaching machine design up until the 1970's. A new generation of faculty has again found these models useful for teaching and research. Also a number of scholars, engineers and artists have begun to travel to Ithaca to see the Reuleaux Collection. As a result Cornell University decided to document and make the Collection available on the World Wide Web as part of the US *National Science Digital Library* or NSDL. The kinematic collection comprises 220 models from the Voigt catalog, a dozen models from the Schroeder works and a Robert Willis model made in Deleuil, Paris and 18 Illinois Gear Company models circa 1950.

KMODDL will provide access to over 300 machine mechanisms for teaching, scholarship and design. KMODDL will couple together visual pictures, and quick-time movies, hands-on animation, including haptic 'feel' of forces in mechanisms, mathematics of mechanisms, history of invention, access to rare books in history of machines, and the ability to 3D 'print' some of the mechanisms to produce working models. KMODDL is expected to be accessible on the web in summer 2004 (kmoddl.library.cornell.edu).

KMODDL is a Cornell collaboration between Mechanical Engineering, Mathematics and the University Libraries who are developing technical descriptions for the models, mathematical tutorials, historical material on the mechanisms and their inventors and on-line user activated simulations.

The KMODDL project has resulted in new uses of the models including;
> Rapid prototyping technology for kinematic mechanisms based on CAD models.
> Student interest in history of machines.
> Student exposure to forgotten machine topologies such as rotary engines.
> Interaction with local high school teachers interested in mathematics of
>> mechanisms.
> Use of Reuleaux models by architecture faculty for creative design course.
> Public interest in history and engineering of machines and mechanisms.

The Cornell KMODDL team plans to use the web to integrate both these textual and artifact collections on the history of machines and mechanisms. Many historic books will be digitized and referenced in the relevant web portal page of each of the kinematic models. Each of the models will contain descriptive and historical text. To see the models in action, we have also created animations of many of the models. A special goal was to provide an opportunity for the web site user to interact with the models. Examples of web-based animations of machine and kinematic models are common on the web. One example is the Kyoto University Museum collection of 19 Voigt-Reuleaux models, which were modeled in a CAD program and animated with the multi-body dynamics code ADAMS. (See Shiroshita et al, 2001.[19].) Cornell has recently created a website which contains a sampling of Reuleaux models with a few animations at www.explore.cornell.edu. What cannot be experienced with a web collection however is the physical handling of the models. When students and visitors get to play with the Reuleaux models, the first thing they want to do is turn the crank and see them move. To this end, we have created the opportunity for web users to reproduce a few of the models through the use of rapid prototyping technology.

Inventors

In addition to technical descriptions of the mechanisms, KMODDL will include over 40 short biographies about inventors associated with many of the mechanisms. For many of the Voigt models, Reuleaux provides the name of the inventor. Some of the names included in the biographies are, Bramah, Cartwright, Chubb, Denison, Evans, Graham, Maudslay, Mudge, Oldham, Pappenheim, Roberts, Root, Tschebychev, Ulhorn and Whitworth. In Addison there are biographies on Redtenbacher, Willis, Reuleaux, Watt and da Vinci.

Rare and Historical Books in the History of Machines-On Line

The goal of KMODDL is not only to show pictures of historical mechanisms but to link historical text-based materials with the technical artifacts. Thus Franz Reuleaux's major books *The Constructor*(1854-1893) and *Kinematics of Machinery*(1875,1900) have been scanned and can be accessed on the web as part of KMODDL in their English translations. Each of the Reuleaux Voigt model web pages will have links to references in the Reuleaux books. In addition books of Robert Willis (1842,1870), [20], F. Redtenbacher, [21], W. Rankine ,A. Kennedy, [22], Martin Grübler (1917) [23], will also be scanned and accessible on line. For the first time readers in the history of kinematics and mechanisms will be able to read these famous works anywhere in the world without having access to large libraries.

Besides theoretical works in 19th century kinematics, KMODDL will provide access to some of the most famous "theatre of machines" pictorial atlases such as Agostino Ramelli (1588),[24], Jacobus Strada (1617), Solomon de Caus (1624), Vittorio Zonca (1656), Georg Böckler (1661), Jacob Leupold (1724),[25], Lanz and Betancourt (1817), [26], and J.A. Borgnis (1818).

Mathematical Tutorials in Kinematics

Along with technical descriptions of the models, are short tutorials on the mathematics of 19th century kinematics and mechanisms. Some of these already written include: the four bar mechanism, Peaucellier's inversor and straight-line mechanisms, spherical geometry of the universal joint, Grübler's mobility criterion, centrodes, the Reuleaux triangle and curves of constant width, the Kennedy three centers theorem and the Roberts-Chebyshev theorem.

Videos and Simulation of Kinematic Models

The KMODDL team (see below for names) was inspired by several existing websites on the history of mechanisms. One of these is the University of Kyoto Museum site that describes their Reuleaux models. For each model they constructed a CAD model and provided an animation of the mechanism. In KMODDL we have implemented three levels of motion. For many of the models we have QUICKTIME videos of the motion of the actual Cornell Reuleaux models. For another set of models, we made QTVR videos that allow the user to move the model with a mouse as well as zoom into the model. A third system is a simulator designed by Professor Hod Lipson of Cornell to allow the user to design a linkage and move it with a mouse.

3D Printing of Reuleaux Kinematic Models

To document the Reuleaux models, CAD drawings of several mechanisms were made using both Pro Engineer and Solid Works software packages. The 3D drawings are then converted to stereolithography format files. This format maps surfaces into a mesh of triangles which can be used as input to rapid prototyping software for a 3D printer, [27].

A rapid prototyping technology, called fused deposition modeling or FDM, was used in this study to reproduce several Reuleaux-Voigt kinematic models. The system was manufactured by Stratasys, (Model FDM 2000). The process creates a sequence of thermoplastic layers from a filament wound coil that is heated and extruded through a nozzle. The x-y planar location of the nozzle is controlled by information from the stereolithography file of the CAD model. In order to create functioning mechanisms, a second, water soluble release material is placed in the gaps between the movable parts. The basic material is named ABS or Acrylonitrile-Butadiene-Styrene. This system has been developed by Professor Hod Lipson of Cornell University, who is also part of the KMODDL project.

The FDM produced copies of the Reuleaux models are remarkably visually true to the originals as shown in a comparison in Figure 6. The models are fairly robust to use and move. The cost and time to produce one is a fraction of that necessary to manufacture a traditional copy in iron and brass. However the time to complete a model from the CAD code is fairly long. A half scale model of the slider crank took approximately 6 hours in the FDM machine.

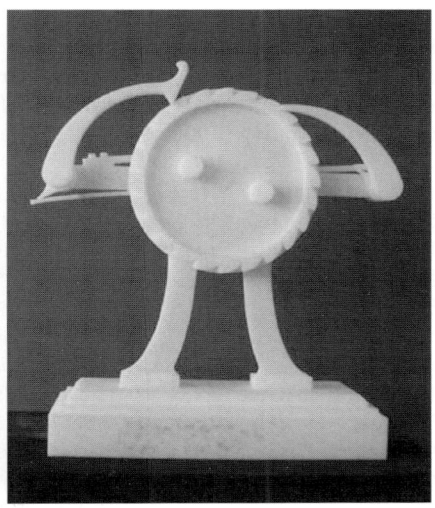

Figure 6. Left-Reuleaux-Voigt ratchet model. Right- Computer-printed model.[27].

INTERNATIONAL KINEMATIC MODEL COLLECTIONS

Over the past five years the Author has visited the sites of a dozen different kinematic collections and in several cases was given access to photograph historical models and collections. These sites include, Aachen, London, Cambridge, Denmark, Rome, Florence, Milan, Berlin, Dresden, Hannover, Karlruhe and Munich. Two of the most important collections we were able to document on digital camera were the 60 models of the Reuleaux collection in the Deutsches Museum, some of which came from Reuelaux's personal collection in Berlin after he died and the Redtenbacher collection of near 100 models many of which appeared in his book, *Die Bewegungs Mechanismen* (1866). The Redtenbacher models are historically important because they likely inspired his student Franz Reuleaux to copy and expand the Karlsruhe collection. This collection also has a model used by the automobile inventor Karl Benz who was also a student of Redtenbacher.

We plan to incorporate the 160 models from Munich and Karlsruhe into the KMODDL website. Many of the Deutsches Museum models are different from the Cornell Reuleaux-Voigt models and therefore compliment each other. It is hoped that the Cornell KMODDL will encourage other institutions to put their collections on the web. We hope too be able to create a link between as many kinematic collections in the world as possible creating an international network of artifacts relating to kinematics and the history of mechanisms. A short descriptions of some of the collections known to the author follows. (See Appendix and Acknowledgements for list of curators and institutions.)

RWTH-Aachen Technische Hochschule: One of the largest modern collections of kinematic models consisting of over 200 models is that in the Institut für Getriebetechnik und Maschinendynamik under Professor Günter Dittrich. Many of the models are motorized and also contain a number of spatial linkages. The Institute has a very nice description catalog (1995) of many of the models and their mathematical principles.

Boston Museum of Science, Boston, Mass.: This unique collection of over 200 motorized models was built in the 1930's and has its roots in the kinematic classification tables of Lanz and Betancourt (1817) and Borgnis (1818). In 1868, a New York patent attorney Henry T. Brown, published a catalog of 501 mechanisms with little documentation, [28]. Many of these figures are remarkably similar to the above books of Lanz and Betancourt and Willis. In the 1930's, a hobbyist named W.M. Clark designed and built a set of motorized panels of these models that were exhibited at the 1933 Century of Progress Exhibition in Chicago, [29]. The models were then exhibited at the Newark Museum, Newark New Jersey, for many years until recently when they were put into storage. A similar set was built for the Boston Museum of Science where around half are on public display.

Deutsches Museum, Munich: The complete collection includes over 100 models and includes around 60 original Reuleaux models sent from Berlin shortly after the time of his death. There are also a number of models originating at the Technical University of Munich, an a fine display of biomechanical kinematic models by Dr Thilo from Riga inspired by Reuleaux's second volume of kinematics published in 1900, [31]. However only about 30 models are on display. The rest are stored in crates in the basement of the museum. The Archive of the Deutsches Museum also contains the archival papers of Franz Reuleaux that have recently been cataloged.

Technische Universität Dresden: This collection has some very fine wooden models by Schubert (see Mauersberger [6]), as well as a number of Reuleaux-Voigt models, Schröder models and locally built models and many modern models. Professor Karl-Heitz Modler has made some fine modern models are made of plastic and are suitable for use with an overhead projector to demonstrate the motions of the mechanisms. The overall collection numbers around 200 models, [6].

Fondazioni Scienza e Tecnica, Florence Italy: This institution has a large collection of 19th century Schröder models of mechanisms and complete steam engines along with a physical cabinet of physics and chemistry models. Dr. Paolo Brenni is the Curator.

Universität Hannover: Another fine modern collection of kinematic models of over 200 models is in the Institut für Getriebetechnik under Professor Reinhard Braune. This collection also contains about 18 Reuleaux models found in the ruins of the university after WWII. Most of these models are of the straight-line mechanisms.

Cornell University, Ithaca New York: As mentioned above this collection contains the largest known extant set of 220 Reuleaux-Voigt models purchased in 1882 by Cornell's first president Andrew D. White. The collection also includes around 12 Schröder models, a Willis model made in Paris, 18 Illinois Gear Co. gear models circa 1950, There are also four model engines, three steam engines and a model gas engine circa 1900. One notable group of models are six full scale 1950's era Chevrolet engines and transmissions cut-away to reveal the inner mechanisms.

Kyoto University Museum: This collection of about 60 models contains 19 Reuleaux-Voigt models and 2 Peter Koch models, [19]. The museum also has provided a web site with CAD animations as well as photos of the original models.

University of Porto, Portugal: This university has 113 Reuleaux-Voigt models, one of the largest extant Reuleaux collections.[30], were acquired circa 1890.

Figure 7 Left-Reuleaux-Voigt model; Kyoto University Museum
Right- Reuleaux model in the Deutsches Museum, Munich.

Science Museum, London; Musee des Art et Metier, Paris: Both public museums purchased Schröder 19th century kinematic models and have a few on public display. There are a number of models in storage at the Science Museum in London in the museum annex. The Museum of Science and Industry in Chicago has some gear models from the 1933 Century of Progress Exhibition.

SUMMARY

A team at Cornell University has built a digital library and virtual museum of kinematic models based on the work of Franz Reuleaux (KMODDL) that is planned to grow into an international resource for the history of kinematic models and mechanisms. The KMODDL team is also using computer-aided modeling tools and rapid prototyping technology to document, preserve, and reproduce in three dimensions, historic machine mechanisms. This technology will allow researchers to share three-dimensional models over the web from mechanism collections around the globe. This research is part of the United States National Digital Science Library project that will make available on the web, science, engineering and mathematics tools for teaching and research. (www.nsdl.edu)

REFERENCES

1.. Moon, F.C. (2003) "Franz Reuleaux: Contributions to 19th century kinematics and the theory of machines" *Applied Mechanics Reviews 56(2),* pp.261-285.
2 .Johnson, W.J. (1963) *Christopher Polhem,* Trustees of Trinity College, Hartford, Conn.
3 . Ferguson, E. "Technical Museums and International Exhibitions"
4 . Moon, F.C. (2003) "Robert Willis and Franz Reuleaux: Pioneers in the Theory of Machines" *Notes and Records Royal Soc. London,* pp.209-230.
5 . Redtenbacher, F. (1866), *Die Bewegungs Mechanismen*
6 . Mauersberger, K. (1997), "Die Getriebemodellsammlung der Technischen Universität Dresden", *Wissenschaftliche Zeitschrift der Technischen Universität Dresden* **46** Heft 3.

7. Thurston, R. (1873) *Machinery and Manufacturers with an Account of European Manufacturing Districts*, report on the Vienna Exhibition published by The US Department of State.pp. 367-8.

8. Gradenwitz, A. (1908), "The Mutual Relations of Geometry and Mechanics and Prof. Reuleaux's Mechanical Movements", *Scientific American,* March 21, 1908, pp. 204-205

9. Voigt, G.(1907) *Kinematische Modelle nach Professor Reuleaux, Verzeichnis I,II.* Berlin.
10.Schröder,J. (1899) *Illustrationen von Unterrichts – Modellen und Apparaten,* Polytechnisches Arbeits-Institut, Darmstadt.
11. Koch, P. (1912) *Katolog II, Kinematik*, Peter Koch Modellwerk, Cöln-Nippes.
12. Deleuil (1865*), Catalogue des Instruments de Physique, de Chemie, d'Optique et de Mathematique*, Paris
13. Reuleaux, F., (1876), *Kinematics of Machinery; Outlines of a Theory of Machines*, A.B.W. Kennedy, Transl., MacMillan and Co., London.
13. Walker, F.A. editor, 1878, *International Exhibition 1876*, United States Centennial Commissioner Reports and Awards, Group XXVIII, J.B. Lippincott & Co., Philadelphia, 1878.
14. Kennedy, Alex. B.W., (1876b), "The Berlin Kinematic Models," *Engineering*, Vol. 22, p. 239-240, London.
14. Reuleaux, F. (1900), *Kinematics of Machinery*, Vol. 1, 2 .
17. Ferguson, E.S., (1992), *Engineering and the Mind's Eye*, MIT Press, Cambridge, Mass.
18. Ferguson, E.S., (1977), "The Mind's Eye: Nonverbal Thought in Technology," *Science*, Vol. 197, No. 4300, pp. 827-836.

19. Shiroshita,S., Kumamoto,H. Nishihara,O., and Jing,D. (2001) "Constructing a Virtual Museum of Machine Mechanism Models Imported from Germany During Japanese Westernization for Higher Education: 3D Animations Based on Kinematics and Dynamics" Presented at *Museums and the Web 2001*.
   [http://www.archimuse.com/mw2001/papers/Shiroshita.html]
20. Willis, Robert, *Principles of Mechanisms*, London, 1841
22. Kennedy, A.B.W., (1886), *The Mechanics of Machinery*, MacMillan and Co., London.
23. Grübler, M. (1917) *Getriebelehre,* Verlag von Julius Springer, Berlin
24. Ramelli, Agostino, (1588) *Le Diverse et Artificiose Machine*, Paris.
25. Leupold, J. (1724), *Theatrum Machinarum,* Leipzig.
26. Lanz ,P.L., Betancourt, A.(1808) *Essai sur la Composition des Machines,* Paris.
27. Lipson, H. Moon, F.C. (2003) "3D-Printing the History of Mechanisms", Cornell University Report, Mechanical and Aerospace Engineering.
28. Brown, H. T., (1868), Editor, *Five Hundred and Seven Mechanical Movements*, NY.
29. Clark, W.M. (1943) *A Manual of Mechanical Movements,* Garden City Publishing Co. Garden City, NY.
30. Vaz Guedes,M. "Ciencia ou Tecnica:Uma Coleccao de Instrumentos Didacticos da Faculdade de Engenharia da Universidade do Porto", *History of FEUP-*Universidade do Porto 2000. (In Portuguese.)
31. Thilo,O. (1901), "F. Reuleaux, Kinematik in Tierreiche" Sonderabdruck aus dem "Biologischen Centralblatt" Bd.XXI Nr.16. 15. August 1901. (514-528)

ACKNOWLEDGEMENTS

The Author was fortunate to have had access to model collections at the Deutsches Museum in Munich, The Science Museum in London, Universität Hannover, Technische Universität Dresden, Universität Karlsruhe, and the Hauch Physike Cabinet in Sorø Denmark.

The Author is also grateful to the Humboldt Stiftung of Germany for financial support to visit the Reuleaux archives of the Deutsches Museum in Munich. The work in Munich would not have been possible without the support from my friend and colleague Professor Franz Kollmann formerly with the Technische Universität Darmstadt. Of particular help at the Deutsches Museum were Dr. Wilhelm Füssl, head of the Archiv, Karl Allwang, Curator for Mechanical Engineering, Manfred Spachtholz, Museum engineer and Mr. Sebastian Remberger.

Also of help were Prof. Reinhard Braune of the Institut für Getriebetechnik, Universität Hannover, Prof. Jorg Wauer, Universität Karlsruhe, Prof. Karl-Heinz Modler, and Dr Klaus Mauersberger, Technische Universität Dresden, Michael Wright and Ben Russel of the Science Museum in London, Prof. D. Severin, of the Technische Universität Berlin, Dr S. Shiroshita, Kyoto University Museum, Jørgen F. Andersen of Sorø Academy, Denmark, Dr. Paolo Brenni, Fondazione Scienza e Tecnica, Florence, Prof. Christoph Glocker, ETH, Zurich, Prof. Paulo M S T de Castro, Universidade do

Porto, Dr. Sule Oygur, Newark Museum, Prof. Gary Kinzel, Ohio State University and the late Dr. Frantisek Peterka, Inst. Thermomechanics, Prague.

The KMODDL participants include Prof. Hod Lipson, Mechanical and Aerospace Engineering, Prof. David Henderson and Dr Daina Taimina, Mathematics, John Saylor, Kizer Walker, Ron Rice, University Library, Cornell University.

Appendix: Table of Collections of Kinematic Mechanisms

Location	Institution	Models	Vintage	Source
Aachen, Germany	RWTH-Technische Hochschule	300	modern	
Berlin, Germany	Technische Universität	40	modern	
Boston, Mass., USA	Boston Museum of Science	150	1940's	Clark/Brown
Cambridge, UK	Cambridge University	40	19th-20th C.	
Chemnitz, Germany	Technische Universität	?	modern	
Columbia, Penn. USA	Nat. Clock and Watch Museum	80	17th-20 C.	Escapements
Columbus, Ohio, USA	Ohio State University	50	1950's	Illinois Gear Co.
Denmark	Hauck Foundation	?	18th C.	
Dresden, Germany	Technische Universität	120	19th-20th C.	
Florence, Italy	Fondazione Scienza e Tecnica	100	19th c.	Schröder
Hannover, Germany	Technische Universität	20	c.1880	Reuleaux
Hannover	Technische Universität	200	modern	
Ithaca, NY, USA	Cornell University	220	c.1882	Reuleaux/Voigt
Ithaca, NY, USA	Cornell University	20	c.1869	Schröder
Karlsruhe, Germany	Üniversität Karlsruhe	100	c.1866	Redtenbacher
Kyoto, Japan	Kyoto University Museum		c. 1890	Reuleaux/Voigt
London, UK	Science Museum	20	19th c.	Schröder
London, UK	Victoria and Albert Museum			Escapements
Milan, Italy	Science Museum		20th c.	L.da VInci copy
Munich, Germany	Deutsches Museum	100	19th c.	Reuleaux
New York, USA	IBM	?	1950-1970	L.da Vinci copy
Newark, NJ, USA	Newark Museum	200	1930's	Clark/Brown
Paris, France	Musee des Arts et Metier	?	19th c.	Schröder
Portugal	University of Porto	113	c.1890	Reuleaux/Voigt
Prague	Technical University	23		Schröder
Riga, Latvia	Technical University	?		Schröder?
Rome, Italy	University	20		

8. Automata and Robots

THE HISTORY OF THE CREATION AND DEVELOPMENT OF HAND-OPERATED BALANCED MANIPULATORS (HOBM)

Vigen Arakelian
Department GMA, INSA-Rennes
20 Avenue des Buttes de Coësmes CS 14315
35043 Rennes Cedex, France
e-mail : vigen.arakelyan@insa-rennes.fr

ABSTRACT - This paper deals with the historical aspects of the creation and the evolution of manual balanced manipulators. Some interesting solutions devoted to the design of these systems are presented as well as some basic results obtained in the field of optimal balancing.

KEYWORDS: Hand-operated manipulator, balancing, manual control, pantograph

INTRODUCTION

The creation of a new class of mechanical systems called manual balanced manipulators (HOBM) was a logical stage in the field of the improvement of the means for the mechanization of production. Manual balanced manipulators are used in such conditions as when the application of hoisting machinery is not efficient and automation by using industrial robots is expensive and consequently unjustified.

We will attempt initially to define what are manual balanced manipulators, what are the characteristics of their design and what is the role of these systems in the mechanization of production.

The creation and industrial application of manual balanced manipulators (HOBM) began in 1964 when Robert A. Olsen suggested a new original balanced assembly [1, 2] (Fig.1, 2). This invention was the basis for the development of the manual balanced manipulator proposed by Reizou Matsumoto in 1975 [3] (Fig.3) which is the prototype of the modern HOBM (Fig.4). In the suggested solution the following property of the pantograph is used: if three points (see Fig. 3: points 5, 6 and 7) of the pantograph lie on the same straight line A-A, then any path traced by point 5 is reproduced by point 7 (for fixed point 6) and any path traced by point 6 is reproduced by point 7 (for fixed point 5).

Thus, if points 5 and 6 are located in the vertical and horizontal guides, the displacement of point 7 in the horizontal plane may be realized by the horizontal displacement of point 6 combined with the rotation of the system about the vertical axis (for fixed point 5). In a similar manner the vertical displacement of point 7 may be

348

realized by the vertical displacement of point 5 (for fixed point 6). The interesting characteristic of such a system is the following: if the pantograph system is balanced about axis through point 5 (see Fig.3), the displacements of the manipulated object in the horizontal plane may be realized without effort (the only resistance being due to friction in the joints) because the gravitational forces are always perpendicular to the displacements.

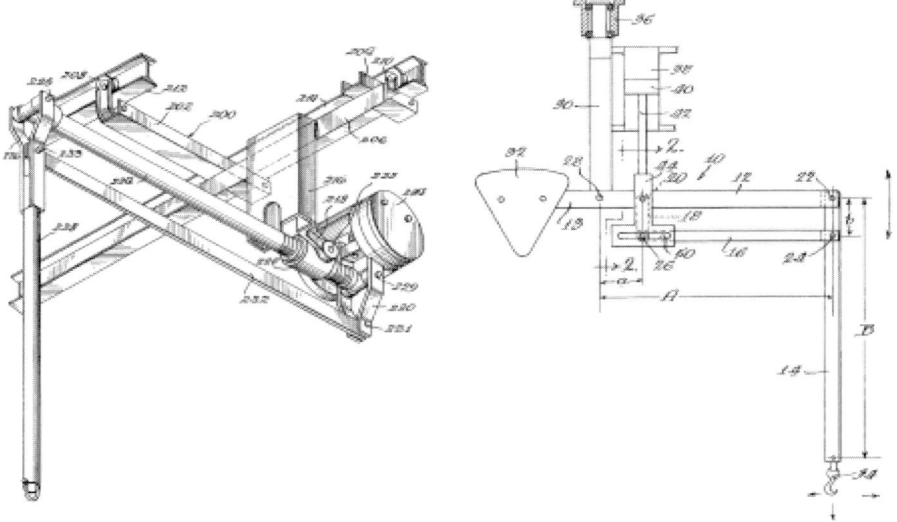

Fig.1: Patent US 3 134 340.   Fig.2: Patent US 3 259 352.

Thus the displacements in the horizontal plane may be realized manually. For the vertical displacement a suitable driver may be used connected to point 5, which can balance the weight of the manipulated object or produce the vertical displacement directly.

We will now try to define just what is a HOBM. One can say that it is a handling system with a simple mechanical actuator in which the manipulated object in any position of the workspace is also balanced. Such a state of constant balance allows displacements of heavy objects to be achieved manually.

Which are the advantages of these manipulators relative to industrial robots?

Primarily it is the simplicity of their construction and the low cost price. They have a very simple command system, a great weight-carrying capacity (up to 2500 kg) and a very large workspace (in the horizontal direction the gripper can reach to approximately 3000mm).

This allows the HOBM to be used for moving workpieces between machine tools very smoothly because it is controlled manually. The implantation of HOBM in existing production is very simple without the need for important additional surfaces, special auxiliary devices or essential reorganization of the production lay-out.

The pantograph application allows the transformation of the small displacements of the points A and B (see Fig.5) into the large displacements of the point G, that gives a large working space to the manipulator.

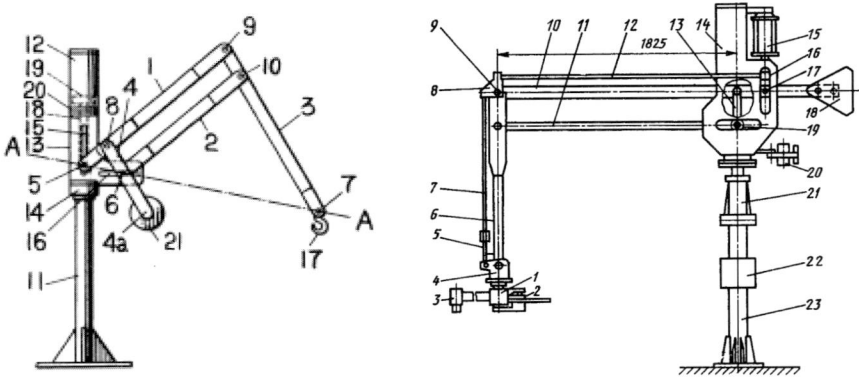

Fig.3: Patent US 3 883 105.　　　　　　　　*Fig.4. HOBM «KCH-160» [24].*

The term "balanced manipulator" shows that in the operating procedure of these systems is very important to achieve an accurate mass balancing.

Many publications [4-27] have examined the design methods and the balancing of HOBM. It was shown that for the balancing of these manipulators it is necessary to apply to the link 1 (see Fig.5) a balancing moment $M_{bal} = M \sin \varphi$, where M is a constant moment depending to the sizes and masses of the pantograph links.

The general approach for determination of balancing conditions was proposed by the study of the motion of the centre of mass of the pantograph actuator.

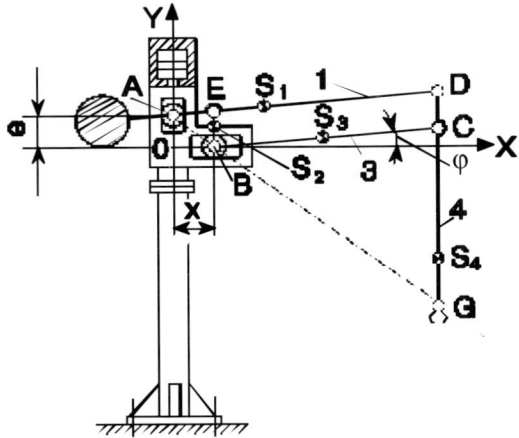

Fig.5: HOBM with pantograph actuator.

The motion of the common centre of mass of the moving links of the pantograph mechanism (Fig.6) along axis OY, is determined by the expression:

$$y_S = \frac{m_{1A}y_A + m_{1E}y_E + (m_{1D} + m_{4D})y_D + m_2 y_2 + m_3 y_3 + m_{4C}y_C + m_{4G}y_G}{\sum\limits_{i=1}^{4} m_i}, \quad (1)$$

where,

$y_2 = l_1 \sin\varphi_1 + (l_2 - r_2)\sin\varphi_2 + e\cos\theta;$

$y_3 = l_1 \sin\varphi_1 + l_2 \sin\varphi_2 - r_3 \sin\varphi_3 + e\cos\theta;$

$y_A = e\cos\theta;$

$y_C = l_1 \sin\varphi_1 + l_2 \sin\varphi_2 - l_3 \sin\varphi_3 + e\cos\theta;$

$y_D = l_1 \sin\varphi_1 - l_3 \sin\varphi_3 + e\cos\theta;$

$y_E = l_1 \sin\varphi_1 + e\cos\theta;$

$y_G = l_1 \sin\varphi_1 + l_2 \sin\varphi_2 - l_3 \sin\varphi_3 - l_4 \sin\varphi_4 + e\cos\theta;$

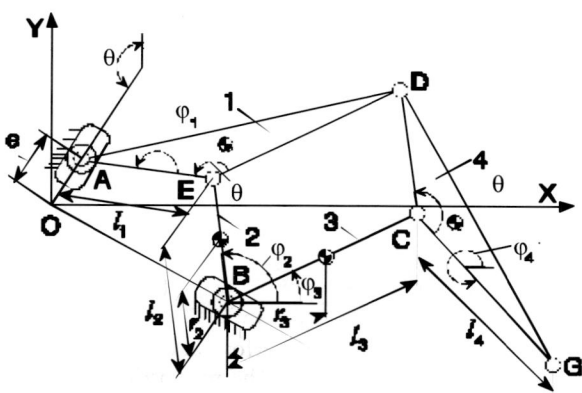

Fig.6: Pantograph mechanism (general type).

So that the operator can move the manipulated object horizontally, overcoming only the forces of friction, the total work done by the forces of gravity of the elements of the actuator must be equal to zero. For this purpose, the centre of mass of the pantograph mechanism must only move in the horizontal direction (or must be fixed), i.e. $\alpha_1 = \alpha_2 = \alpha_3 = \alpha_4 = 0$.

Fig.5 represents a HOBM with planar collinear pantograph which is a special case of the skew pantograph (Fig.6) when $\theta = \pi$.

For such a pantograph mechanism, equation (1) can be presented by the expression:

$$y_S = \left\{ \alpha \sin \varphi + \left(\sum_{i=1}^{4} m_i - \beta \right) e \right\} \Big/ \sum_{i=1}^{4} m_i, \qquad (2)$$

where

$$\alpha = m_1 r_1 + m_{2E} l_1 + m_3 r_3 + m_{4D}(l_1 + l_3)$$

$$\beta = m_2 (l_2 - r_2)/l_2 + m_4 r_4 / l_2 + m_3.$$

In this case, to balance the pantograph actuator, it is necessary that the condition $\alpha = 0$ is satisfied. This may be accomplished by the installation of a counterweight.

A further requirement is the invariability of the loads on the driving system due to the weights of the movable links of the actuator. This must be independent of the position of the manipulated object in the workspace of the manipulator.

In this case together with the condition $\alpha = 0$, it will be advisable to add the condition:

$$\beta = \sum_{i=1}^{4} m_i$$

$$\text{(since } x_S = (\alpha \cos\varphi + \beta x)/\sum_{i=1}^{4} m_i \text{)}.$$

When these conditions are achieved, the centre of mass of the actuator of the manipulator coincides with the centre of the joint B (fig.5) and during vertical displacement of the manipulator gripper, the effort on the driving system is constant and proportional to the magnitude of the load. The proportionality factor is equal to the similarity factor of the pantograph mechanism.

It is obvious that this balancing condition may be achieved by a counterweight mounted on link 1, but it should be noted that there are also balancing schemes based on the application of springs (Fig.7).

Practical applications show that the operation of the manual balanced manipulator is easier when the mechanical actuator contains only revolute pairs. In this case, the frictional forces are low and, consequently, the manual effort exerted by the operator is less.

Fig.8 shows a version of a manual balanced manipulator with only revolute pairs [27].

352

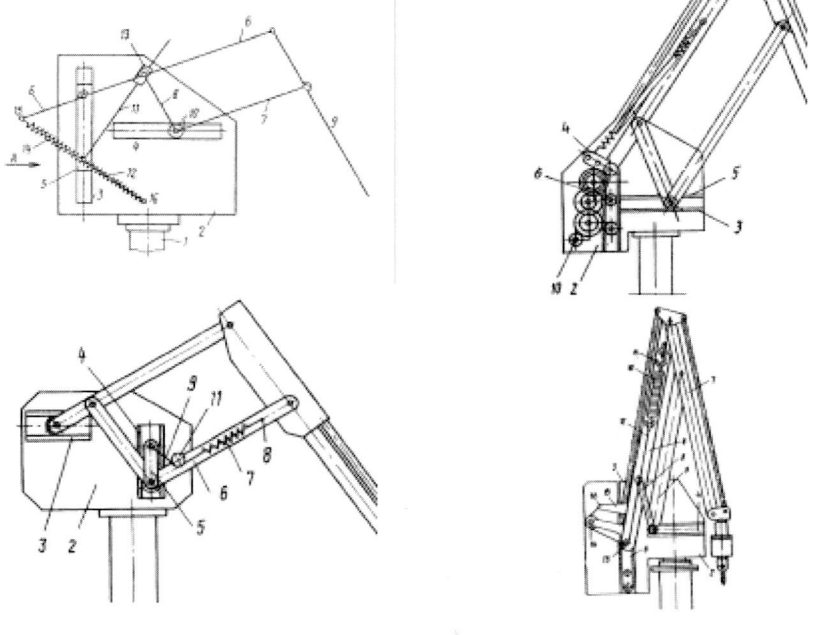

Fig.7: Balancing schemes with springs [11, 22-24].

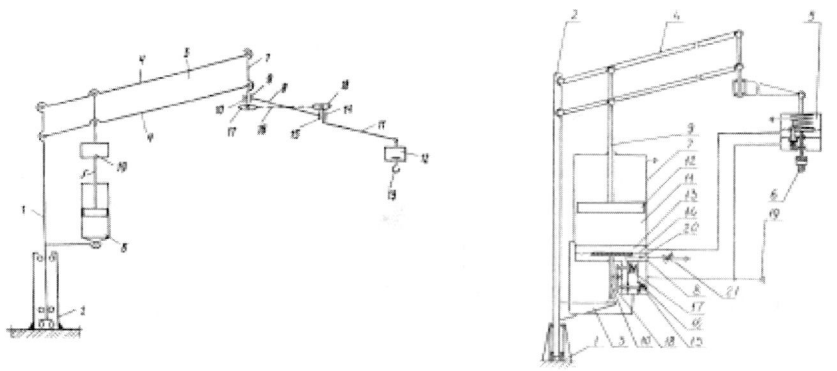

Fig.8: HOBM with parallelogram mechanism [26, 27].

Fig.9: Grippers of manipulator «dalmec» (Italy).

354

Fig.10: Some examples of HOBM applications («dalmec» Italy).

It is obvious that for wide applications of these manipulators they should be equipped with different grippers. These devices for gripping are designed for each specific application and the sizes and forms of the objects to be manipulated. Fig.9 illustrates a range of different grippers (pneumatic and mechanical) while Fig. 11 shows different types of load hooks.

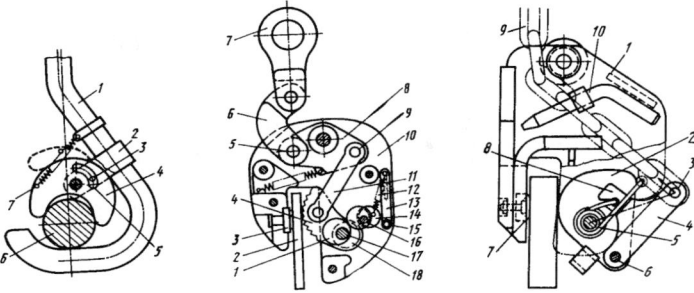

Fig.11: Load hooks of HOBM [24].

The design of the drive system in manual balanced manipulators also determines the construction of the manipulator, its functional capacity, as well as its reliability and speed. In industry one can find manipulators with electric, pneumatic and hydraulic drives. For manipulators having up to 150 kg load capacity, one uses electric motors or pneumatic drives; if the load capacity is up to 500 kg one uses electric motors. With regard to manipulators with very great load capacity (up to 2500 kg) essentially one uses hydraulic drives.

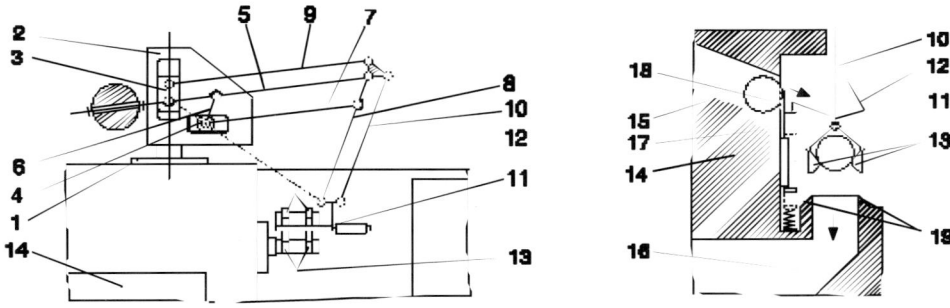

Fig.12: HOBM without driving system [9].   *Fig.13. Device (14) [9].*

It should be noted that the applicability of hand-operated balanced manipulators in the different fields is very wide; loading and unloading of machine tools, presses and other production equipment; displacement of heavy objects during machine assembly; movement of foundry ladles; handling of heavy objects as, for example electric motors, various devices, etc.; as well as handling of objects during storage. Fig.10 represents some examples of their application.

Fig.12 shows a type of hand-operated balanced manipulators without driving system designed to assist the loading and unloading of a machine tool [9]. The need for a driving system is avoided by the balancing method, in which the system parameters take into account the mass of the effective load equal to one half of the masses of the blank

356

part and the machined part. In other words, the manipulator is always loaded either by the blank part or the machined part. Fig.13 shows the loading and the unloading device of the manipulator in addition to the input of the blank and the ejection of the machined piece.

In conclusion therefore it may be seen that these manipulators have found a broad application in several fields of industry where it is necessary to carry out mechanization of heavy manual work. The production of HOBM can be found in several countries. Well known examples are «Dalmec» (Italy), «Balaman» (Japan), «Conco-Balancer» (USA), «Auto-Balancer» (Germany), «Muscle» (Switzerland), «KCH» (Russia), «MPB» (Bulgaria), «RMS» (Czech Republic), «MY» (Motoda / Japan), «HBL» (Japan), «SA» (Marol / Japan). The design methods for HOBM continue to develop and new solutions are constantly being reported. The development of design methods is directed particularly towards the improvement of safety of operation and weight minimization in their construction. Also of great promise is the creation of new actuators based on the application of path-generating mechanisms.

ACKNOWLEDGEMENTS - The author thanks Dr. Mike Smith for the advice while editing of the manuscript.

REFERENCES

1. Olsen R. A.. Balanced assembly. Patent US 3 134 340, May 26, 1964.
2. Olsen R. A.. Loading balanced assembly. Patent US 3 259 352, July 5,1966.
3. Reizou Matsumoto. Load handling equipment. Patent US 3 883 105, May 13, 1975.
4. Hank F.. Mashinenmarkt. 82, 1976, pp.1376-1378.
5. Krsnak H.G., Howe M.J.. Fördern und Heben. 15(26), 25, 1976.
6. Mechanisieren von Handhabungen mit Synchon-Manipulatoren. Maschinenmark. 84(41), 1978, pp.809-810.
7. Arakelian V. Equilibrage des manipulateurs manuels. Mech. and Mach. Theory, 33(4), 1998, pp.437-442.
8. Arakelian V. Balanced manipulator. Patent SU 1673432, 30 August, 1991.
9. Arakelian V. and all. Robotic technological complex. Patent SU 1530432, June 16, 1986.
10. Djavakhian R. and all. Balanced manipulat or. Patent SU 1468739, July 19, 1987.
11. Djavakhian R., Djavakhian N. Balanced manipulator. Patent SU 1521579, February 1, 1988.
12. Djavakhian R. and all. Manipulator. Patent SU 1828803, May 21, 1990.
13. Moor B.R., Akouna H.M. Vertical counter balanced test head manipulator, Patent CN 1408065T, April 2, 2003.
14. Haaker Lester. Balanced articulated manipulator. Patent GB 1097800, January 3, 1968.
15. Kakihara Kiyoaki. Manual work assistant manipulator control system.Patent JP 11277473, October 12, 1999.
16. Bittenbinder W.A. Lifting device for manual manipulat or. Patent DE 4342716, June 22, 1995.
17. Gvozdev Y. Manipulator. Patent SU 1777993, November 30, 1992.
18. Gvozdev Y. Manipulator boom counterweighing mechanism. Patent SU 806405, February 23, 1981.
19. Popov M.B., Tyurin V.N. Balanced manipulat or. Patent SU 1664544, July 20, 1989.
20. Popov M.B., Tyurin V.N. Balanced manipulator. Patent SU 1491698, November 24, 1987.
21. Popov M.B., Tyurin V.N. Balanced manipulator. Patent SU 1484698, November 24, 1987.
22. Vladov I., Danilevskij V. Balancing manipulator. Patent SU1423367, September 15, 1999.
23. Vladov I., Danilevskij V. Loading manipulator. Patent SU83113, April 23, 1981.
24. Belyanin P.N. Balanced manipulators. Ed. Mashinostroynie, 1988, 263p.
25. Mishkind C.I. and all. Manipulators with manual control and automatic weight balancing for engineering industry. Moscow, 1981, 74p.
26. Derkach G., Umantsev A. Balanced manipulator. Patent SU 1546408, 28 February, 1990.
27. Derkach G. Balanced manipulator. Patent SU 1826956, July 7, 1993.

DEVELOPMENT OF WALKING MACHINES; HISTORICAL PERSPECTIVE

Teresa Zielinska
Department of Power and Mechanical Engineering, ul.Nowowiejska 24
Warsaw University of Technology, 00-665 Warsaw, Poland,
e-mail: teresaz@meil.pw.edu.pl

ABSTRACT - The paper discusses the early history of machines and mechanisms related to the design of walking machines. Selected facts concerning the theory and the first ideas of mechanisms later used in the design of walking machines are presented.

KEYWORDS: History of Walking Machines, Mechanism Design

INTRODUCTION: FIRST WALKING MACHINES

From the ancient times people have been fascinated by legged locomotion. This fascination was reflected in mythology. According to the books of Iliad, written by Homer (VIII c. BC), one of the Greek gods built different walking devices. Some of them were human-like. In the old scripts from India we can find descriptions of mechanical elephants. In Egyptian pyramids archaeologists identified a wooden dog (toy) dated from the XX c. BC (Wolovich W.A. (1987)). Mimicking human or animal motion started with the decoration of water organs and water clocks with moving figures. The precursor of this type of devices was Ctesibius, who worked in Alexandria circa 270 BC. His student, Philo of Byzantium, wrote (circa 200 BC) the Mechanical Collection in which he described his teacher's inventions. The continuation of these works can be found in Hero's of Alexandra (I century AD): Treatise on hydraulics, Treatise on pneumatics, and Treatise on mechanics. He can be named the precursor of entertainment robots, due to his theaters with moving figures (Rosheim M.E. (1994)).

In the III c. AD in Sichuan province (China) a wooden walking machine Mu Niu Lu Ma was built under the supervision of a Chinese officer Zhu Ge-Liang (Shaoping Bai (2000); Yan H-S. (1999)). Mu Nu Liu Ma was used as a wheelbarrow for transportation of food supplies needed by the army. The machine was able to cover a distance of 10km in a day in an undulating terrain carrying a load of 200-250kG. (in Chinese MU means wooden, MA means horse, NIU means cow, in free translation it is *device powerful as horse and fast as cow*) was designed under the supervision of a Chinese officer Zhu Ge-Liang in the frame of preparation for the war against Wei kingdom located in central part of current China. Mu Nu Liu Ma was used as a wheelbarrow for transportation of food supplies needed by the army. The machine was able to cover a distance of 10km in a day in an undulating

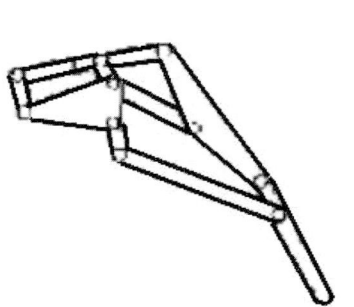

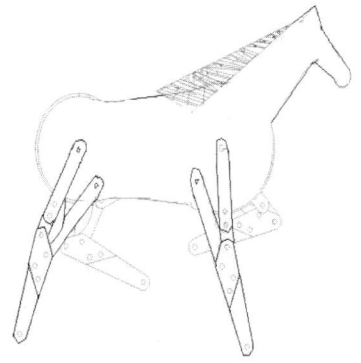

Figure 1: *Model of the leg build of 10 links*

Figure 2: *Scheme of running wooden horse*

terrain carrying a load of 200-250kG. This story was recorded but no one of the authors gave an information about the design details. Probably, thanks to the complex mechanism incorporating wooden gears, the machine - when pushed - transferred its legs in a sequence similar to that of a cow or a horse when moving slowly. This story fascinated many researchers interested in walking machines who tried to reconstruct the device. Reconstruction done by Wan Jian from Xinjiang Institute of Technology (XX c) is one of the most well known. Works on Mu Niu Liu Ma copies were also taken in Taiwan. The investigations resulted in several prototypes resembling by view the horses or cows.

The Wang Jian prototype is probably the most complex one, each leg consist of 10 links – Fig. 1. After work of Wang Jian, Chiu-Chengping from Taiwan elaborated proposition of Mu Niu Liu Ma with similar leg structure. Fig. 2 illustrates the legs positions during motion of proposed mechanism. The size and proportions of the mechanical components is chosen on this way that the leg-end trajectories are similar to the trajectories observed during walk of animals – Fig. 3. There are also models with simpler leg structures. For example Chen-Paihung suggested the leg design with 4 links (Fig. 4). Models build by Shen-Huanwen and Kwang-Kai (Fig. 5) uses similar idea. In Mu Niu Liu Ma reconstructions the key question is how worked the mechanism powering the sequence of legs motion. This problem was studied by Wang Jian and others. In several models the proper leg movement was obtained by careful choice of lengths of leg links and the special design of mechanical connections between them.

Thanks to the complex mechanism incorporating wooden gears, the machine - when pushed - transferred its legs in a sequence similar to that of a cow or a horse when moving slowly. The detailed design (size of the mechanical components, the assembly details) is not know. This story was recorded but no one of the authors gave an information about the design details. Probably, thanks to the complex

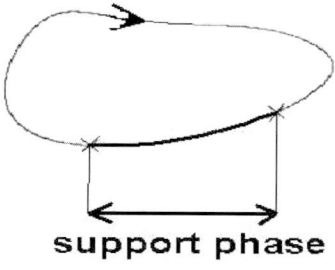

support phase

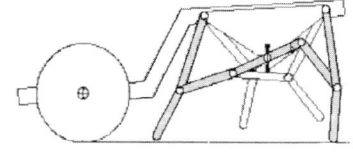

Figure 3: *Leg-end trajectory*

Figure 4: *Scheme of the design with 4-link legs (Chen-Paihung)*

mechanism incorporating wooden gears, the machine - when pushed - transferred its legs in a sequence similar to that of a cow or a horse when moving slowly. This story fascinated many researchers interested in walking machines who tried to reconstruct the device. Reconstruction done by Wan Jian from Xinjiang Institute of Technology (XX c) is one of the most well known. Works on Mu Niu Liu Ma copies were also taken in Taiwan. The investigations resulted in several prototypes resembling by view the horses or cows (Fig. 1).

In the XII entury AD Badi'as- Zaman Isma'il bin ar-Razzaz al-Jazari designed a figure which upon manual empting of a water basin automatically filled it again with water (Rosheim M.E. (1994)). He also described in his book many other devices, some of his own devise. In those gadgets actuation was due to the force of gravity transferred by levers or hydraulics to the limbs of figures. Leonardo da Vinci (1452-1519) used spring mechanisms as actuators in his ingenious machines. In the XVI and XVII c., as the precision mechanics capabilities improved, dolls dancing and/or playing diverse musical instruments were designed by many watch-makers (e.g. Juanelo Torreano, Tukob Bullman, Christof Margraf). As we move through the ages towards the more recent times examples of toys or machines with manipulation or locomotion abilities become more abundant. Frequently the propulsion of the machine was produced by the motion of the wheels and not legs, but an outward appearance of the motion resembled walking.

An excellent example of mechanical dolls are the mechanisms built in the XVIII c. by Swiss watch-makers: Pierre Jaquet-Droz, Jean Frederic Leschat, Henri Jaquet-Droz and Henri Millardet (Chronicle (1992); Rosheim M.E. (1994)). Those dolls were programmable by exchange of pegs pushing cams. The dolls were capable of drawing and writing. As they were programmed, what was written or drawn could be changed. Very complex gears, cams and levers inside their bodies were powered by spring mechanisms. Droz brothers miniaturized the mechanical components

360

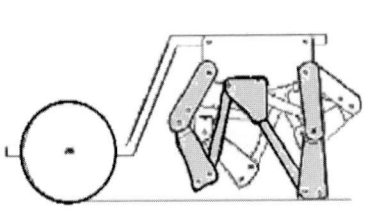
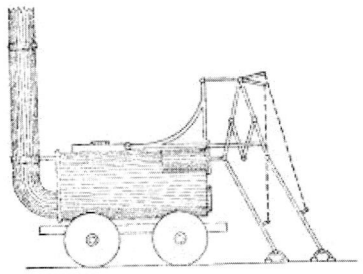

Figure 5: *Another version of Mu Niu Liu Ma (according to Shen-Huanwen and Kwang-Kai)*

Figure 6: *Steam engine with legs*

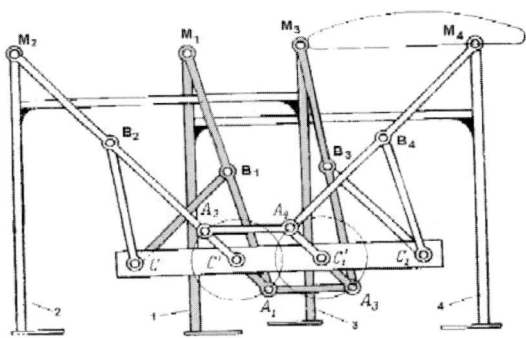

Figure 7: *Drawing of feet-walking machine; copy from Tchebychev works*

(Fig. 8). In their dolls they often applied mechanizm transferring the motion by chains and teeth wheels. Until now this mechanism in large size, was used in milling machines, steam engines and wall clocks

In Japan (XVII-XIX c.) the famous Karakuri wooden dolls were produced in a significant number. The height of those devices was 35 cm. They were propelled by spring mechanisms built using whale bones. The doll was capable of moving forward holding a tea-cup. Once the cup was removed the doll stopped. Upon replacing the cup the doll turned around and moved away. Shoji Tatsukawa, using the design manual Karakurizui written by Yorinao Hosokawa in 1796, reconstructed such a doll (Rosheim M.E. (1994)). In the design of bigger machines, elaborated from the XVII till the XIX century, legs were used as prerequisite for propulsion. In 1814 Levis Gompertz proposed a "square wheel" consisting of 4 feet attached around a square frame capable of motion without inducing the vertical

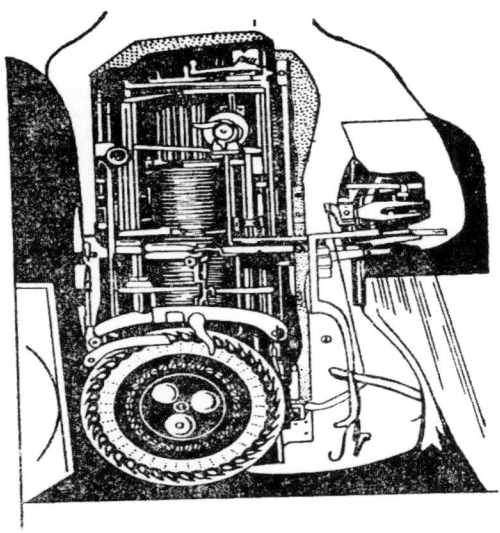

Figure 8: „Writer" designed by Droz brothers: view of the mechanizm (Milonov Ju.K. (1936))

oscillation of the axis. This idea was later modified and used in Pedrail - many feet attached around a frame. The device was used in a tractor for motion over soft terrain (Thring M.W. (1976)). In such a case feet are superior to wheels as they do not compress the soil in the direction of tractor's motion, thus reduce considerably resistance to motion. An excellent example of a vehicle supported by wheels, but powered by legs, is the so called Blueprint vehicle (XVIIIc.) (the name of the designer is not known) (Berns K. (2002)). In the early steam engine vehicles it was difficult to initiate the motion due to low friction between the wheels and rails (at that time the friction phenomena were not well identified). To overcome the problem with accelerating, extra legs were added to push or stop the vehicle. The devices designed at the break of the XVIII and XIX c. by Branton (Milonov Ju.K. (1936)) (Fig. 6) are examples of such steam vehicles. In 1893 L.A.Rygg patented the design details of the Mechanical Horse (Shin-Min Song and Waldron K.J. (1989)). In the same year G.Moore introduced the idea of The Steam Man (Thring M.W. (1976)). In the years 1821-1894 P.L.Tchebychev elaborated the design details of Stopochodjaszczaja Machina (Feet Walking Machine, Tchebychev P.L. (1955)) which similarly to the Chinese Mu Nu Liu Ma, when pushed transferred the legs in the same sequence as a horse or a cow. P.L.Tchebychev used parallelograms in his design. He underscored that the leg- end trajectories produced by his machine in relation to its body are similar to those of animals. XX c. was marked by an extensive development of diverse walking machines, especially past the fifties when first computer controlled machines appeared. As the resent

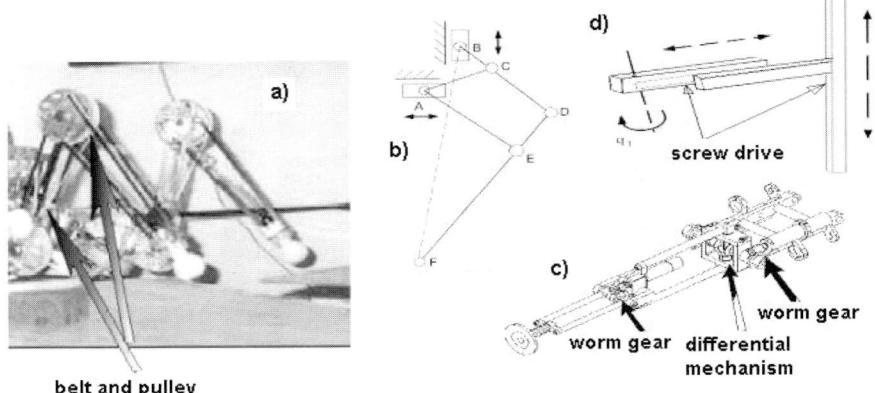

Figure 9: *Walking machine leg structures (with the data from Berns K. (2002)): a) leg with a belt and pulley (picture of LAURON I, b) a pantograph leg (e.g. ASV), c) a differential mechanism placed in the hip joint (e.g. LAVA), d) a screw drive in a leg (e.g. MELWALK)*

history of walking machines is very rich and is well described in several books (e.g. Shin-Min Song and Waldron K.J. (1989); Todd D.J. (1985)) or web pages (Berns K. (2002)) we shall not include it in this paper.

SHORT HISTORICAL OVERVIEW OF MECHANISM AND MACHINES USED IN WALKING MACHINES

The further text is an attempt to relate the currently used design solutions, utilised in walking machine legs, to the history of development of mechanisms. Fig. 9 presents the most frequently used leg structures applied in multi- legged machines. Usually revolute connections of many bars are used (Shin-Min Song and Waldron K.J. (1989)), e.g. the four-bar linkage presented in Fig. 10. Pantographs are also popular. Motion transfer is realised by: belts and pulleys, worm gears or differential gears. In conjunction with electric motors toothed-wheel gears are used for the reduction of angular velocity and to increase the torque.

If linear motion is desired screw and nut mechanisms are employed. The design of ankle joints is usually based on spherical connections (Fig. 11) or Cardan universal joints (Berns K. (2002)). There are also ankle joints that use several plates with revolute joints (Fig. 12, Tsukagoshi B.H. et al. (1997)). Parallel linkage passive ankle systems (e.g. ASV, Shin-Min Song and Waldron K.J. (1989)), which are a modification of the solution used in drafting machines, are also used. Summarising, the list of basic mechanisms employed in walking machines includes: pulleys, belt drives, toothed wheel gears, worm gears, differential gears, pantographs, four-bar linkages, spherical joints and universal joints.

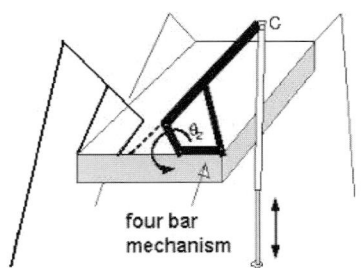

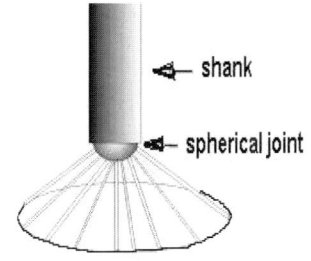

Figure 10: *Leg with a four bar mechanism*

Figure 11: *Spherical connection in the design of walking machine ankle*

Now let us take a closer look at the history of their invention. Pulleys with ropes were used in antiquity. Herodotus (Vc. BC) mentions them in a passage of his Histories describing the construction of Egyptian pyramids (Fig. 13). Moreover, they are depicted in Assyrian relieves (VIIIc. BC) (Kucharzewski F. (1924)). In the IVc. BC Aristotle mentions in his Problemata Mechanicea "iron or metal wheels transferring motion to each other", what suggests that at his times simple gears were already known (Kucharzewski F. (1924), Fig. 14). On the sidelines, let us note that Problemata Mechanicea is believed to be the first book mentioning mechanics in its title. A more detailed description of pulleys appears in the works of Archimedes (III century BC). He also developed the theory of compound pulleys - he constructed, among others, the polyspatos - a kind of a multiple pulley in which several ropes run in parallel over several rolls (Chronicle (1992)).

Belt drives were already known to the Chinese in the I c. BC. The proof of that can be found in Yang Xiong's The Dictionary of Dialects (15 BC) which describes the belt drives used in machines producing silk threads (Temple R. (1991)). Screws were known probably in Assyrian times. Certainly Archimedes (287-212 BC) employed them in his machines for launching ships. He used a system of pulleys and screws for that purpose. Unfortunately the blueprints of this mechanism have not survived till our times.

Detailed descriptions of mechanisms with toothed wheels (gears) are presented in Hero's On Lifting Loads (Ic. AD) (Kucharzewski F. (1924), Fig 15). This work also describes screws as wedges wound around cylinders. Moreover, he underscores that the action of a screw is obtained by its rotation and not by hammering as is the case with wedges. Hero also describes how with the help of a screw and a toothed wheel fixed to a shaft a considerable load can be moved. This brings to our mind a worm gear, in which a crank turning a screw imparts motion to a toothed wheel. Hero's writings also contain the description of a pantograph used

364

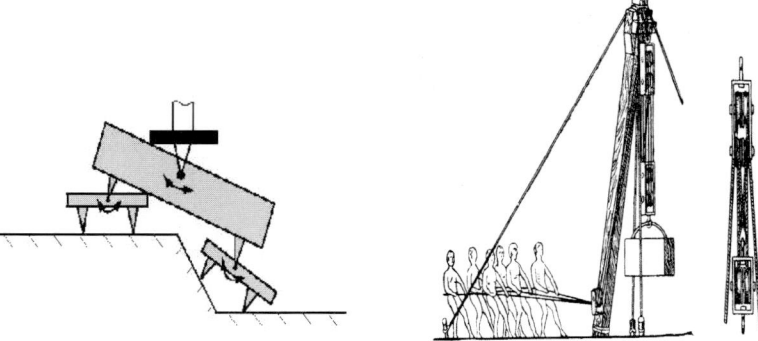

Figure 12: *Walking machine foot: plates with revolute joints*

Figure 13: *Egyptian crane (Milonov Ju.K. (1936))*

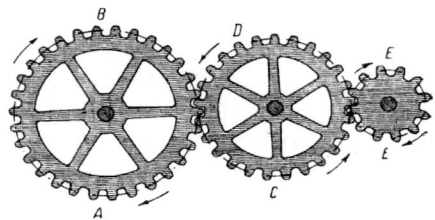

Figure 14: *Transfer of motion by the system of teeth wheels (according to Aristotle description, Milonov Ju.K. (1936))*

for copying drawings. It should be noted that Hero of Alexandria is treated as the first engineer, because he was the first to document his machines with detailed drawings. His drawings are very precise, convincing us that the machines really existed. Until his times the descriptions of mechanisms had been very general, thus imprecise, and the habit of illustrating them had not existed.

In Europe Philo of Byzantium (260-180 BC) described something that he called a cross joint (Chronicle (1992)), i.e. a Cardan universal joint. This type of mechanism was used in China in the II century BC in lamp holders (Temple R. (1991), Fig. 16). The earliest Chinese reference to this mechanism is made in the Poem About Beautiful Ladies composed about 140 BC by Sim Xiangru. The Cardan universal joint got its name from Girolamo Cardano (1501-1576), who personally did not claim that he invented it, but described it in detail in his work De Subtilitate (1550). Differential gear was used in China in the II century AD in the so called carts pointing south. A nephrite figure regardless of the trajectory of motion of the cart always pointed its finger to the south. This was possible due to the system of gears containing a differential gear connected to the wheels of the

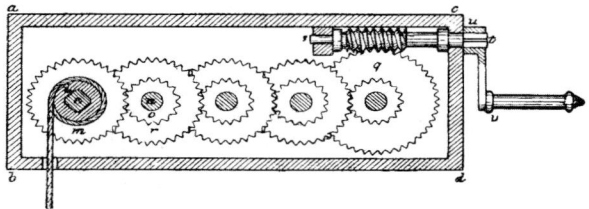

Figure 15: *Mechanism designed for lifting the loads (reconstruction according to Heron works, Milonov Ju.K. (1936))*

cart. It is probable that this type of gear existed earlier. According to a chronicle, dated 500 AD, the south pointing cart was built on emperor Zhou's order (beginning of the first millennium BC), to enable a legation to return home in low visibility. Fully confirmed is the information that differential gears existed in 80 BC in Greece (Temple R. (1991)). Many works of ancient authors have been forgotten. After stagnation in the middle ages, progress in mechanics was marked by Leonardo da Vinci (1452-1519), who reinvented the worm gear, and described the universal joint. He experimented with many motion transmission mechanisms (Kucharzewski F. (1924), Fig. 17). He also noted that every motion is subjected to resistance, thus he concluded that to sustain motion an engine is needed. It is difficult to ascertain when exactly did multi-link devices with planar revolute joints appear. Nevertheless, it is pretty certain that they were used in conjunction with levers in the times that pyramids were constructed, because texts mentioning such devices exist (e.g. Herodotus in Vc. BC wrote about such mechanisms). According to the blueprints of the already mentioned Chinese machine they existed in the IIIc. AD. It is difficult to ascertain when the spherical joint appeared. The above brief overview shows that many simple mechanisms that are currently used in walking machines appeared in antiquity. Neglecting the problem of actuation, the modern designs of walking machines could have appeared over a thousand years ago. However the lack of appropriate motors hindered the development of walking machines.

Up till the XIX c. mechanisms had been designed on the basis of a flash of genius and the experience gained by trial and error. This was due to the lack of an adequate theory. The development of such a theory was impeded by the prevailing misconception established by Aristotle (384-322 BC) that the natural state of a body is its immobility and thus bodies remain motionless or try to attain that state by themselves (due to natural motion). Otherwise for a body to move a force has to be applied to it (causing unnatural motion). Only in the XIV century this claim was disputed. It was Galileo Galilei (1564- 1642) in his Dialogue Concerning the Two Chief Systems of the World - Ptolemaic and Copernican, published in 1632, who finally asserted that for a body to move with a constant velocity no force is required, thus he laid down the foundation for the formulation of the principle of inertia. He claimed that the state of rest is indistinguishable from the state of uniform motion in a straight line, hence formulated the principle of relativity. The investigations of motion were inseparable from astronomy. Initially scientists aspired just to describe the motion of planets without studying the reason for their behaviour. Nicolas Copernicus (1473-1543) postulated the heliocentric system, which was a geometrical construction considerably simplifying the description of the motion of planets in comparison with Claudius Ptolemy's (II AD) epicycles specific to the geocentric system. Later Johannes Kepler (1571-1630), using detailed astronomical data collected by Tycho Brahe (1546-1601), improved the Copernican model and developed his laws of motions of planets, in which he postulated elliptic trajectories - still geometrical concepts. Rene Descartes (1596-1650) showed the relationship between geometry and algebra by creating analytic geometry, hence reducing geometrical considerations to the manipulation of algebraic equations. Galileo established the relationship between: time, velocity and distance travelled by a body in a uniformly accelerated motion. The considerations of astronomers and Descartes treated trajectories as static entities - imaginary static traces left by moving bodies. It was Isaac Newton (1643-1727) who combined Galileo's kinematic considerations with geometrical models produced by astronomers and through the introduction of the method of fluxions (differential calculus) and by postulating gravitational force formulated his three laws of motion and the law of gravitation. Those were contained in Philosopiae Naturalis Principia Mathematica published in 1687. This book is the foundation of mechanics, i.e. the science investigating the interactions of bodies through forces. It combines both dynamics (theory of causes of motion) and kinematics (theory of the process and effects of motion). But it was due to Gottfried Wilhelm Leibniz (1646-1716), the independent co-inventor of the differential calculus, that this calculus received its current notation. It must be mention that in 1675 G.W. Leibniz constructed a computing machine (Fig. 18). The first such devices appeared earlier, in break of XVI and XVIIc., but Leibniz machine was the most complex and minaturized, mechanical structures utilized there were later applied in first commercially available mechanical computing machines (from 1878) (Milonov Ju.K. (1936)). Leonhard Euler (1707-1783) combined Newton's method of fluxions and Leibniz's differential calculus into mathematical analysis and described it in Mechanica (1736). He formulated analytical mechanics in his Theory of the Motions of Rigid Bodies (1765). Later he formulated variational principles to determine the optimal ship design. Finally in Theoria Motus Corporum Solidorum (1765) he decomposed the motion of a solid into translational and rotational motion and introduced Euler angles. Meanwhile, in 1743, in Traite de dynamique, Jean Le Rond d'Alembert showed how to reduce a dynamics problem to a one of statics. To do this one has to balance all the external forces and a fictitious force equal to the force of inertia. In 1756 Joseph-Louis Lagrange (1736-1813) applied the calculus of variations to mechanics. His foundations of dynamics were based on the principle of least actions and on kinetic and potential energy. In 1788 he summarised in his Mecanique Analytique all the results obtained in mechanics since Newton. He introduced generalised coordinates and provided us with a method of generating second order differential equations of motion for a system with any number

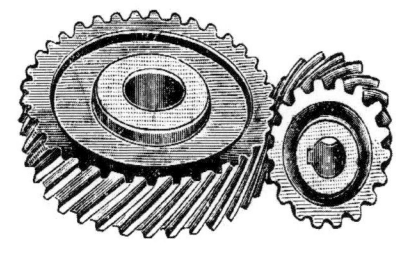

Figure 16: *Lamp holder*

Figure 17: *Leonardo da Vinci worked on the gears improvement; he invented helical gears, Milonov Ju.K. (1936)*

of degrees of freedom. In 1843 William Rowan Hamilton (1805-1865) discovered quaternions. The works of Leonard Euler, Michael Chasles (1793-1880), stating that any motion of a rigid body can be reproduced by a single translation followed by a single rotation about some axis, culminated in the formulation of the theory of screws in 1876 by Robert Stawell Ball (1851-1913) (Duffy J. (1999)). All those mathematical developments were well suited to the description of motion of point masses or rigid bodies. With the emergence of a multitude of mechanisms in the early XIX century, spurred by the invention of the steam engine due to Thomas Newcomen (1663-1729) and its improvement by James Watt (1736-1819), mathematicians were faced with the problem of finding a framework for categorising them. Moreover, in the XIXc. scientists- engineers started to believe that machines should rather be designed using mathematical principles than by trail and error. In 1854 Pafnuty Lvovitch Tchebychev (1821-1894) published his Theorie des mecanismes connus sous le nom de parallelograms in which his polynomials appeared for the first time. He later studied mechanisms converting rotary into translational motion and even mechanical calculating machines. But it was Franz Reuleaux (1829- 1905) who classified mechanisms on the basis of how they constrained motion. He created a library of mechanisms that could be combined to create complex machines. In 1864 he came upon the idea to describe machines as kinematic chains linked by pairs of geometric constraints (Moon F.C. and F. Reuleaux (2000)), thus inventing the notion of a kinematic pair. Towards the end of the XIX c. many scientists concentrated on the theoretical aspects of kinematics, e.g. in 1892 Lorenzo Allievi (1856-1941), Ceccarelli M. (1999) wrote Kinematics of Planar Couplers, in which he summarized the knowledge about planar mechanisms that had been accumulated until his times and proposed a classification of elementary mechanisms. Due to complex kinematic structures, difficult dynamics and problems with the generation of stable gaits of walking machines, control of those devices is of paramount importance. As friction hinders the transmission of motion in complex mechanisms the problem of their actuation becomes vital. Only at the turn of the XVIII c. steam engines or clockwork mechanisms were proposed as a means of propulsion in walking machines (e.g. in the Mechanical Horse). It cannot be denied that the current intensive development of walking machines should be attributed to two factors: firstly, the invention of an electric motor, and secondly, the rapid development of control systems sustained by the ever increasing computational power and miniaturisation of computers. This brings us to the XIX c. In 1829 Michael Faraday (1791-1867) elaborated the idea of a motor which consisted of a static magnet and a conductor rotating around it. This led M.H.Jacobi (1801-1874) to build in 1834 the first electric motor capable of doing useful work. The advent of miniaturised digital computers, in the later half of the XX c. enabled the creation of walking machines. The influence of the developments in the fields of electronics and computer science on the design of walking machines cannot be overestimated. The history of electronics and computers in particular is well documented, e.g. Ligonniere R. (1987). Nevertheless, for the sake of completeness, a few milestones have to be included in this overview. Both the evolution of technology (hardware) and mathematical foundations were important. The former is marked by the invention of the transistor in 1946 in Bell Telephone Laboratories by John Bardeen (1908-1991) and Walter Brattain (1902-1987) supplemented by the theory of p-n junctions and the theory of transistor operation formulated (1948) by William Shockley (1910-1989). This led in 1958 to the design of the first integrated circuit by J.S. Kilby (b.1923) from Texas Instruments. At that time Robert Noyce (1927-1990) invented the diffusion method of connecting electronic elements (Temple R. (1991)). The introduction of the so called planar technology in 1961 by Fairchild Semiconductors enabled the mass production of integrated circuits. Finally, in 1972 Intel Corporation introduced to the market its first microprocessor - the 4-bit Intel 4004, substituted by the 8-bit Intel 8008 in 1972, and the famous 8-bit Intel 8080 in 1974. By no means this summarises all important technological developments, but the bounds imposed on conference papers require briefness. The technological developments have been paralleled by the evolution of the theory of computations. Again, only just a few milestones can be cited. Gottfried Wilhelm Leibniz in the years 1679-1702 worked on the binary representation of numbers.

The foundations for the formal logic were laid down by the seminal publications of Augustus de Morgan (1806-1871) in 1847 and 1860, and George Boole (1815-1864) in 1847 and 1854 (O'Connor J.J. and Robertson E.F. (2003)). In 1937 Allan Turing (1912-1954) working on the theory of computability defined an abstract machine (Turing machine) containing a finite state control automaton and an infinite length tape divided into cells which can hold symbols. This abstract machine was later instrumental in the classification of formal languages (Noam Chomsky, b.1928) and thus the development of high level computer programming languages. In 1938 Claud Shannon (1916-2001) in his Ph.D. thesis showed how to use the binary

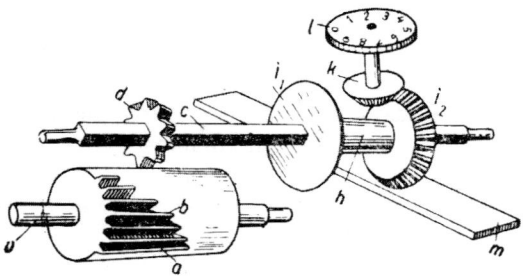

Figure 18: *Structure of Leibniz computing machine, Milonov Ju.K. (1936)*

number system and relays to construct arithmometers. In 1945 John von Neumann (1903-1957) pointed out that there is no contradiction between the representation of data and programs operating on that data, thus both can be stored in the same memory. This led to the design of computer architectures with programs and data residing in the same memory, i.e. the vast majority of modern computers. All the above developments led to the present state of walking machine design. It is anticipated that towards the middle of the XXI century we shall be able to design biped running robots capable of outperforming human soccer players (Burkhard H.D. et al. (2002)).

CONCLUSIONS

The study of the history of engineering developments is by no means an easy task, especially its early stages, as diverse sources quote different chronology. In the above overview we tried to verify the dates in several sources (not all cited). We believe that the knowledge of history besides enriching can also influence the design of modern machines by reminding of certain solutions. Before the II World War at Warsaw University of Technology the history of mechanics was lectured to the students as a separate subject. The lecture dealt not only with historical facts but also with the design of simple mechanisms and the evolution of theoretical knowledge associated with mechanics. This led to a better understanding of the subject, because the students learned how modern designs have been arrived at. Mechanics was presented as an evolution of engineering thought and not as a set of disjoint modern solutions. The reintroduction of such lectures should be reconsidered once again.

Acknowledgment: The research work and preparation of the manuscript was supported by statutory funds of Warsaw University of Technology.

REFERENCES

Berns K., Walking Machines Ctlg. (www.fzi.de/ids/WMC)

Burkhard H.D, Duhaut D., Fujita M., Lima P., Murphy R., Rojas R., The Road to RoboCup 2050, IEEE Robotics and Automation Magazine, vol.9, no.2, 2002, pp.31-38

370

Ceccarelli M., Cinematica Della Biella Piana by Lorenzo Allievi in 1895. 10th World Congress TMM. Finland, 1999. pp.37-42.

Chronicle of Technical Sciences (in Polish). Chronicle, Warsaw, 1992.

Duffy J., A Short History of Screws. 10th World Congress TMM, Finland, 1999. pp.48-56.

Kucharzewski F., Mechanics in its Historical Development (in Polish), Bibljioteka Polska, Warsaw, 1924.

Ligonniere R., Prehistoire et Histoire des Ordinateurs. Editions Robert Laffont, Paris, 1987.

Milonov Ju.K., Etudes on History of Mechanism Kinematics (in Russian). NTI Moscow, 1936.

Moon F.C., F.Reuleaux, Contributions to 19th C. Kinematics and Theory of Machines. Cornell University, February 2002.

O'Connor J.J., Robertson E.F., The MacTutor History of Mathematics Archive. University of St. Andrews, Scotland, 2003. http://www-history.mcs.st-adrews.ac.uk/history

Rosheim M.E., Robot Evolution: The Development of Anthrobotics. John Wiley, New York, 1994.

Shaoping Bai, The Kinematics of the terrain-adaptive locomotion of walking machines. Ph.D.diss., 2000. NTU Singapore

Shin-Min Song, Waldron K.J., Machines that Walk, The MIT Press, 1989.

Tchebychev P.L., Selected Works (in Russian). Izdatelstvo Akademii Nauk, Moscow, 1955.

Temple R., The Genius of China. 3000 years of Science, Discovery and Invention, Multimedia Books Limited, 1991.

Thring M.W., Some experimental walking machines. 2nd RoManSy, Ed.Morecki A, Kedzior K., PWN Warsaw, 1976.

Todd D.J., Walking Machines: An introduction to legged robots, Kogan Page, 1985.

Tsukagoshi B.H., Hirose S., Yoneda K., The Quadruped Walking Robot on a Steep Contruction Site Supported by Wires. Proc. New Approaches on Dynamic Walking and Climbing Machines. 8th ICAR USA, 1997, pp.31-43.

Wolovich W.A., Robotics: Basic Analysis and Design. Holt, Rinehart and Winston, Australia, 1987.

Yan H-S., A Design of Ancient China's Cattle Machines - ChuKo Liang's Wooden Ox and Gliding Horse. 10th World Congress TMM, Finland, 1999. pp.57-62.

BIPED ROBOTS: THE STATE OF ART

Carlos André Dias Bezerra
Dept. of Mechanical Engineering. Federal University of Ceara. Campus do Pici. Bloco 714. 60455-760. Fortaleza-Ceará. Brazil.
cadbe@ufc.br.

Douglas Eduardo Zampieri.
School of Mechanical Engineering. Dept. of Computational Mechanics. State University of Campinas. Postal Number 6122. 13083-970 Campinas, SP. Brazil.
douglas@fem.unicamp.br.

ABSTRACT - *Researchers from all over the world have studied biped robots. The robots have an inherently unstable structure and whose modeling and control can be carried out by different techniques. The aim of this study is to present a critical literature review of the main types of biped robots developed until the present date, including Brazil. According to their particular features presented in the literature, robots will be treated concerning dimensions, degrees of freedom or system of control. Advantages and disadvantages of moving robots when compared to leg robots will be presented. This comparison will also be done taking into account other types of robots. The evolution of biped robots also will be presented including those ones for entertainment purposes. Mathematical modeling for determination of equation of motion, Zero Moment Point, and application of control theory in order to allow a response for a standard input is also showed.*

KEYWORDS. Biped robots, state of art, robotics review.

INTRODUCTION

The aim of this review is to keep the knowledge on biped robots up to date, in order to allow the knowledge of the main research carried out through and in progress in the world-wide context of this area. The main objective of this work is the presentation of the biped robots that have most contributed for the state of art and the application or development of new theories of control or locomotion. A brief review of the main robots developed in Brazil is also presented.

The methods of locomotion on the ground include tracks, wheels and legs. The legged motion has the peculiar characteristic of discontinuous contacts with the ground, which allows to avoid obstacles, walk in an uneven ground and climb stairs. While wheel robots require a continuous path of support, legged robots can cross in a terrain that is discrete and discontinuous. As an example, humans can walk not only on flat surfaces but also on a rough and uneven ground such stairs, ladders, ramps and even on an unknown terrain containing obstacles.

To develop human-like walking robots, many researches on the biped locomotion robot have been archived (Hasegawa et al., 2000). The researches agree that biped robots can easily adapt to several types of terrain. In theory, the biped robots can move on unknown environments such as is the human being live due to their anthropometric forms. And so on they can also be made to realize different tasks in all environments where a human being can move (Sardain et al., 1998).

It is hoped that biped robots can to operate in a human environment with greater efficiency than other mobile robots due to his mobility provide by his legs. In the future biped robots can be used to complete tasks which are too dangerous or difficult for human beings such as: industrial tasks, people guidance, transportation on uneven terrain, rescue or operation on dangerous zones. They can also be used to help handicapped people to walk or as a prosthetic device to replace human legs, among others examples. One another potential of the biped robots would be the substitution of a part of the human work such as domestic tasks or to help aged, because the bipedal gait can access any environment where the human being can go.

Therefore, as the bipedal gait is the best one for locomotion in the human environment, the technology for its reproduction is one of the most important for the construction of an android (Fujimoto, 1998).

BIPED ROBOTS

The early studies of biped robots (brs) had emphasis just in locomotion and not in their real application in the industrial environment. However with the advance of research, some authors considering that the biped robot (br) had two legs, similar to the human being, is the most appropriate to move itself in environments destined to the human beings.

This finding had guided studies in the substitution of the operators by brs in dangerous operations and/or where a too high mobility was demanded. Two of the most difficult challenges in the study of brs are the support their structure above their waist and how to provide an efficient and steady gait.

The biped robots can perform static and dynamic walking. The so called static balance or static walking refers to a system which stays balanced by always keeping the center of mass (COM) of the system vertically projected over the polygon of support formed by the feet during the motion (Raibert, 1986). Dynamic balance or dynamic walking refers to a system that the COM may leave the support area formed by the feet for periods of time. The ground projection of the COM is a static criterion of walking.

Balancing of a human in motion does not result from the fact that the foot soles have non-zero areas or the ground projection of COM, but it results from the complicated coordination of the body parts. The human being can walk in special shoes with pinpoint supports. When the supporting point is modeled as a hinge joint, the point is called zero moment point.

In 1973, Vukobratovic presented the first studies on the control and bipedal locomotion (Vukobratovic and Stepanenko, 1973; Vukobratovic, 1973). He who was created the term Zero Moment Point (ZMP), that it is termed as the point in the ground where the robot's total moments at the ground is zero. The ZMP has been used as stability criterion. If the ZMP is inside the support area the walking is considered dynamically stable, because the foot can control the robot's posture. The ZMP criterion cannot be applied to biped robots that do not continuously keep at least one foot on the ground or that do not have active ankles joints. In the robotics literature Center of Pressure (COP) is called ZMP.

Sardain and his coworkers (1998) classified in three categories the biped robots: static walkers, whose projection of COM is always inside the polygon that demarcate the one or two soles in support phase. Second, dynamic walkers, whose ankles are actuated in order that the COP is situated in the support base. Third, purely dynamic walkers, whose ankles are inactuated, or that have purely and simply no feet. In this case, the polygon of contact between the robot and the ground is reduced to a single point during the single support phase, and to a single segment during the double support phase.

Goswami (1999) introduced the term Foot-Rotation Indicator point (FRI), which is a point on the foot/ground contact surface where the net ground reaction force would have to act to keep the foot stationary. The position of the FRI outside the foot/ground contact surface indicates the direction of the rotation and the magnitude of rotational moment acting on the foot. The FRI point may leave the support area as opposed to the ZMP, and indicates the severity of instability of the gait as well.

The above concepts are extensively used in biped robot literature as bases from new approaches and application of control theory in robotic locomotion.

HISTORY

The first studies in brs locomotion were concentrated in the theoretical analysis of the legged locomotion and the conceptual project of walking machines, among them the so called "Ideal Walking Machine", developed by Frank (1968). This "walking machine", besides being a theoretical model of bipedal locomotion, served as reference for construction of brs. In recent years, research has been concentrated in the study of robots able to perform a steady locomotion. Some of main brs and their respective researchers will be presented respectively to their invention.

Ichiro Kato and his colleagues from the Department of Mechanical Engineering of the School of Science and Engineering of the University of Waseda in Tokyo constructed the first brs. During 1966 and 1967, an artificial lower-limb mechanism WL-1 (WASEDA LEG - 1) was constructed to analyze the locomotion of lower limbs. In 1968, a mechanical model of lower limbs WL-3 (WASEDA LEG - 3), actued by an electro-hydraulic servo-actuator and controlled by a master-slave method was built. It has a human-like movement with swing and stance phase, and it was also able to stand up and sit down. In 1969, a br with anthropomorphic appearance was built and was called WAP-1 "Pneumatically-Activated Pedipulator". It was constituted by a mechanical structure activated by artificial rubber muscles. WAP-1 was able to move through a "playback" control of its artificial muscles.

In the following years, Kato and his colleagues had developed a series of brs, among them the man-like robot, WABOT-1 (1973). It was first br proportional to the human being. It had a member controlling system, a viewing system and a conversational one, allowing the communication with a person in Japanese and the measurement of distances and directions by using external sensors, artificial ears, eyes and mouth. This robot was static steady in all the phases of the walking cycle, keeping its gravity center inside the area of its feet support. Besides walking, it was able to catch and to carry objects with its hands, which had tactile sensors. It was estimated that WABOT-1 had mental age of a 1 year and half child.

WABOT-1 was the result of the junction of different archetypes, previously constructed. Thus, it was composed by superior members of the WAM-4 and inferior members of the WL-5 (Kato et al., 1974). In the next years, Kato and his collaborators developed a type of locomotion for brs called of locomotion "quasi-dynamic". The main characteristic was that the phase support transition foot-to-foot occurred very fast without double support phase (Kato et al., 1983).

Miomir Vukobratovic from Belgrade school of robotics developed a scientific acknowledged on the theory and practice of robotics, locomotion and active rehabilitation systems. Ichiro Kato, founder of Robotics in Japan said about him: "... Since 1975 we can speak about world recognized Yugoslav robotic school headed by professor M. Vukobratovic... The robotic research in Japan is very influenced by the initial work of Prof. Vukobratovic, whose several monographs were published in 1975, 1979 and 1985" (Vukobratovic, 2000). The another concept from Vukobratovic is the semi-inverse method (Vukobratovic and Juricic, 1969; Vukobratovic and Stepanenko, 1972 and Juricic and Vukobratovic, 1972) that statement: "The conditions of dynamic equilibrium with respect to coordinate frame located at the Zero Moment Point give three relations between generalized coordinates and their derivatives. As the whole system has n degrees of freedom (n>3), the trajectories of the (n-3) coordinates should be prescribed as to ensure the dynamic equilibrium of the entire system (trunk motion including arms if the biped robot is in question). If there are some supplementary zero-moment-points (like the passive joints of biped arms) then for every additional ZMP another three equilibrium conditions are available. Thus, when applied to the problem of investigating the dynamics of biped systems, the motion of links is partly known, while the unknown moments are equal to zero. Vanishing of the given moment results from the equilibrium conditions about the supporting point (ZMP) and about the joints of passive links". Vukobratovic wrote: "ZMP concept was born thirty years ago. Because of its biological justification and biomechanical validity I strongly believe in its permanence and further successful application in legged locomotion robotics generally".

Miyazaki and Arimoto (1980) carried out theoretical studies in a seven - degree of freedom (DOF) br able to dynamically walking in some phases of the march, the projection of its center of gravity (COG) was out of the feet area. The controlling technique was based in the principle that bipedal locomotion can be divided in two ways: fast and slow. They proposed reduced order models in which, by supplying a reference signal function to each controller and using this function for each step, a steady locomotion was obtained. In 1981, they had constructed a br called IDATEN (Miyazaki and Arimoto, 1981).

In 1982, Takanishi and his colleagues presented the WL-10R (Waseda Leg - 10 Refined). This is a milestone in biped robotics, marked by the shift from static to dynamical walking.

In 1984, Mita and his collaborators had simulated and constructed br CW-1 (CHIBA-WALKER-1) with 7 links (two feet, two lower legs, two upper legs, and one hip), six DC motors (15W) fixed at six joint, and with 5 DOF. They had considered a new controlling method using the linear optimal state regulator, which stabilizes the stance supported by legs as a commanded attitude. They applied the modern control theory (control LQR and the theory of the optimal regulator) to stabilize the phase of simple support. They also presented simulated and experimental data of the locomotive motion. The CW-1 height was of 75.5 cm, weighed 15.3 kg and it was able to perform a 30 cm step in 1s, in the dynamic steady way (Mita et al., 1984).

BIPER brs were developed in the University of Tokyo by Kimura, Mitsuishi, Shimoyama and Miura (Miura and Shimoyama, 1984). They developed five kinds of bipedal robots named BIPER - 1, 2, 3, 4 and 5. All of then were statically unstable but could perform dynamically stable walk with proper control. The BIPER-1 and BIPER - 2 could walk only sideways. BIPER-3 could walk sideways, backward and forward. It was shaped with three links (pelvis and two legs), actuated by four DC engines situated in his pelvis, being two to move the legs in frontal plane and two to move them forward. BIPER-3 did not have a torque in the ankle, or in the knee and presented a point contact between the leg and the ground. Its weight was of approximately 2 kg, and a height of 30.8 cm. A characteristic of this br was the necessity of a succession of steps to keep it stand up, because it would fall if both feet were kept in contact with the ground. The 7 DOF BIPER- 4 robot had nearly the same DOF as human legs and had eight DC motors mounted at each joint. The BIPER - 5 was similar to BIPER - 3 except by the computer mounted on it.

Katoh and Mori (1984) constructed 4 DOF robot called BIPMAN (BIPedal walking MAchiNe). This br was constituted of two massless telescope legs, two feet and one pelvis. The legs did not have knees but they could be extended or contracted. They studied the control method of a dynamic biped locomotion by using a dynamical system having a stable limit cycle. The properties of this dynamical system were utilized to design a control method in the single support phase and double support phase. This control method was formulated as a parameter control by making use of the bifurcation dynamical system. They utilized the couple van der Pol's equations as the fundamental dynamical system to study the control method. BIPMAN was programmed to track the reference signals given by a microcomputer and was able to walk a step. It weighed 36.5 kg and performed a 0.50 m step for a leg length of 1 m.

Raibert used symmetry to extended his studies with the uniped or monopod robot called "CMU hopper", to analyze brs of one, two and four legs and to derive a global characteristic from the system for the bipedal run (Kajita et al., 1992). He used a model of a structure composed by multiple links to do the planning of foot positioning and developed a desacoupled controlling system of for body attitude, body height and foot positioning. The control law was gotten considering dynamic characteristics of a leg with spring. Using this control law the br could vary its velocity in real time.

In 1985, Jessica Hodgins and Marc Raibert developed a two legged machine with 4 DOF able to jump 44 cm up to 67 cm. The br was actuated by hydraulically actuators, being able to run with a frontally inclined posture keeping a dynamic balance similar to Miura and Shimoyama br (Raibert, 1984; Hodgins, 1988 and Hodgins and Raibert, 1990) with velocity of 4 m/s.

In 1986, using ZMP concept and semi-inverse method, which were elaborated by Vukobratovic (Vukobratovic et al., 1975 and Vukobratovic and Stokic, 1975), Kato and his coworkers were the first who realized the dynamic walking compensation with a 2 DOF trunk in the hydraulic biped robot WL-12 (WASEDA LEG - 12).

Furusho and Masubuchi constructed the KENKYAKU-1 br from the plane model of five links, resulting in a body, two upper legs and two lower legs, and a hip. Its knee joints were driven by four DC servomotors. They adopted a local feedback at each joint as the lower level control of dynamical walking br. To the upper level control, they used the reduced order model to correct the reference inputs to the local feedback. The reference signal to each local controller was stored in its higher level (memory of a minicomputer) as the walking pattern (Furusho and Masubuchi, 1986). The br was able to walking with 0.8 m/s in a walking cycle of 0.4 – 0.5s. Later, with addition of two more links, this br became KENKYAKU-2, with 40 kg weight and 1.10 m height (Furusho and Masubuchi, 1987).

In 1988, Hodgins (1988) also studied dynamic locomotion in irregular land using a modified version of the controlling method by Raibert for step variations of the of br. Raibert developed his work in the MIT Leg Lab, that currently has researchers acting in two lines: elastic actuators project and algorithm of control for the virtual locomotion (Pratt, 2000).

In 1988, Sano and Furusho (1988) constructed BLR-G1 br and then the BLR-G2 (Furusho and Sano, 1990), with nine links and 8 DOF. It was composed of a hip and two legs with knees and ankles. Knees presented 1 DOF each and the ankle possesses 1 DOF in flexing-stretching plus 1 DOF in lateral rotation each. It was 25 kg weight and 97 cm height. It walked along a straight line on flat horizontal ground with a velocity of 0.18 m/s. In 1989, Sano and Furusho developed the quadruped robot COLT 3 (Sano and Furusho, 1989).

In 1990, McGeer designed a dynamic br with 1 DOF, without hip and with only two fine legs articulated one in another, weighing 3.5 kg, 0.50 m of height and capable to develop 0.46 m/s in a surface inclined for low of 2.5 degrees. He studied steady passive locomotion in this inclined surface. This robot was able to generate a passive standard of march without the use of any type of active control, actuated only by gravity (McGeer, 1990).

Zheng and his collaborators (Zheng and Sias, 1988 and Zheng and Shen, 1990) studied the dynamic locomotion using SD-1 (without ankles) and SD-2 (with ankles) brs, with nine links and 8 DOF. The method of dynamic analysis of br consisted on the division of the model system in simple subsystems. The SD-2 had four joints that allowed movement in the frontal plane and other four joints, that allowed movement it in the sagittal plane and was able to pass from a surface plane to an inclined one, however in the static way of locomotion. The SD-2 used force sensors under the foot that allowed to the recognition of the transition between a plane surface and another inclining of up to 10 degrees. The SD-2 mass was 9 kg and it was 0.9 m high. It could move at 0.13 m/s.

Takanishi and his collaborators (Takanishi et al., 1990) developed the WL-12RIII (Waseda Leg-12 Refined III) with 9 DOF. It was capable of going up steps and inclined surfaces. It had seven links including a trunk. Moreover, the trunk had 2 DOF and moved in the frontal plane as an inverted pendulum, allowing the robot to pitch and roll relative to the forward direction. It weighed 100 kg and was 170 cm height. The actuators were hydraulic rotary jacks situated close to articulations. This biped could climb a staircase with 2.6 s/step and walk on a slope with 1.6 s/step. The control system was able to change the ZMP in order to achieve dynamical walking on even or uneven terrain. Yamaguchi et al. (1993) added the yaw-axis movement to the trunk motion to WL-12RIII, creating the WL-12RV.

Kajita and co-workers (Kajita et al., 1992) developed simple linear differential equations for an ideal biped robot with massless legs, in which COG was moved horizontally. Such model was used to control MELTRAN br of 4 DOF that walked in a dynamic way. It was 2.5 kg weight and 0.40 m height and developed a velocity of 0.3 m/s. In the following years, Kajita and Tani (1991) developed an algorithm for locomotion in rough land.

Gruver and collaborators (Gruver et al., 1993) developed a br of 12 DOF able to dynamic walk, composed of one pelvis and two legs. The br had five links and 4 DOF in the frontal plane and seven links and 6 DOF in the sagittal plane (hips had 3 DOF, knees had 1 DOF and ankles had 2 DOF). It was able to change direction, due to the 2 DOF vertical axes of its hips. The br weighed 57 kg and was 92 cm height. It presented 0.71 m/s velocity on the straight-line floor.

Miller (1994) modified the SD-2 with the addition of knees. The resultant br was able to realize static walk and had five rotating joints for each leg, in a total of 10 DOF. It carried through strategies of control based on artificial neural network of type CMAC (Cerebellar Model Arithmetic to Computer), without using a detailed model of the kinematics and dynamics of a br. The br weighed 6.8 kg and was 104 cm height. Moreover, it used eight polymer film force sensors for the determination of the center of application of force in the foot.

Grishin (Grishin et al., 1994) and collaborators constructed a dynamic br with 2 DOF. Made up of one hip and two telescope legs without ankles, it actuated by two DC motors. It was 7.5 kg weight, 0.75 m height, with velocity of 0.3 m/s and actuators in the hip and each thigh.

Yamaguchi and collaborators (Yamaguchi et al., 1995) developed a br that walked in the dynamic way and it was capable of adapt in the inclined surfaces, through the use of a material that absorbed impact to get information on the land.

Kajita and Tani (1996) presented an experimental study on MELTRAN II br with 6 DOF, where each leg was shaped as a parallel mechanism and set in motion by two DC motors of 11W, assembled in pelvis. The ankle joint was driven by a 6.4W DC motor on each leg. It was capable to move themselves and to transpose a 3.5 cm height box in a velocity of 20 cm/s. They had restricted the displacement of cg to a rectilinear trajectory, thus simplifying, the dynamic model and its control.

Italian researches Figliolini and Ceccarelli designed and published several papers about the ElectroPneumatic Walking Robot EP-WAR2 (1997) by improving the previous prototype EP-WAR (1995). The br was able to walk, turn and climb stairs. The br was provided of logical sensors in order to avoid five different types of static obstacles. The EP-WAR2 was provided of suction-cups to walk along straight paths, turn right and left, and climb stairs. A programmable logic controller was able to carry out the br control during the walk helped by an external sensors system.

Honda Motor Co. has developed several brs since 1986, beginning with E0 up to E6 in 1996. At this year P1 was created from E6. P1 was the first prototype of a humanoid robot, 1915 mm height and 175 kg weight. It was able of turning a switch on and off, gripping the door knob, and carrying something. The P2 android robot had two legs, trunk, two arms, head and was presented in December 1996 (Gomi et al., 1992; Hirai,

1997, Hirai1, 1997 and Hirai et al., 1998). Its inferior members had 12 DOF: 3 DOF for hip, 1 DOF for knee and 2 DOF for ankle. This br had five links in the frontal plan that conferred 4 DOF and seven links in the sagittal plan, giving 6 DOF. The height of the robot was of 182 cm and the mass 210 kg. It was able to move at 0.83 m/s in plan surfaces, to go up and to go down stairs, floor in an (uneven) inclined surface of up to 10 percent. Moreover it could walk of side, turn and to walk stops backwards (Sardain et al., 1998). The P3 br was fully autonomous two-legged humanoid and was presented in September 1997, standing 1600 mm and weighed 130 kg.

In 1997, Zheng and collaborators used a neural algorithm of non supervised learning to allow SD-2 br to go up inclined surfaces. The control system of the eight motors of the joints was PI type and based on neural controller that stored different standards of march in an adapting unit. The adapting unit was responsible for the variation of the joint trajectories, allowing it to walk in unknown surfaces (Salatian et al., 1997).

Shih and Chiou (1998) studied the control of the movement of BR-1 (BIPED ROBOT-1), able to walk in the static way on an inclined land. The BR-1 had 7 DOF, with two telescope legs and a moving mass in pelvis to keep the projection of cg inside of the area of support feet. It weighed 42 kg, height of 80 cm, with a foot of 20 cm x 15 cm and used seven 80 W DC servomotors.

In 1998, Sardain et al. designed an anthropomorphic biped robot called BIP that was able to achieve dynamic walking gaits. Four French Laboratories carried out the BIP project. The BIP robot weighted 105 kg, had two legs and one trunk (42 kg), 180 cm height, and was actuated by brushless DC motors. The 15 DOF BIP br was able to walk forward with extension/flexion in the sagittal plane with the ability to changing direction given by the trunk and pelvis (Espiau and Sardain, 2000). A two axis inclinometer was used in order to determine the gravity vector component and ultrasonic sensors were installed on his legs for reconstructing the ground profile (Azevedo, 2001).

In 1998, the Johnnie br was developed on the Institute of Applied Mechanics of the Technische Universität München (Johnnie, 2003). The latest Johnnie's structure resembles the human locomotor apparatus. A total of 17 joints were included in the robots structure. The br was about 40 kg and was 1,80 m height. Each leg incorporates six driven joints. Three of them are located in the hip, one actuates the knee and another two drive the ankle joint (pitch and roll). The upper body was equipped with a rotational DOF about the body vertical axis. Two arms with 2 DOF each were employed to compensate the total momentum about the body vertical axis. The joints were driven by DC motors, and encoders measured joint angles and velocities. Johnnie has force sensors in the feet to measure the ground reaction forces (Gienger et al., 2001).

In 1998, The Multibody Mechanics Group of the Department of Mechanical Engineering at the Vrije Universiteit Brussel (VUB) - presented LUCY, a lightweight biped, which was able to walk in a dynamical stable way restricted to move only in the sagittal plane and powered by an antagonistic pair of Pleated Pneumatic Artificial Muscles on each legs. The robot weighted about 30 kg and was 150 cm tall (Verrelst et al., 2003).

In 1999, Albert and his coworkers (Albert et al., 1999) presented the BARt-UH (Bipedal Autonomous Robot - Universität Hannover) br. Each leg was composed of three joints, the hip joint, the knee and the ankle resulting in a total of 6 DOF. BARt-

UH was able to perform statically (climb stairs) and dynamically stable walking. Later, BARt-UH received a stereo vision in order to detect obstacles and to calculate the characteristics of stairs as well as new feet with force sensors in order to measure the ground reaction forces. Nowadays it is able to perform autonomous navigation in a structured but a priori unknown environment with obstacles, ditches and step traces. The successor BARt-UH II was born in 2002 (Gerecke et al., 2002).

Yamaguchi and collaborators (Yamaguchi et al., 1999) developed a humanoid robot with 41 DOF called WABIAN-R. The robot achieved stable walking using inverted pendulum acting like an inverted pendulum in his upper body.

In 1999 the Shadow Project in the United Kingdom built the prototype bipedal robot Shadow Walker. It was a wooden leg-skeleton actuated by twenty-eight air muscles (14 on each leg) commanding eight joints and enabling a total of 12 DOF. It stood 160 cm tall, its 'upper body' consisting of the control valves, control electronics and computer interfaces. Under each foot there were five force sensors, two at the toe, one in the center, and two across the heel. These provided a fairly good indication of both contact with the ground, and the distribution of the COM of the robot (Shadow, 2003).

Hasegawa (2000) and collaborators proposed a hierarchic evolutionary algorithm to generate locomotion, considering energy optimization. The algorithm was able to generate a natural movement considering the ZMP. They applied this algorithm in a br weighed 24 kg, 1.20 m height, with 13 joints (13 DOF) and 13 DC motors. This br walked with a step of 30 cm in 5s and was capable to follow a definitive reference trajectory.

In 2000, the human-like biped robot called WABIAN-RII (WAseda BIped humANoid robot-Revised II) was presented (Lim and Takanishi, 2000). It had 6 DOF legs, two 10 DOF arms, a 4 DOF neck, 4 DOF in the eyes and a torso with a 3 DOF waist. Its height was about 1.84 m and its total weight was 127 kg. The researches presented a follow-walking control for the biped robot that selects and generates switchable unit patterns, based on the action model for human-robot interaction. They also presented an emotional walking to the biped robot, which expresses emotions by parameterizing its motion. In the experiments, three emotions such as happiness, sadness and anger are considered. The experimental schemes where based on preset walking patterns of the lower-limbs, waist and head and were planned according to the motion parameters determined by emotional simulations. The emotional walking patterns were commanded to the WABIAN-RII using the control program. The experimental dynamic walking presented a step time of 1.28 s/step and the step width of 0.15 m/step. The emotional walking was evaluated by ten undergraduates as two steps (not agree or agree).

The Nishiwaki and his coworkers presented the H6 and H7 humanoid robots constructed by University of Tokyo. The H6 br was 1370 mm height, and has a total of 35 DOF. Its weight was 55 kg. It had a strong and light structure since aircraft technologies were applied to the body frame. H6 and H7 can walk up and down 25 cm high steps and can also recognize pre-entered human faces (Nishiwaki et al., 2000).

Recently, Honda Motor Co. launched a br called ASIMO (Advanced Step in Innovative Mobility) with height 1.20 m, width of 0.45 m, with 43 kg weight. This robot was constructed for entertainment purposes and a new walking technology (i-WALK) allowed ASIMO to walk continuously while changing directions, and gave the robot even greater stability in response to sudden movements. (Hirose et al., 2001).

Another robot developed for entertainment was SDR-X3 (Sony Dream Robot), made by Digital Creatures Laboratory, Sony, in Tokyo. SDR-3X was presented in November of 2000 and had mass of 5 kg, 50 cm of height and 24 DOF (Sony, 2000). On March of 2002 was presented a br with 38 joints called the SDR-4X. (Ishida, 2003). On September of 2003, Sony announced the QRIO (from "Quest for Curiosity") br similar to the SDR-4X. It can walk and recover from falls. His movement control method was based on the ZMP and it was able to walk on an uneven surface, even if the slope suddenly changes. QRIO can take a step in the direction it was pushed to keep from falling over, when pushed by an external force, or it can immediately ceases all body motion avoiding the fall. If its actions not prevent a fall, it assumes an impact position and the control system commands the actuators to relax slightly avoiding the shock of the fall. After a fall, it turn itself face up, and recover a stable positions (Sony, 2004).

The Ministry of Economy, Trade and Industry (METI) of Japan had launched a humanoid robotics project (HRP) (Inoue et al., 2001). A humanoid robotics platform (HRP-1) and HRP-2 where developed.

On December of 2002, the Intelligent Systems Institute (IST) of the National Institute of Advanced Industrial Science and Technology (AIST) developed a br capable of working on highly advanced tasks with human workers and moving flexibly on a rough terrain by means of a remote control through a communication network. The Humanoid Robot HRP-1S (Humanoid Robotics Project - 1S) was able to operate a forklift and a backhoe by using a remote control. The Honda R&D hardware was provided with control software developed by the AIST.

The br HRP-2 was developed to create a robot that can walk on a rough terrain, stand after falling and work with human beings. The robot was 154 cm tall and weights 58 kg. It has a body similar to humans by eliminating a backpack for electronics installation. The HRP-2 was made with feminine size and it was expected to be 1500 mm height and 60 kg weight (Kaneco et al., 2002).

In the recent years the Robot World Cup (RoboCup, 2003) (Kitano and Asada, 2000) has attempted to foster AI and intelligent robotics research on soccer game. The biped robot teams must perform a football game and must include: design principles of autonomous agents, multi-agent collaboration, strategy acquisition, real-time reasoning, robotics, and sensor-fusion. RoboCup is a task for a team of multiple fast-moving robots under a dynamic environment, it also offers a software platform for research on the software aspects of RoboCup. It is stated on RoboCup's site that "The ultimate goal of the RoboCup project is by 2050, develop a team of fully autonomous humanoid robots that can win against the human world champion team in soccer".

Brazilian researchers have published studies related to practical and theoretical biped robots. Pinto and Bevilacqua (1991) developed a br in the Federal University of Rio de Janeiro, built in aluminum and driven by four step motors. It had legs of 28 cm of height with a thigh of 14 cm of length and 42 cm height. With mass of 7.5 kg approximately, his torque was 0.7 Nm without reductions. This robot was modeled with 5 DOF and was able to perform steps in the sagittal plan. It had forward control and was able to follow a standard walking sent to its motors. Dutra (1993) published a series of works on modeling and simulation of mechanical structures with legs. He is currently working in formulation of coordination and synchronization movement system using nonlinear oscillators. Schemas and collaborators shaped and simulated a uniped robot using two types of control algorithm (Schemas et al., 1999). A comparison between them was carried out, having as performance and index the spent energy. This model was based on the "CMU hopper" developed for Raibert. The Laboratory of Active Systems and Mechatronics of State University of Campinas (UNICAMP/FEM/DMC) modeled, simulated (Campos et al., 2001), projected and constructed (Bezerra, 2002) two biped robots called RB-1 and RB-2 in the period of 2000 to the 2001. The brs were able to carry march in the static steady way in the sagittal plan using a forward control. Both had seven links, 45 cm height and were actuated by servomotores of 14 Kgf.cm of torque each. The RB-1 had 6 DOF and the RB-2, 8 DOF, being capable to carry through curves using sensors. Gonçalves and Zampieri (2003) used a recurrent neural network (RNN) to determine the trunk motion for a br, based on the ZMP criterion, to plan a stable gait for 10 DOF br that has a trunk like as inverted pendulum. So a RNN is trained to determine a compensative trunk motion that makes the actual ZMP get closer to the planned ZMP.

MODELING OF BIPED ROBOTS

Determination of kinematic and dynamic models for biped robots can be carried out in different ways. Taking the biped robot RB-1 as an example, constituted of seven bodies, so called right foot - link 1, right leg - link 2, right thigh - link 3, pelvis - link 4, left thigh - link 5, left leg - link 6 and left foot - link 7, Fig. 1(a). L_i is the length of each robot link, a_i is the localization of the COG of each link and, A, B, C, D, E, F, G, H, I and J are the geometric centers of each joint. All the six joints between the links have rotation of 1 DOF each, not considering the friction between them. An independent actuator moves each joint. Then, the position of point I in relation to the point A is given by (Campos et al, 2000):

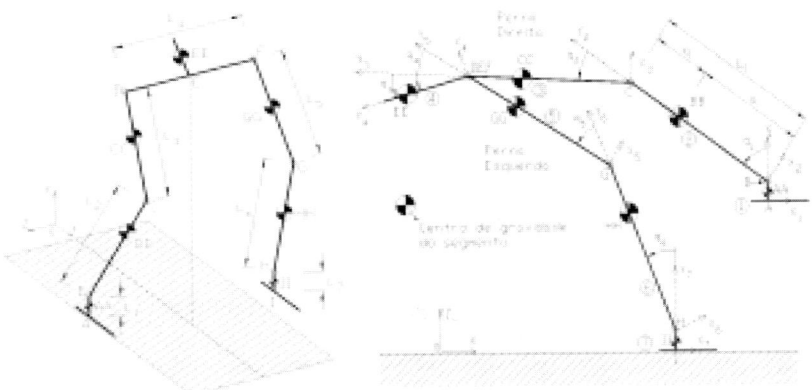

Figure 1. (a) 3D Biped Robot view　　　　*(b) Moving Frames (Mita et al, 1984)*

$$x_{AI} = -L_2 \operatorname{sen}(q_1) - L_3 \operatorname{sen}(q_1 + q_2) + L_5 \operatorname{sen}(q_1 + q_2 + q_3 - q_4) + L_6 \operatorname{sen}(q_1 + q_2 + q_3 - q_4 - q_5)$$
$$y_{AI} = L_1 + L_2 \cos(q_1) + L_3 \cos(q_1 + q_2) - L_5 \cos(q_1 + q_2 + q_3 - q_4) - L_6 \operatorname{sen}(q_1 + q_2 + q_3 - q_4 - q_5) - L_7$$
$$z_{AI} = -L_4$$

$$(1)$$

Considering that both feet remain parallel to the ground during all movement, where q_i ($i = 1..6$) are the relative angles between the robot links, L_i are the dimensions of each link and A and I are the points that represent the right and left foot, respectively. The centers of gravity of each link are represented by two letters (CC). Velocity equations can be derived From Eq. (1). As an example for the center of gravity of the right thigh (CC) represented in the moving frame B3:

$$_{B3}v_{CC} = {}_{B3}v_C + {}_{B3}\omega_3 \times {}_{B3}r_{CCC} + \frac{d}{dt}\left({}_{B3}r_{CCC} \right) = \left\{ \begin{array}{c} -l_2\dot{q}_1 \cos q_2 - a_3(\dot{q}_1 + \dot{q}_2) \\ l_2\dot{q}_1 \operatorname{sen} q_2 \\ 0 \end{array} \right\} \quad (2)$$

Where in Eq. (2), $_{B3}v_c$ is the linear velocity of point C in the moving frame B3, $_{B3}\omega_c$ is the angular velocity of the right knee in the moving frame B3, $_{B3}r_{ccc}$ is in the distance of point C to point CC in the moving base B3, ℓ_2 is the length of link 2, a3 is in the distance of the center of gravity of link 3 in relation to point C, and is the angular velocity of the generalized coordinate. Equation (3) gives the acceleration of point CC in the moving frame B3 can be obtained from equation 3 (Campos et al, 2001), where $_{B3}a_c$ is the linear acceleration of point C in the moving frame B3, $_{B3}\omega_3$ is the angular velocity of link 3 in the moving frame B3, and, ($i = 1..6$).

$$_{B3}a_{CC} = {}_{B3}a_C + {}_{B3}\omega_3 \times {}_{B3}\omega_3 \times {}_{B3}r_{CCC} + {}_{B3}\dot{\omega}_3 \times {}_{B3}r_{CCC} + 2\,{}_{B3}\omega_3 \times \underbrace{\frac{d}{dt}\left({}_{B3}r_{CCC} \right)}_{=0} + \underbrace{\frac{d^2}{dt^2}\left({}_{B3}r_{CCC} \right)}_{=0} \Rightarrow$$

$$(3)$$

$$_{B3}a_{CC} = \left\{ \begin{array}{c} l_2\left[-\ddot{q}_1 \cos q_2 - \dot{q}_1^{\,2} \operatorname{sen} q_2\right] - a_3(\ddot{q}_1 + \ddot{q}_2) \\ l_2\left[+\ddot{q}_1 \operatorname{sen} q_2 - \dot{q}_1^{\,2} \cos q_2\right] - a_3(\dot{q}_1 + \dot{q}_2)^2 \\ 0 \end{array} \right\}$$

Dynamic Modeling
The dynamic modeling can be carried out by different methods. The most used ones, suitable for brs, are the methods of Lagrange, Newton-Euler and Hamilton principia, respectively. Most researchers treat the problem of nonlinear dynamic modeling by the method of Lagrange, once external reactions are not need for the attainment of the motion equation. Moreover, Lagrange method presents advantages when applying to br control. Using the Lagrange method (Eq. (4)):

$$\frac{d}{dt}\{\frac{\partial L}{\partial \dot{q}_i}\} - \frac{\partial L}{\partial q_i} = Q_i \qquad (i = 1....7) \qquad (4)$$

Where L is the lagrangean, defined by the difference between kinetic and potential energy and Qi is the vector of motor torques. After the substitution of robot parameters in Eq. (4), motion equation is obtained under matricial form

$$[P]\ddot{q} + [H]\dot{q}^2 + [G] = [E]u_i \qquad (5)$$

In Eq. (5), the [P] matrix contains the masses and moment of inertia of each link, [H] is the matrix that contains velocity dependent terms, [G] is the matrix with the gravitational terms and [E] is the matrix of transformation between the relative and absolute generalized coordinates and ui is the drive torques (u_B, u_C, u_D, u_F, u_G and u_H).

Zero Moment Point Determination
As viewed above the ZMP has been widely used as an index of stability and motion analysis to the biped robots. In dynamic motion analysis its is analogous to the projection of the COM on the ground plane. As COG should be kept within the supporting area for static balance, ZMP should be kept within the supporting area for dynamic balance.
The distribution of ground reaction force under the supporting foot is quite complicated, in general, it can be replaced with an equivalent pair of force and moment that act on a single point, within the foot sole. Here the moment depends on the choice of the point. The point P can be defined as ZMP such that the moment vanishes to zero.
The ZMP can be even further extended to the double support phase. In this case the support area is the convex hull of two supporting feet. The distribution of ground reaction force is a little peculiar; the pressure is zero everywhere except for the two sole areas. As the foot pressure changes during a motion, ZMP draws a certain trajectory. It translates from the heel to the tiptoe of the supporting foot in a walking step.

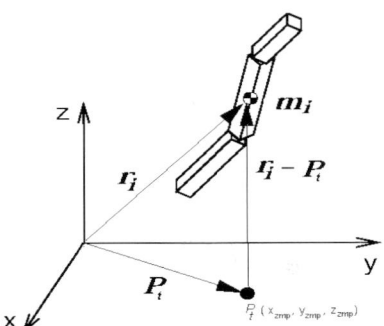

Figure 2. ZMP localization

Since that is not easy to measure the distribution of ground reaction force over time, The ZMP trajectory is calculated according to inverse dynamics. In the motion of an articulated biped robot, each body segment contributes to the moment at the foot.

According to D'Alambert's principle and Figure 2, the moment contribution at point P_t is given by :

$$m_i(\mathbf{r}_i - P_t) \times (-\ddot{\mathbf{r}}_i + \mathbf{g}) \tag{6}$$

Where m_i is the mass of the i-th body segment, $\mathbf{r}_i$ is the position of the mass center of i-th body segment. So, the ZMP can be obtained by solving the following equation for P_t:

$$\sum_i m_i(\mathbf{r}_i - P_t) \times (-\ddot{\mathbf{r}}_i + \mathbf{g}) = 0 \tag{7}$$

The solutions for Eq. (7) are an infinite number. Is possible to obtain a unique solution by fixing z value of P_t to zero, which corresponds to constraining P_t into the ground plane. The solution P_t (x_{zmp}, y_{zmp}, 0) is given by

$$x_{zmp} = \frac{\sum_i m_i(\ddot{z}_i - g_z)x_i - \sum_i m_i(\ddot{x}_i - g_x)z_i}{\sum_i m_i(\ddot{z}_i - g_z)} \tag{8}$$

$$y_{zmp} = \frac{\sum_i m_i(\ddot{z}_i - g_z)y_i - \sum_i m_i(\ddot{y}_i - g_y)z_i}{\sum_i m_i(\ddot{z}_i - g_z)} \tag{9}$$

Where x_i ; y_i and z_i are the position of the COM of i-th body segment and $\ddot{x}$, $\ddot{y}$ and $\ddot{z}$ are the acceleration of the COM of i-th body segment.
The ZMP trajectory is calculated from Equations 8 and 9 at each time.

BIPED ROBOT CONTROL

Furusho and Masubuchi (1987) classified the studies about dynamic biped locomotion systems in two classes according to their aims. The first one uses the human motion data to discover walking characteristics. The second one aims to analyzing biped motion using the torque control at each joint to find the control law of human walking and to derive the control algorithms of biped robots. Three main approaches were used in control theoretic studies about biped robots. One approach is to simplify the control system design by decoupling the high order linearized dynamics of the biped locomotion into independent low order subsystems (Gollidary and Hemami, 1977). The second approach is to use a dynamic equation to specify the ideal trajectory by modifying the joint torques (Vukobratovic and Stepanenko, 1972). This method requires much computation and the real-time control of br becomes difficult. The third approach is the hierarchical control. This control is based on physiological studies of human walking control having a hierarchical structure and that at lowest level there exist some local feedback between muscles and spinal cord (Furusho and Masubuchi,

1987). Miyazaki and Arimoto (1980) have first used a hierarchical control strategy and a model reduced method for biped walking.

The most robots studied by the above groups had slow and quite primitive motion when compared to human motion ability. Their work mainly focused on developing a stable control algorithm. The study of biped robots has being relatively slowly improved due the complexity of dynamic models and to difficulties in the movement control, since it is subject to dynamic instabilities to remain in balance (Zheng and Shen, 1990; Goddard et al., 1992; Wang et al., 1992; Shih and Gruver 1992; Kajita et al., 1992; Miller, 1994; Shih and Chiou, 1998; Zhang, 1998 and Sardain et al., 1998).

Other robotics researchers (Raibert, 1986; Shih, 1996; Pratt et al., 1997; Fukuda et al., 1997; Fujimoto and Kawamura, 1998; Dasgupta and Nakamura, 1999) have studied dynamics and motion control of brs realizing robust locomotion by analyzing the reactive force at the foothold. Most of these works aims to maintain ZMP within the stable region by compensating upper body motion. Fukuda et al. (1997) obtained ZMP trajectory by using sensors at each sole and the joint motion was determined by employing the recurrent neural networks to maintain ZMP in the supporting area. Park and Rhee (1998) determined the ZMP trajectory using fuzzy logic on the leg trajectories. The trunk and swing leg motion were compensated to stabilize the locomotion based on this trajectory. They concluded that a moving ZMP increases the locomotion stability and guarantees natural motion rather than a fixed ZMP. Dasgupta and Nakamura (1999), using human motion captured data proposed an algorithm to produce feasible walking motion of a robot. They set a desired ZMP trajectory referring to the human data, and then corrected the motion of a selected joint so that the resulting motion approximately matches the desired ZMP trajectory. They used the joint motion represented by Fourier series and formulated an optimization problem to determine the unknown coefficients.

In the most studies mentioned, a particular technique is proposed, adjusted to that model, as much for the linearization of motion equation for robot control in real time.

CONCLUSION

A bibliographic review on the main biped robots developed for scientific and entertainment purposes in last the 4 decades was presented, including Brazilian researchers. Mathematical modeling of a biped robot, determination of ZMP and the main studies on the theory of control was briefly presented. The subject was not depleted and it was not possible to include photos or figures and to mention the several homepage on this subject. However, two great repositories related to robots with legs can be accessed in http://www.fzi.de / ids / WMC / preface / walking machines _katalog.html and also in the website dedicated to those interested in the subject http://www. androidworld.com/prod01.htm.

ACKNOWLEDGEMENT
The authors would like to thank Fapesp for support (Project 9907715-6).

REFERENCES
Albert, A., Gerth, W., Hofschulte, J., and Schermeier, O. " Echtzeitsystem für einen zweibeinigen Roboter". Workshop über Realzeitsysteme, PEARL99, Springer-Verlag, 1999, pp. 69 - 78.

Azevedo, C. "On the interaction between the human and the robot in bipedal walking". Proceedings of the 4th International Conference on Climbing and Walking Robots CLAWAR 2001, 24-26 Sep. Karlsruhe. Germany.

Bay, J.S. and Hemami, H. "Modeling of a neural pattern generator with coupled nonlinear oscillators". IEEE Transactions on Biomedical Engineering. 34(4). 1987. 123-145.

Bezerra, C. A. D. "Desenvolvimento de um robô bípede autônomo para locomoção em ambiente desestruturado". Tese de Doutorado. Unicamp - SP. 2002.

Campos, D.C., Bezerra, C.A.D, Zampieri, D.E., Mendeleck, A. "Modelagem e simulação dinâmica de um robô bípede". XVI Congresso Brasileiro de Engenharia Mecânica, COBEM 2001. Uberlândia. MG, 2001.

Dasgupta, A. and Nakamura. Y. " Making feasible walking motion of humanoid robots from human motion capture data". In Robotics and Automation '99 Proceedings, volume 2, 1999, pages 1044–1049.

Dutra, M.S., Kinematical Analysis of a Six-Legged Walking Machine. In DINAME Dynamical Problems in Mechanics, Angra dos Reis-RJ, 1997.

Espiau, B., Sardain, P. "The Anthropomorphic Biped Robot Bip2000", Proceedings of the IEEE International Conference on Robotics and Automation, San Francisco, CA-USA, 2000, pp. 3997-4002.

Figliolini G., Papa L., "Biped walking electropneumatic robot", 4th International Workshop on Robotics in Alpe-Adria Region, Portschach, 1995, pp. 245-248.

Figliolini G., Ceccarelli M., "Mechanical Design of an Anthropomorphic Electropneumatic Walking Robot", Eleventh CISM-IFToMM Symposium on Theory and Practice of Robots and Manipulators. Wien, Springer-Verlag, 1997, pp.189-196.

Figliolini G., Ceccarelli M., "Gait analysis of EP-WAR2 Biped Robot for Walking and Climbing Stairs", Proceedings of the 4th International Conference on Climbing and Walking Robots ClaWar2001, Professional Engineering Publ., London, 2001, pp. 827-834.

Figliolini, G., Ceccarelli, M. and Di Cocco, V. "Walking stability of EP-WAR2 biped robot". Proceedings of the Fourth European Workshop on Advanced Mobile Robots. EUROBOT '01. September 19-21, 2001. Lund, Sweden.

Frank, A.A., "Automatic control systems for legged locomotion machines", PhD Thesis University of Southern California. 1968.

Fujimoto, Y. and Kawamura. A. "Simulation of an autonomous biped walking robot including environmental force interaction. IEEE Robotics and Automation Magazine, 1998, pages 33–42.

Fukuda, T., Komata, Y. and Arakawa. T. " Stabilization control of biped locomotion robot based learning with gas having self-adaptive mutation and recurrent neural networks. In Robotics and Automation '97 Proceedings, volume 1, 1997, pages 217–222.

Furusho, J. and Masubuchi, M. " Control of a dynamic biped locomotion system for steady walking". Journal of Dynamic Systems, Measurement and Control. (108), 1986, 111-118.

Furusho, J. and Masubuchi, M. "A theoretically motivated reduced order model for the control of dynamic biped locomotion". Journal of Dynamic Systems, Measurement and Control. Transaction of ASME. (109), 1987, 155-163.

Furusho, J. and Sano, A. "Sensor-based-control of a nine link biped". International Journal of Robotics Research 9(2), 1990, 83-98.

Gerecke, M., Albert, A., Hofschulte, J., Strasser, R., Gerth, W. " Towards an autonomous, bipedal service robot". Tagungsband 10 Jahre Fraunhofer IFF, 24.-28.06.2002, Magdeburg, S. 2002, 163-168.

Gienger, M., Löffler, K. and Pfeiffer, F. " Towards the Design of Biped Jogging Robot," Proceedings of IEEE International Conference on Robotics and Automation, 2001, pp. 4140-4145.

Goddard, R.E., Zheng, Y.F. and H. Hemami, H. "Control of the heel-off to toe-off motion of a dynamic biped gait". IEEE Transaction on Systems, Man, and Cybernetics. 22(1) , 1992, 92-102.

Gollidary, C. L. and Hemami, H. " An approach to analyzing biped locomotion dynamics and designing robot locomotion controls". IEEE Transaction on Automatic Control. AC 22-6, 1977, pp. 963 – 972.

Gomi, H., Kumagai, T., Hirose, M. and Nishikawa, M. "Articulated structure for legged walking robot", U.S. Patent 5159988. 1992. Nov. 3.

Gonçalves, J.B. and Zampieri, D.E. "Recurrent Neural Network Approaches for Biped Walking Robot Based on Zero Moment Point Criterion". Journal of the Brazilian Society of Mechanical Sciences and Engineering. Vol. XXV (1).2003, 69-78.

Goswami A., "Postural Stability of Biped Robots and the Foot-Rotation Indicator (FRI) Point", The International Journal of Robotics Research, Vol. 18, No. 6, 1999, pp. 523-533.

Grishin, A.A., Formalsky, A.M., Lensky, A.V. and Zhitomirsky, S.V. "Dynamic walking of a vehicle with two telescopic legs controlled by two drivers". International Journal of Robotics Research. 13(2). 1994, 137-147.

Hasegawa, Y., Arakawa, T. and Fukuda, F. "Trajectory generation for a biped locomotion robot". Mechatronics. (10), 2000, 67-89.

Hasegawa, Y., Arakawa, T. and Fukuda, F. "Trajectory generation for a biped locomotion robot". Mechatronics. (10), 2000, 67-89.

Hirai, K. "The Honda Humanoid robot". Proceedings of IEEE-RSJ International Conference on Intelligent Robots and Systems. 1997, 499-508. Grenoble, France.

Hirai1, K. "Current and Future Perspective of Honda Humanoid Robot". In Proceedings of the IEEE/RSJ Int. Conference on Intelligent Robots and Systems IROS, (Grenoble, France), 1997, pp. 500 - 508.

Hirai, K., Hirose, M., Haikawa, Y. and Takenaka, T. " The Development of Honda Humanoid Robot," In Proceedings of the IEEE International Conference on Robotics and Automation, (Leuven, Belgium), 1998, pp. 1321 - 1326.

Hirose, M., Haikawa, Y., Takenaka, T. and Hirai, K. "Development of Humanoid Robot ASIMO". In IEEE/RSJ International Conference on Intelligent Robots and Systems (IROS) - Workshop 2, Maui, Hawaii, 2001.

Hodgins, J. "Legged robots on rough terrain: experiments in adjusting step length". Proceedings of IEEE International Conference on Robotics and Automation. 1988, 824-826.

Hodgins, J. and Raibert, M. "Biped gymnastics". The Int. Journal of Robotics Research. 1990. Vol. 9 (2).

Inoue, H., Tachi, S., Nakamura, Y., Hirai, K., Ohyu, N., Hirai, S., Tanie, K., Yokoi, K. and Hirukawa, H. "Overview of Humanoid Robotics Project of METI," Proceedings of the 32nd International Symposium on Robotics, 2001, pp. 1478-1482.

Ishida, T. "A small biped entertainment robot SDR-4X II". Computational Intelligence in Robotics and Automation. Proceedings of IEEE International Symposium on Publication Date: July 16-20, 2003 On page(s): 1046- 1051 Volume: 3.

Johnnie, 2003. http://www.amm.mw.tu-muenchen.de/index_e.html. Accessed 01.12.2003.

Kajita, S. and Tani, K. "Study of dynamic biped locomotion on rugged terrain". Proceedings of the IEEE International Conference on Robotics and Automation. 1991, pp. 1405-1411.

Kajita, S. and Tani, K. "Experimental study of biped dynamic walking". IEEE Control Systems Magazine. 16(1) , 1996, 13-19.

Kajita, S., Yamaura, T. and Kobayashi, A. "Dynamic walking control of a biped robot along a potential energy conserving orbit", IEEE Transactions on Robotics and Automation". 8(4), 1992, 431-438.

Kaneko, K., Kajita, S., Kanehiro, F., Yokoi, K., Fujiwara, K., Hirukawa, H., Kawasaki, T., Hirata, M. and Isozumi, T. "Design of Advanced Leg Module for Humanoid Robotics Project of METI". Proceedings of 2002 IEEE International Conference on Robotics and Automation. Washington. DC. May 2002.

Kato, S. Ohteru, H. Kobayashi, K. Shirai and A. Uchiyama, "Information-power machine with senses and limbs," First CISM-IFToMM Symposium on Theory and Practice of Robots and Manipulators, Springer-Verlag, 1974.

Kato, T., Takanishi, A., Ishikawa, H. and Kato, I. "The realization of quasidynamic walking by a biped walking machine", Fourth CISM-IFToMM Symposium On Theory and Practice of Robots and Manipulators". PWN, Warsaw. 1983, pp 341-351.

Kitano H., Asada, H. " The RoboCup humanoid challenge as the millennium challenge for advanced robotics", Advanced Robotics 13(8), 2000, 723-736.

Katoh, R. and Mori, M. "Control method of biped locomotion giving asymptotic stability of trajectory". Automatica. 20(4), 1984, 405-414.

Lim, H and Takanishi, A. "Emotion-based Walking for a Biped Humanoid Robot", CISM-IFToMM International Symposium on the Theory and Practice of Robots and Manipulators, CD-ROM proceedings, 2000.

McGeer, T. "Passive dynamic walking", The Int. Journal of Robotics Research, vol. 9, no. 2, April 1990.

Miller-III, W.T. "Real time neural network control of a biped walking robot". IEEE Control Systems. Feb. 1994, pp. 41-48.

Mita, T., Yamaguchi, T., Kashiwase, T. and Kawase, T., "Realization of high speed biped using modern control theory". International Journal of Control. 40 (1), 1984, 107-119.

Miura, H. and Shimoyama, I. "Dynamic walk of a biped". Int. Jour. of Robotics Research. 3(2). 1984, 60-74.

Miyazaki, F. and Arimoto, S., "A control theoretic study on dynamical biped locomotion". Journal of Dynamic Systems, Measurement, and Control, vol. 102, 1980, pp. 233-239. December.

Miyazaki, F. and Arimoto, S. "Implementation of a hierarchical control for biped locomotion". Proceedings 8th IFAC Control Science and Technology Congress. 1981, pp 1891-1896.

Nishiwaki, K., Sugihara, T., Kagami, S., Kanehiro, F., Inaba, M. and Inoue, H. " Design and Development of Research Platform for Perception-Action Integration in Humanoid Robot: H6," Proceedings of International Conference on Intelligent Robots and Systems, 2000, pp. 1559-1564.

Park, J. H. and Rhee, Y. K. "ZMP trajectory generation for reduced trunk motions of biped robots". In Proceedings of IEEE/RSJ International Conference on Intelligent Robots and Systems (IROS'98), pages 90 – 95. IEEE/RSJ IROS, October 1998.

Pinto, F.A.N.C. and Bevilacqua, L. "Dynamical analysis of the biped machine". IV Simpósio International Sobre Sistemas Dinâmicos da Mecânica, Pouso Alto, MG. 1991.

Pratt, J. , Dilworth, P. and Pratt., G. "Virtual model control of a bipedal walking robot. In 1997 International Conference on Robotics and Automation, 1997.

Pratt, G. A. "Legged Robots at MIT: What's New Since Raibert". IEEE Robotics and Automation Magazine. 7(3), 2000, 15-19.

Raibert, M.H. "Legged Robots That Balance". "MIT Press". Cambridge, MA. 1986.

Raibert, M. H. "Hopping in legged systems—modeling and simulation for the two-dimensional one-legged case", IEEE Transactions on Systems, Man, and Cybernetics, vol. SMC-14, no. 3, May/June 1984.

RoboCup, 2003. http://www.robocup.org/02.html. Accessed in 12.12.2003.

Salatian, A.W., Yi, K.Y. and Zheng, Y.F. "Reinforcement learning for a biped robot to climb sloping surfaces". Journal of Robotic Systems. 14(4), 1997, 283-296.

Sano, A. and Furusho, J. "3D steady walking robot with kick-action". Proceedings of USA-Japan Symposium on Flexible Automation. 1988. Vol.II.

Sano, A. and Furusho, J. "Dynamically stable quadruped locomotion - a pace gait in the colt-3". Proceedings of the 20th International Symposium on Industrial Robots. 1989.

Sardain, P., Rostami, M. and Bessonnet, G., "An anthropomorphic biped robot: dynamic concepts and technological design". IEEE Transactions on Systems, Man and Cybernetics, Part A: systems and humans. 28(6), 1998, 823-838.

Shadow Group. Homepage. http://www.shadow.org.uk/projects/ biped.shtml #Anchor - Prototype - 44591. Accessed in 10.11.2003.

Shih, C. L. " Analysis of the dynamics of a biped robot with seven degrees of freedom". In *Robotics and Automation '96 Proceeding*s, volume 4, 1996, pages 3088– 3013.

Schammass, A., Valente, CMO and Caurin, GAP. "Modelling and control of a simulated one legged robot". XV Congresso Brasileiro de Engenharia Mecânica. Aguas de Lindoia. São Paulo.2000

Shih, C.L. and Gruver, W. "Control of a biped robot in the double-support phase". IEEE Transactions on Systems, Man and Cybernetics. 22(4). 1992, pp. 729-735.

Shih, C.L., Gruver, W. and Lee, T. "Inverse kinematics and inverse dynamics for control of a biped walking machine". Journal of Robotic Systems. 10(4), 1993, 531-555.

Shih, C.L. and Chiou, C.J. "The motion control of a statically stable biped robot on an uneven floor". IEEE Transactions on Systems, Man and Cybernetics, Part B: cybernetics. 28(2), 1998, 244-249.

Sony Corporation. Homepage http://www.sony.net/SonyInfo/News/Press/200203/02-0319E/. accessed on 10.04.2003.

Sony Corporation. Homepage http://www.sony.net/SonyInfo/QRIO/top_nf.html. Accessed on 10.12.2003.

Takanishi, A., Naito, G. , Ishida, M. and Kato, I., "Realization of plane walking by the biped walking robot WL-10R", Robotic and Manipulator Systems, 1982, pp. 283-393.

Takanishi, A., Lim, H., Tsuda, M. and Kato, I., "Realization of dynamic biped walking stabilized by trunk motion on a sagittally uneven surface", IEEE International Workshop on Intelligent Robots and Systems, IROS, 1990.

Verrelst, B.& , Van Ham, R., Vanderborght, B., Lefeber, D. & Daerden, F. "Lucy: a walking robot". Proceedings of the 6th National congress on theoretical and applied mechanics, Ghent, May 2003, paper no. 073 (CD-ROM).

Vukobratovic M. and Juricic D., "Contribution to the Synthesis of Biped Gait", IEEE Trans. on Biomedical Engineering, Vol. 16, No 1, 1969.

Vukobratovic, M. and Stepanenko, Y. "On the stability of anthromorphic systems". Mathematical Biosciences. Vol. 15. No. 1, 1972, pp. 1 - 37.

Vukobratovic M., Hristic D. and Stojiljkovic Z., "Development of Active Anthropomorphic Exoskeletons", Medical and Biological Engineering, Vol. 12, No. 1, 1975.

Vukobratovic M., and Stokic D., "Dynamic Stability of Unstable Legged Locomotion Systems", Mathematical Biosciences, Vol. 24, No. 1/2, 1975.

Vukobratovic, M., Borovac, B., Surla, D. and Stokic, D. "Scientific fundamentals of robotics 7. Biped locomotion: Dynamics stability, control and application". Springer - Verlag. 1990.

Wang, H., Lee, T.T., and W.A. Gruver, W.A. "A neuromorphic Controller for a three link biped robot". IEEE Transaction on Systems, Man, and Cybernetics. 22(1), 1992, 92-102.

Yamaguchi, J., Takanishi, A. and Kato, I. "Development of a biped walking robot compensating for three-axis moment by trunk motion", Proc. of the 1993 IEEE/RSJ International Conference On Intelligent Robot and Systems (IROS), pp. 561-566, July 1993.

Yamaguchi, J., Takanishi, A. and Kato, I. "Experimental development of foot mechanism with shock absorbing material for acquisition of landing surface position information and stabilization of dynamic biped walking". Proceedings of IEEE International Conference Robotics and Automation. 1995, pp 2892-2899.

Yamaguchi, J., Soga, E. Inoue, S. and Takanishi, A. "Development of a bipedal humanoid robot – Control method of whole body cooperative dynamic biped walking". Proceedings on IEEE International Conference of Robotics and Automation. 1999, pp 368-374.

Zheng, Y.F. and Sias, F., "Design and motion control of practical biped robots". "International Journal of Robotics and Automation". 3(2), 1988, 70-78.

Zheng, Y.F. and Shen, J. "Gait synthesis for the SD-2 biped robot to climb sloping surface". IEEE Transactions on Robotics and Automation. 6(1), 1990, 86-96.

THE EARLY DEVELOPMENT OF REMOTE TELE-MANIPULATION SYSTEMS

Robert Bicker[†], Kevin Burn[‡], Zhongxu Hu[†], Watcharin Pongaen[†], Abdouslam Bashir[†]
[†] School of Mechanical & Systems Engineering, Stephenson Building,
University of Newcastle upon Tyne, NE1 7RU, UK
e-mail: robert.bicker@newcastle.ac.uk
[‡] School of Computing, Engineering and Technology,
University of Sunderland, Sunderland SR1 3SD, UK
e-mail: kevin.burn@sunderland.ac.uk

ABSTRACT - Humans have been practicing 'remote manipulation' for many centuries. Our early ancestors would have used sticks to manipulate food on an open hearth fire without getting themselves, or the food burnt, The blacksmith subsequently developed special tools, called tongs, to manipulate and heat the workpiece in the hot embers and during forging. However, the key developments in remote manipulation and handling took place in the middle of the last century during the pioneering days of the nuclear industry, once the extreme radiation hazards to humans became apparent. From relatively simple mechanical devices there now exist extremely sophisticated computer controlled tele-robotic systems and virtual environments that allow humans to plan and execute complex tasks that would otherwise be impossible to undertake because of the nature of the environment, or the extreme distances involved. Having chronicled the history of remote handling technology from its early development in the nuclear industry to state of the art tele-robotic systems, mention is also made of developments in other application areas, such as surgery, military, space and undersea exploration.

KEYWORDS: Remote handling, master slave manipulators, tele-operation, tele-robotics

INTRODUCTION

Tele-manipulation involves the extension of the human operator's sensing and manipulative capability into a remote, often hostile and unstructured environment, in order to perform a wide variety of tasks that demands a high degree of manual dexterity, judgement or intelligence which only humans can provide. Historically, tele-operation, which means literally 'performing work from a distance' can be said to have its roots in some of our most primitive tools that have been available for centuries. Blacksmiths tongs are a crude but effective example of an early remote handling tool, allowing the blacksmith to extend his reach into a hostile environment (the forge), whilst still enabling him to retain control of the position and orientation of the workpiece whilst working the white hot metal (Vertut & Coiffet, 1984). The handling of radioactive chemicals during the pioneering days of the nuclear industry required extreme caution and led to the introduction of glove boxes to eliminate any direct contact with the highly toxic elements. With increasing radiation levels, direct handling, even gloves, was not safe, and simple mechanical tong devices were subsequently developed to provide the operator with improved protection whilst situated behind a relatively safe biological shield comprising a lead lined enclosure, or hot-box. However because direct handling is no longer possible, manual dexterity becomes seriously impaired due to the complex relationship between the operator, the remote manipulator system and the task.

During the latter half of the last century it became necessary to develop truly remote handling systems that allowed dexterous manual operations to take place, not only in the nuclear environment but also for undersea and space exploration, in order to extend the operators reach over significant distances; or where the working space is too restricted for manual operations, or where the operator does not possess sufficient reach or power, or other physical attribute. An example of a restricted workspace is to be found within the brain in which a surgeon uses micro-manipulators to perform delicate surgical operations (Matsushima, 1981, 1984), whereas the GEC Handyman (Mosher, 1986) is an example of an exosketal device designed to enhance the operator's own capabilities - in this case both strength and reach.

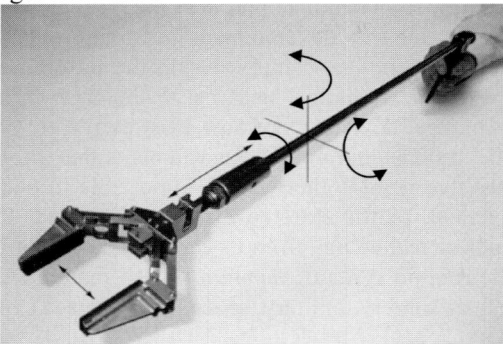

Figure 1 Mechanical tongs with 4 degrees of freedom, and pinch gripper

REMOTE HANDLING IN THE NUCLEAR INDUSTRY

As both toxicity and radiation hazard levels increased, it became necessary to further remove the operator from the task. This led to the introduction of simple mechanical tong devices of the type shown in Fig. 1, which the operator manipulates as a pair (with one in each hand) whilst viewing the task directly through a glass window. The tong itself is usually mounted in a spherical lead bearing that permits rotation about three axes, and translation along the axis of the tong shaft. Having only four degrees of freedom (DOF), there is a consequential and significant loss of dexterity, and the lack of articulation prohibits over reaching obstacles; nevertheless trained operators are still able to accomplish quite complex and precise manipulations. The robust mechanical design of the tong is simple with the pinch gripper operated via a simple push rod located inside the shaft, and is fitted with a pistol type handle with spring return. In order to improve the range and dexterity of the handling tasks that can be performed the grippers are interchangeable from a magazine positioned inside the hotbox. The tongs are usually fitted with a protective gaiter to provide an adequate atmospheric seal back through the bearing housing.

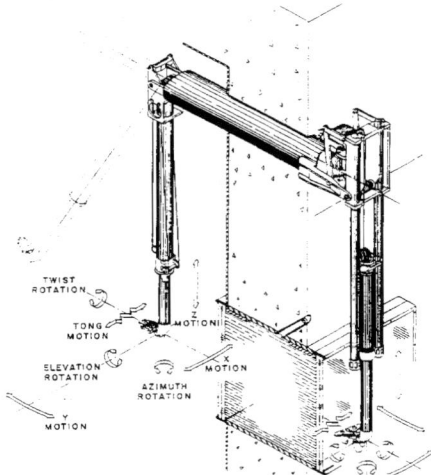

Figure 2   MSM developed at the Argon National Laboratories (Goertz, 1954)

Further developments in nuclear physics in the early 50's created more dangerous radiation hazards, and with more stringent legislation designed to limit exposure of technicians to 'safe' radiation levels, thicker and more effective biological shields were designed which required handling devices which could be used to carry out tasks at a distance of several metres. The first generations of master-slave manipulators (MSM's) were developed at the Argonne National Laboratory (ANL) specifically for use in what were called hot-cells, as shown in Fig. 2 (Goertz, 1952). By using an ingenious mechanical arrangement of pulleys, belts and/or cables the end effector could be provided with six DOF, which enabled the operator to manipulate the 'master arm' and in so doing produce a corresponding motion of the identical 'slave arm'. The MSM is located above the operator's head, in a 'through-tube' and can be easily withdrawn for maintenance. The design was based on a two DOF shoulder with pitch and roll, and a third telescoping axis to improve reach. The wrist design incorporates a two DOF differential gearing arrangement coupled to an azimuth twist. Having six DOF the MSM was thus capable of reasonably unrestricted dexterous manipulation, within its working envelope. In time, the working envelope was further enhanced, by motorising the shoulder (pitch) axis to offset and extend the forward reach. Although relatively complex mechanisms, when well maintained and used by skilled operatives these devices can be used to undertake highly dexterous and extremely precise tasks. A more compact all revolute design of MSM utilizes an articulated elbow joint, not only helps to reduce the overall height of the device (and the hot-cell) but also provides improved obstacle avoidance over the telescoping design (PAR systems). The simplified schematic shown in Fig. 3 illustrates the pulley-cable arrangement for the lower 3 DOF.

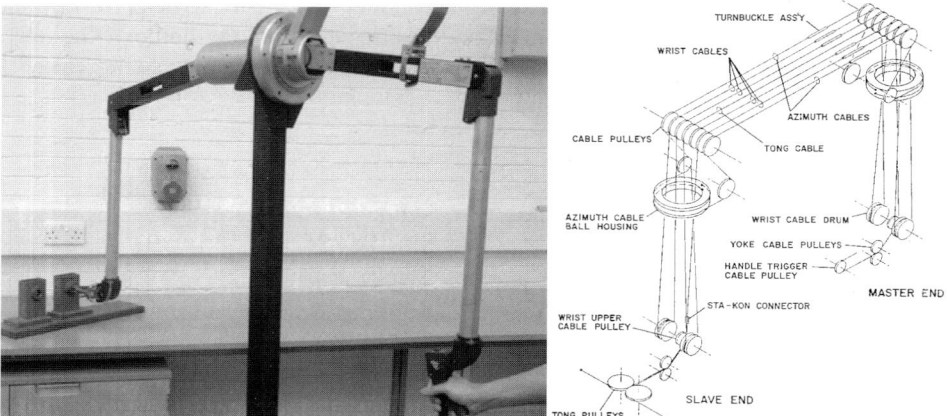

Figure 3 PAR Systems Tru-Motion Mini MSM - arrangement of joints and cabling

One of the major benefits of using mechanical MSM's is the direct one-to-one kinematic correspondence between master and slave motions, thus enabling the operator to produce co-ordinated 6 DOF movements of the end-effector with little mental effort. However another important advantage is the intrinsic force-feedback of the mechanical transmission, which allows the operator to sense the magnitude of the forces and torques applied at the slave arm. Although a fairly crude form of force reflection, the forces and torques felt at the master arm are modified by dynamic effects such as columbic and viscous friction, backlash, link inertia and the inherent flexibility of cables and tapes. It still nevertheless provides the operator with an important sense of 'feel', as well as detecting contacts from unseen objects within the workspace, thus making the operator aware that further movement might well result in damage to either the MSM or the object in the hazardous environment. Tele-operators that enable the operator to sense the forces and torques in this way are commonly referred to as bilateral systems. Fig 4a illustrates the mechanical coupling between a single DOF system with the input (on the left) and the output (on the right) which can be rotated without impedance, unless a disturbance torque is applied at the output, in which case the operator will 'feel' the reaction, and the shaft will wind-up (Goertz, 1954).

DEVELOPMENT OF ELECTRICALLY LINKED TELE-MANIPULATORS

Although mechanical MSM's are adequate for a certain range of tasks, and in situations where it is safe to work in relative proximity to 'hot' areas, they have a number of limitations. Their main disadvantage is the immobility of the master arm relative to the slave, so that where radioactivity in the workspace is of an intermediate level, the operator's safety can only be ensured by extending the physical size of the device and by providing thicker shielding. However, even in situations where this is practical, there is an inevitable reduction in performance of the tele-operator system, due to increased flexibility in the drive cables coupled with the link mass/inertia, and a further deterioration of the operator's vision with increased window thickness.

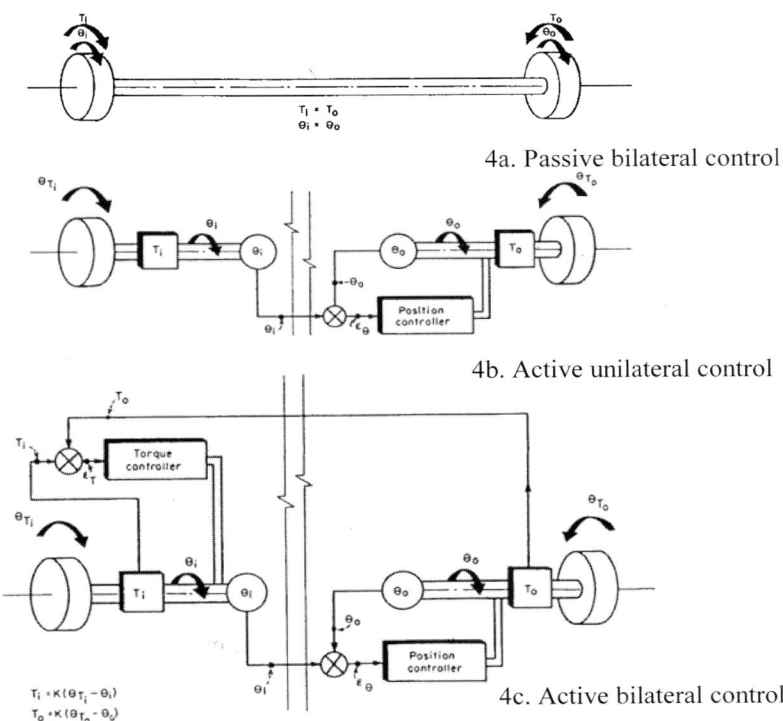

4a. Passive bilateral control

4b. Active unilateral control

4c. Active bilateral control

$T_i = T_o$
$\theta_i = \theta_o$

$T_i = K(\theta_{T_i} - \theta_i)$
$T_o = K(\theta_{T_o} - \theta_o)$

Figure 4 Master- Slave control modes (Goertz, 1954)

Further increases in both radiation and toxicity levels, particularly in applications such as in-reactor repair and maintenance, precluded the use of direct 'passive' mechanical systems, and led directly to the development of electrically linked 'tele-manipulators', where the operator could drive electric motors via a bank of switches or joysticks to actuate the remotely sited slave arm joints, whilst viewing the scene indirectly through a television screen. The 'active' unilateral position servo-control concept is clearly illustrated in Fig. 4b. However, such developments created significant problems in that by driving only one joint at a time this removed the luxury of the kinematic correspondence previously available in the MSM, as well eliminating the kinaesthetic force feedback. Furthermore it also brought about a dramatic increase in mental concentration as well as significantly increasing the time needed to complete the task in hand. The development of the general-purpose servo manipulator by Goertz subsequently restored both the kinaesthetic bilateral coupling and axis correspondence between master and slave arms, whilst simultaneously providing a firm foundation for significant advances in remote systems technology (Goertz, 1954). His replica servo-manipulator design incorporated servomotors driving all 12 axes of the master and slave by initially employing a position-position bilateral control scheme, in which the position error between master and slave joints is used to actuate the slave motor, and the slave to master position error is used to backdrive the master. A more advanced and sensitive

In remote manipulation where the master controls are situated at some distance from the slave arm the field of view may be partially obscured, often through thick plate glass windows, which can diminish the operator's visual perception of the scene. Force feedback is an intrinsic feature of a mechanically linked master-slave manipulator, its fidelity, as perceived by the operator is directly related to the efficiency of the transmission. Since force and tactile sensory feedback are essential for dexterous manual manipulations it can be readily appreciated that this sensitivity will be seriously impaired when factors such as compliance, backlash and friction exist in the direct mechanical transmission of the master-slave system. This is further compounded when the system becomes more complex, i.e. in electrically linked tele-operator systems, which is more likely to degrade the quality of this type of feedback.

It is an unfortunate, though inevitable consequence in using a tele-operator system that overall performance is degraded and a penalty is paid, partly in the form of extra time to accomplish a task, and partly in the increased likelihood of mistakes. This is due to the fact that no electro-mechanical manipulation system has, or is ever likely to have, the degree of coordination and dexterity found in a skilled human being. Thus, although tele-operation will always remain less efficient and more mistake prone than comparable skilled manual operations, it is advantageous to look for improvements, to reduce the performance degradation, minimize the penalty, and reduce operator fatigue, particularly when circumstances demand the use of such systems.

In view of the complex operator-machine and machine-task interactions it is not surprising that the criteria by which the performance of remote manipulator systems was assessed has been largely qualitative, often using task completion time to carry out a specific task, or set of tasks manually, then compared to the same tasks undertaken using a tele-operator system (Sheridan, 1976). Investigators have attempted to analyse dexterity using several different criteria. Simple peg-in-hole type tasks have received considerable attention in the past, and the 'index-of-difficulty proposed by Fitts (1954), and modified to include constrained tasks is often used, particularly where performance is significantly affected by factors such as transmission delay, system dynamics and controller characteristics, as well as the neuromuscular response of the human operator (Ferrell, 1966; McGovern, 1974). Studies have identified that the operator strategy changes according to the extent of the time delay in the communication between the master and slave, as is experienced in operating remote systems in undersea and space exploration, due to the affect on stability of the closed-loop control system, particularly when the operator is in the control loop. With delays of less than 0.3 seconds the operator was found to adopt a predictive or preview behavioural strategy, whilst still maintaining a continuous motion, whereas if the delay is up to 3 seconds the strategy changed to one of a move-and-wait, i.e. move open-loop and wait for a visual and/or force feedback response (Sheridan & Ferrell, 1963).

DEVELOPMENTS IN TELE-MANIPULATOR CONTROL TECHNOLOGIES

Direct geometric correspondence between master and slave arms automatically resolves the problem of the operator having to mentally coordinate the motion of the slave arm end-effector or tool. When restrictions in workspace and size exist it is necessary to make use of alternative strategies, with some limitations in usefulness. Whilst highly trained operators can successfully coordinate the motions of up to three joints independently, fatigue can quickly diminish performance.

Early robotics research, having in part stemmed from the early developments in remote handling technologies had, by the 1970's become a focused area of research in its own right, largely due to the rapid advances being made in computer technology. In terms of robot manipulator control, one of the most important developments was Whitney's Resolved Motion Rate Control (RMRC), (Whitney, 1969). Based on the human's reflex system, which automatically resolves motion of the hands during dexterous manipulations, it was devised to allow coordinated motion control of manipulators and prosthetics, and permitted the motion of the end-effector to move along straight lines relative to a fixed coordinate system. With RMRC all actuators are required to run simultaneously, at different and time-varying rates in order to accomplish steady motion of the tool or end-effector along a specific coordinate, also with the ability to change from fixed world to 'tool' coordinates as well as maintaining the orientation of the tool during translations, or changing orientation whilst 'stationary'. RMRC was found to be a particularly useful control strategy for deploying the remote manipulator, when the relative scaling between the master and slave is large, or in remotely operated vehicles, which have a relatively unrestricted workspace.

RMRC is not suitable for bilateral force reflecting systems because of the incompatibility between velocity and force, however, it can be easily implemented as a unilateral controlled using a pair of 3-axis proportional joysticks to provide the position and orientation control of the end-effector, respectively. Rate controlled multi-DOF joysticks can provide a more natural correspondence between the operator's hand and the slave manipulator, and have been the subject of much research in the UK nuclear industry during the 80's, aimed specifically at in-reactor deployment of large redundant manipulators (Owen, 1986). Other novel rate controller designs were also proposed during this time, based on parallel configurations, such as the 9-string joystick, and the Stewart platform (Venkata et al, 1981; Siva et al, 1988).

The implementation of force feedback strategies in tele-operators employing dissimilar master and slave arms is of particular interest, and extended the scope of so-called generalized controllers developed at JPL (Bejczy & Handlykken, 1980; Bejczy & Brooks, 1981). The force reflecting universal hand-controller concept provides kinaesthetic human-machine coupling, via a 6 DOF command device which is backdrivable and can, in principle be integrated to any type of dissimilar slave arm. The kinematics of the slave arm are related to the master arm joint position variables using mathematical transformations implemented in real-time using a microprocessor, as is the force/torque feedback back to the operator via the master arm, again resolved in real-time, to give the operator a sense of 'feel' regarding the reactions and task interactions at the slave end-effector. The concept is computationally very intensive compared to RMRC, but nevertheless provides great flexibility, and is relatively straightforward, as shown diagrammatically in Fig. 7. An experimental 3-axis generalized force reflecting system was developed at Newcastle in the 80's, with similar and dissimilar slave arms to evaluate performance, see Fig. 8 (Bicker,1989).

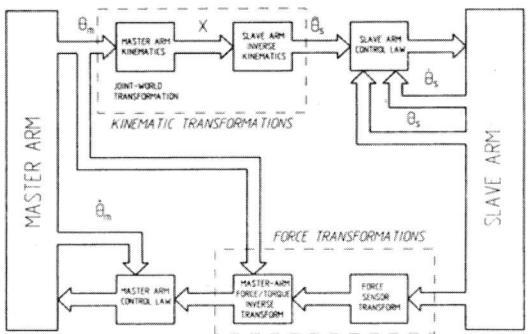

Figure 7 Schematic implementation of Generalized Control

With the availability of high performance microprocessors, applications in real-time control of tele-operation flourished. High level supervisory control schemes were also developed at this time, which permit the computer to assist the operator in providing additional status information and enabling semi-autonomous operation via pre-programmed sub-tasks to be performed under the supervision of the operator. In addition to improving the speed, reliability and cost of the system, these control schemes greatly reduce the operator's workload, however, the system relies on the ability to be able to switch smoothly between manual and autonomous control modes.

Figure 8 Three-axis force reflecting generalized tele-manipulation

TELE-ROBOTICS

Computer assisted tele-manipulation was an important step in the evolution of specialist remote handling systems, however, the need to develop robust remote handling devices led directly to the concept of tele-robotics, where an industrial robot is adapted to operate as a 'slave' manipulator, with the potential for pre-programmed autonomous operation. Having reliability and tolerance to radiation, these slave robots opened up further avenues for cost-effective research, as an alternative to developing very expensive advanced tele-manipulators with only limited market potential. The classification of the most important tele-robotic control modes is illustrated in Fig. 9, and ranges from the 'straightforward' manual tele-operation, through to fully autonomous (or traded) control, in which it can be observed that the operator's role varies from full control to that of an observer, during traded control. Significant development work in this area was undertaken at JPL as part of their ongoing flight tele-robotic control program (Backes & Tso, 1990; Zimmermann et al, 1992).

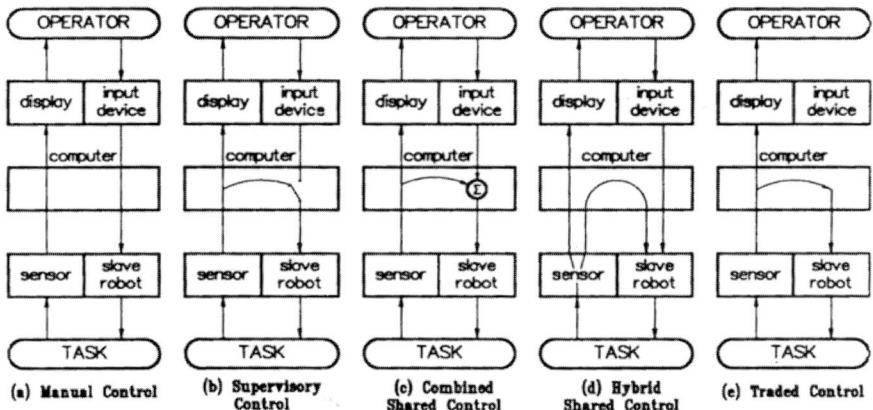

Figure 9 Classification of tele-robotic control modes (Zimmerman & Backes, 1992)

Work on developing a prototype tele-robotic system at Newcastle was begun in 1987, as shown in Fig. 10 (Burn, 1993). Three different rate controllers were evaluated, i.e. a prototype joystick, an isometric force/torque stick, and a Puma 260 robot, which was also implemented as a unilateral position controller, with Puma 560 and 760 'slave' robots. This study subsequently led to the development of a very sophisticated 6-axis force reflecting tele-robotic system using re-engineered Puma 260 and Puma 760 robots as the master and slave arms, respectively (Sin Ming Ow, 1997). The implementation relied heavily on a multi-processor architecture, but included most of the advanced control schemes illustrated in Fig. 9, above. These enabled one or more axes to be controlled using either rate and position control (unilaterally and bilaterally), as well as autonomous force control, simultaneously. The Newcastle tele-robotic system is illustrated in Fig. 11. By bringing together the demand for increased performance of tele-operator systems with sophisticated control strategies developed primarily for industrial robots this allowed the operator to act in a more supervisory capacity. Fully autonomous operations in remote and/or hostile environments only become feasible when it is possible to anticipate all possible scenarios that may occur during task execution. However, semi-autonomous operation becomes realistic with the operator-in-the-loop, who is capable of intervening at any time when the likelihood of a problem arises.

402

Figure 10 – Newcastle prototype tele-robotic system with rate controlled joystick.   *Figure 11 - Newcastle advanced 6-axis force reflecting tele-robotic system*

VIRTUAL REALITY AND HAPTIC INTERFACES

The technological push in remotely operated systems is presently being directed towards the concept of virtual reality, or what Sheridan formerly called 'tele-presence' (Sheridan & Furness, 1992), where the operator is 'immersed' within a computer generated virtual environment (VE), in which they can interact with virtual objects within the environment. Even so, VE may still make the user 'feel' isolated within an abstract world, which requires sophisticated equipment to generate a sufficiently accurate CAD model that in turn, demands high-level information, such as light, shadow, texture, etc, to perform the real time interaction. Augmented reality systems have overcome some of these issues by mixing real scenes with the virtual one in order to create an augmented display. This precludes the whole environment being under the control of computer graphics, although the computer-generated scene must be accurately registered with the real environment at all times to ensure that the real and virtual video streams are correctly synchronised. This correlation must be continuously monitored in real time whilst the user moves about in order to prevent motion sickness which is a frequent occurrence in these systems.

Whilst the virtual object may be seen but not felt, this is not sufficient, as most tasks involve touching and manipulating objects. A 'haptic interface' is therefore an essential pre-requisite to achieving the kinaesthetic coupling, and thus increase the level of presence of the user. Several different types of haptic interface have been developed, including a fixed force interface display known as a pen-based master, which allows the user to interact with virtual environments though tools such as pens, pointers or scalpel. The workspace of these masters typically is quite large and they have between 3 and 6-DOF of movement. Iwata has developed a 6 DOF pen base force display, the design of which is based on two 3-DOF arms placed in parallel (Iwata, 1993). The pen shaped handle is connected between the two arms and the force feedback system is integrated with the graphical system to produce reaction forces and torques at the users hand, through which the operator can respond. However, the most popular commercially available force feedback device is the 'PHANTOM' (SensAble Technology Inc., 2000). Whilst the PHANTOM can however only provide a single force vector its excellent design permits very good fidelity, albeit over a relatively low force range. In order to replicate a gripping action on a virtual object by several fingers then multiple PHANTOM devices are required, with the consequence that the working volume may be compromised and this quite restrictive (Burns, 1996).

VR technology is becoming increasingly important in the development of training aids for surgical procedures as well as in the more traditional fields of nuclear engineering. The public imagination is of course stimulated with computer-generated images of autonomous vehicles traversing the surface of distant planets, and of course those laboratories heavily engaged in space exploration rely heavily on VR technology in order to plan and execute their missions using autonomous control, which would otherwise be impossible to undertake because of the extreme distances involved and the consequential transmission delays which otherwise prevent remote control.

CONCLUSIONS

In looking back over a little more than half a century it is evident that great strides have been made in the technological developments associated with remotely operated systems, not only in the nuclear industry but in many other fields. From its early beginnings with the simple tong devices used in the evolving nuclear industry, we can observe the development of highly complex remote systems that are capable of executing tasks as diverse as those undertaken within nuclear reactors, to planetary rovers capable of carrying out experiments on the surface of distant planets, and to the precise manipulative skills of surgeons carrying out brain surgery with the aid of micro-manipulators in real-time over the internet. Within the not too distant future there is the prospect of the user being completely 'immersed' within a virtual environment that is totally hostile, but nevertheless permits a wide range of complex tasks to be undertaken in the relative safety of a laboratory.

REFERENCES

Backes PG & Tsu Kam S, 'UMI: An interactive supervisory and shared control system for telerobotics', Proc of IEEE Conf on Robotics & Automation, Vol 2, pp1096-1101, 1990.

Bashir AM, 'Design and evaluation of a virtual augmented reality system with kinaesthetic feedback', University of Newcastle upon tyne, PhD thesis. To be submitted, 2004.

Bejczy AK, Handlykken M, 'Generalization of bilateral force-reflecting control of manipulators', Proc. of 4th RoManSy, pp300-312, CISM-IFToMM, 1980.

Bejczy AK & Brooks TL, 'Advanced control techniques for tele-operation in earth orbit', Proc of 7th Annual Assoc of Unmanned Vehicle Systems, pp59-74, Ohio, 1981.

Bicker R, 'Force feedback in remote tele-manipulation', PhD, University of Newcastle, 1989.

Burn K, 'Control of a force reflecting tele-robotic system', PhD, University of Newcastle, 1993.

Burns DT, 'Design of a six-degree of freedom haptic interface', Master's thesis, Department of Mechanical Engineering, Northwestern University, August, 1996.

Fitts P, 'Information capacity of the human motor system in controlling the amplitude of movement', J of Experimental Psychology, Vol 47, Part 6, pp 381-391, !954.

Ferrell WR, 'Delayed force feedback', Tran IEEE, J of Human Factors in Electronics,Vol HFE-6,pp24-32,1966.

Flatau CR, Vertut J, et al, 'A compact bilateral servo master-slave manipulator', Proc of 20th Conf. on Remotely Manned Systems, pp296-302, American Nuclear Society, Pasedena [1972].

Flatau CR, 'SM229 – A new compact servo master slave manipulator', Proc 25th Conf on Remotely Manned Systems, pp 169-173, ANS.

Goertz RC, 'Fundamentals of general purpose remote manipulators' Nucleonics, Vol 10, Part II, pp36-42, 1952.

Goertz RC, 'A force reflecting positional servo mechanism', Nucleonics, Vol 10, Part II, pp43-45, 1952.

Goertz, RC et al, 'General purpose manipulators', Nucleonics, Vol 12, Part II, pp45-49, 1954.

Goertz RC, 'Manipulator system development at Argonne National Laboratory', American Nuclear Society, Proc of 12th Int Conference on Remote Systems Technology, pp117-136, 1964.

Iwata H, 'Pen-based Haptic Virtual Environment', Proc of IEEE Virtual Reality Annual international symposium, Seattle, WA, pp287-290, 1993.

McGovern DE, 'Comparison of two manipulators using a standard task of varying difficulty', Proc of ASME Annual Meeting, Paper 74-WA-BIO.4, pp106-110, 1974.

Matsushima K and Nagai A, 'On the remote mini-manipulator Tiny-Mini control – it's arm and gripper', Proc. 11th Symposium on Industrial Robots, Part 11A-6, pp645-653, 1981.

Matsushima K, ' Development of servo manipulators', Proc of International Computer Symposium, Taipei, Taiwan, 1984.

Mosher RS, 'Hardiman to Handyman', Proc of SAE Automotive Congress, Paper 670088, Detroit, 1967.

Owen CKV, 'Joysticks for manipulators, CERL report TPRD/L/2881/N85, 1986.

PAR Systems, 'TRU-Motion master-slave manipulator', Programmed and Remote Systems Corporation, St Paul, Minnesota, 1974.

SensAble Technology Inc, PHANTOM^{TM}, www.sensable.com/products/phantom_ghost/phantom.asp, 2000.

Sin Ming Ow, 'Force control in tele-robotics', PhD, University of Newcastle, 1997.

Sheridan TB & Ferrell WR, 'Remote manipulative control with transmission delay', Trans IEEE, J on Human Factors in Electronics, Vol HFE-4, pp25-29, 1963.

Sheridan TB, 'Evaluation of tools & tasks: reflections on the problems of specifying robot manipulator performance', National Bureau of Standards, Report Sp-459, pp27-38, 1976.

Sheridan TB & Furness TA, 'Presence: Teleoperators and Virtual Environments', Cambridge, MIT Press, 1992.

Siva KV, et al, 'Development of a general purpose hand controller for advanced teleoperation', Proc IMechE Int Symp on Teleoperation & Control, pp277-290.

Venkata T, Gobburu T and Doty KL, 'A novel design principle for microprocessor based 6-DOF manual controllers', Proc IEEE Southeastern 81 Regional Conf, Huntsville, 1981.

Vertut J and Coiffet P, 'Teleoperation and Robotics – Evolution and Development', Robot Technology, Vol 3A, Kogan Page, 1984.

Whitney DE, 'Resolved motion rate control of manipulators and human prostheses', Trans IEEE, pp47-53, Vol MMS-10, No .2, 1969.

Yamamoto M, Inada E, Maeda M et al, 'Remote maintenance equipment for hot-cell facilities', Proc. of 30th Conf. On Remote Systems Technology, pp132-137, ANS, Washington, 1982.

Zimmerman W, Backes PG & Chirikjian G, 'Telerobot control mode performance assessment', Prof of 15th Annual Rocky Mountain Guidance Control Conf, pp305-318, 1992.

Author Index